Alternative Energy Sources

for the centralised generation of electricity

Modern Energy Studies

Consultant Editor: **Professor N H Lipman**

Other books in the series

Heat Pumps
J T McMullan and R Morgan

Stirling Cycle Engine Analysis
I Urieli and D M Berchowitz

Alternative Energy Sources
*for the centralised
generation of electricity*

R H Taylor

Principal Engineer, Technology Planning
and Research Division, CEGB

Adam Hilger Ltd, Bristol

British Library Cataloguing in Publication Data

Taylor, R. H.
    Alternative energy sources for the centralised generation of electricity.
    1. Electric utilities
    2. Power resources
    I. Title
    338.4′562′1312   HD9685.G7
    ISBN 0-85274-476-5

Published by Adam Hilger Ltd, Techno House, Redcliffe Way, Bristol
BS1 6NX.
The Adam Hilger book-publishing imprint is owned by The Institute of
Physics.

Text set in 10/12 pt Linotron 202 Times, printed and bound
in Great Britain at The Pitman Press, Bath

For Caroline and Nicholas

'You see, we should utilize natural forces and thus get all our power. Sunshine is a form of energy, and the wind and tides are manifestations thereof. Do we use them? Oh no!, we burn up wood and coal, as renters we burn up the front porch for fuel.

We live like squatters, not as if we owned the property'.

Thomas Edison, 1916

Contents

Series Editor's Preface

Since the early 1970s it has become recognised that energy is a factor central to the well-being of both the developed and developing world. The present slump affecting much of the world is at least in part due to the financial distortions caused by the succession of dramatic rises in the price of oil over the past decade or so. There has been a growing awareness of the finite nature of the supplies of the main fossil fuels on which we rely, namely oil, gas and coal. It appears that we will exhaust these valuable assets, produced by Nature over millenia, in a matter of a few tens of years in the case of oil and gas and in a few hundred years in the case of coal.

It is therefore not surprising that there has been a considerable surge of interest and activity worldwide in searching for more efficient ways of using existing energy sources, and in developing new ones, including nuclear fusion, solar, wave, tidal and wind energy. Equally important are initiatives to improve the energy efficiency of buildings, vehicles and of industry.

In the UK there is a great deal of research and development going on in universities, industry and in certain of the government-funded research laboratories. Some will argue that UK activity is on too small a scale compared with that of our competitors in other developed countries. Others will argue that the bias in UK funding remains too strongly entrenched in favour of nuclear energy and neglects some of the alternatives such as wind or wave energy.

However, it is not our purpose in this series to take sides in the argument and the subjects presented cover the work of leading experts in the field of energy studies. It follows that many of the leading 'energy contenders' will be fairly represented. The academic level of the books will be permitted to vary according to the tastes of each author, but they will, in general, require of the reader some scientific education.

Professor Norman H Lipman
Consultant Editor

Author's Preface

This book has been written as a general survey of the progress made in developing alternative energy sources. It should primarily be of interest to the engineer or scientist who wishes to be introduced to the subject in fairly broad terms. However, parts of it should also be of interest to the more general reader without a formal engineering training, to students and perhaps to specialists in one subject area of the book who wish to find out about the others. The analysis of the prospects and economics of the various sources should be of interest to the energy planner. To cover such a large field in a short book means that certain important subjects have tended to be dealt with rather superficially. The list of references given for each chapter should allow those who are interested to explore issues in greater depth.

Choosing a title for this book has been very difficult because there is always confusion about what is meant by alternative energy sources, renewable energy, solar energy and other loosely interchangeable terms which are sometimes used to cover the broad range of technologies considered here. Discussion of alternative sources frequently includes such subjects as oil shale technology, solar heating, biomass etc—these, however, are not covered here because the emphasis of this book is on the large-scale generation of electricity. However, fusion, thermoelectric generation, MHD and other technologies are sometimes included as 'alternatives'. Renewable sources include hydro power and geothermal energy (although the purist would argue that the latter is not strictly a renewable energy source). It is therefore left largely for the individual to determine what should be included under these umbrella classifications and in this book I will be discussing wind, wave and tidal energy, ocean thermal energy conversion (OTEC), ocean energy concepts such as the use of salinity gradients and ocean currents, geothermal energy, photovoltaic conversion, solar thermal generation and solar ponds. Hydroelectricity is now such a widely used and well developed technology that it is the major source of generation for some electrical utilities, and it has not therefore been included in the book. Several of the subjects surveyed have no apparent potential in the UK, and whilst the scene in this country is given some prominence, an attempt is made to discuss the sources, wherever possible, in a world context.

Finally, I would like to thank many colleagues and others who have given valuable advice and made helpful comments on the text; in particular David Milborrow, Tony Rockingham, Brian Count, Tony White, Bob Ashbee, Peter Harrison, John Robertson, Vic Newman, Peter Musgrove, Clive Grove-Palmer and Christopher Armstead. I also wish to thank the Central Electricity Generating Board for granting permission to publish this volume in which I have drawn extensively on the knowledge and experience which has been gained by me in my professional capacity and for their permission to reproduce many of the illustrations used in the book. Finally, I should thank my wife and family who have suffered almost as much as I have in producing this volume!

R H Taylor

List of Abbreviations

ACTF	Advanced Component Test Facility
AlCoA	Aluminium Company of America
BHRA	British Hydraulic Research Association
BWEA	British Wind Energy Association
CEC	Commission of the European Communities
CNEXO	Le Centre National pour l'Exploitation des Oceans (France)
CEGB	Central Electricity Generating Board (UK)
DEn	Department of Energy (UK)
DoE	Department of Energy (US)
EEC	European Economic Community
EFG	Edge Defined Film-fed Growth
EPRI	Electrical Power Research Institute
ERA	Electrical Research Association
ERNO	A division of the German company VFW-Fokker
FPUP	Federal Photovoltaics Utilisation Program
GRP	Glass Reinforced Plastic
HDR	Hot Dry Rock (geothermal technology)
HRS	Hydraulic Research Station (as in HRS (Russell) rectifier)
ICE	Institution of Civil Engineers
IEA	International Energy Agency
IEE	Institute of Electrical Engineers
IEEE	Institute of Electrical and Electronic Engineers
ISES	International Solar Energy Society
KMW	Karlstad Meckaniska Werkstad
LASL	Los Alamos Scientific Laboratory
MEL	Marchwood Engineering Laboratory
MHD	Magnetohydrodynamic (generation)
NASA	National Aeronautical and Space Administration
NEL	National Engineering Laboratory
NPL	National Physical Laboratory
OECD	Organisation for Economic Cooperation and Development
OTEC	Ocean Thermal Energy Conversion
OWC	Oscillating Water Column

List of abbreviations

PNL	Batelle Pacific Northwest Laboratory
RMCS	Royal Military College of Science
SEA	Sea Energy Associates
SERI	Solar Energy Research Institute
SPS	Satellite Solar Power Stations
STTF	Solar Thermal Test Facility

1

Introduction

In many parts of the world there is a growing awareness that some alternative energy sources could have an important role to play in the production of electricity by the end of the century. It is probably fair to say that there is a mood of cautious optimism about their long-term potential. However, there are many technical problems which are yet to be overcome, the economics of nearly all of them require to be improved significantly before they can be competitive on a large scale with most conventional sources and some of them present environmental problems. This book has not been written to plead the case for alternative energy sources, but rather in the endeavour to present a balanced view of their technical, economic and environmental status and to explain the progress which has been made and the problems which still remain to be overcome. No attempt is made to argue, for example, that alternative energy sources should replace nuclear power. It is the author's view that in the very long term the world will require all the energy it can get, and sources of various kinds will have an important role to play.

In this short introduction, the background to the present energy considerations in the UK and other developed countries will be sketched and some of the special problems facing developing countries will be mentioned.

1.1 The role of the alternative energy sources in the UK and other developed countries

The future energy situation in Britain is, in the long term, unlikely to be very different from that in most other developed countries. Whilst the benefits of having indigenous oil and gas provide a breathing space, they do not alter the longer term outlook. Demand for energy and, in particular, for electricity will be sensitive to many factors—not least the amount of growth in the economy and the degree to which governments actively pursue the objective of fuel conservation. Whichever view is taken however—be it the official line as exemplified by the arguments in *Energy Paper* 39 [1] or the view of those advocating low-energy strategies as exemplified by the *Leach Report* [2]—the general conclusion is that during the first half of the next century there will be a progressive shortfall between what conventional sources will supply economically and acceptably, and the demand for electricity. These conclusions are reached on the basis of quite different assumptions. In the official view, growth is likely to increase demand significantly in the long term; in other scenarios nuclear power is deliberately constrained because it is believed to be undesirable and even with optimistic projections of the possible energy savings that can be brought about by the more efficient use of energy, a shortfall begins to develop between supply and demand. The argument then reduces to how this 'energy gap' can best be filled.

Coal—which at present supplies about 80% of UK electricity requirements—will probably become increasingly the subject of international trade and costs are likely to follow those of oil as it begins to be required for the production of synthetic oil and gas at the end of the century or shortly after. Furthermore, growth in the UK production of coal may be restricted by environmental objections to the siting of pits and the cost of electricity production from coal may be increased if, as in some other countries, measures are required to reduce sulphur emissions. In the long term, and on a global scale, there exists anxiety in some quarters about possible increase in atmospheric carbon dioxide resulting from the burning of fossil fuels and the impact this may have on climate. Thus, in terms of environmental effects, fuel supplies, and the cost of the electricity which might be produced, it would appear likely that coal will play a diminishing role in electricity generation in the UK and possibly in other developed countries. The use of conventional thermal nuclear plant is less likely to be constrained by the cost of fuel or by the problems of supply until well into the next century, but high-grade uranium ore is a depleting resource and will also eventually be in short supply unless the breeder reactor can be developed to the required standards and at acceptable cost. Furthermore, it would be foolish not to recognise that the strength of opposition to nuclear power is such that there must be a significant chance that its future development will be restricted to less than might be justified on economic grounds. Nuclear fusion offers a long-term prospect of making an important contribution to electricity production but prospects for commercial use are even less certain than those of most of the energy sources discussed in this book, despite the relatively high levels of investment in research in this field.

The argument about the long-term role of alternative energy sources in the UK and some other developed countries thus depends critically on judgements about whether the breeder reactor will be able to provide a platform for a continuing nuclear power programme in the next century.

The UK Department of Energy (DEn) and the present administration in the USA believe that this option should be the main thrust of their energy policy for electricity generation. In this situation, most of the alternative energy source research and development is relegated to the role of exploring possible insurance policies against failure, for one reason or another, of the favoured strategy. It is accepted, however, that some of the alternatives may be able to play a useful (although not an essential) role in particular applications, in adding diversity to supply and in providing quick response to changes in demand patterns which might occur over periods of only a few years. Other countries have decided that they wish to move away from nuclear power as the prime form of generation in the future and are therefore concentrating a proportionately greater part of their research and development budgets for energy on alternative sources. The level of expenditure shown for some developed countries in table 1.1 thus reflects a range of factors—the size and mix of present generating capacity, reliance on energy imports, views on the most likely growth in demand and particularly on the future role of nuclear power. To this should be added the view taken of the need to provide a government stimulus to commercialise these technologies and the degree to which they are seen to be an important future source of exports. It should be noted that the large expenditures on alternatives shown for the USA reflected the policy of the Carter administration that government targets, provision of tax incentives and well-funded programmes for the commercialisation of these sources were essential to stimulate industry; a policy which has largely been reversed by President Reagan.

Among the alternative energy sources which can be used for electricity generation, the UK DEn recently [3] selected wind and tidal energy as those they thought most likely to be providing a substantial contribution to UK electricity production by early in the next century. They suggested that 10 TW h p.a. might be generated from wind energy and about twice this from tidal sources. The degree of enthusiasm shown by governments for the various options is largely determined by geographical factors such as wind regimes, wave climates, geothermal potential and the seasonal variation of solar radiation. However, other factors such as the degree to which a technology is perceived to have export potential, the way that output corresponds with demand and the extent to which other countries have established a technical lead in the area, clearly influence the breakdown of expenditure in the various areas by governments in different developed countries.

1.2 The role of alternative energy sources in developing countries

It is even more difficult to generalise about the role of the alternatives in developing countries than it is in the developed world. Not only are there

Table 1.1 1980 International Energy Agency (IEA) government R, D and D on new energy sources ($Mp.a.). Information taken from *Energy Research, Development and Demonstration in the IEA Countries* 1980 Review of National Programmes OECD (IEA 1981).

	CEC	Australia	Canada	Denmark	Germany	Italy	Japan	Holland	Sweden	UK	US
Wind		0.14	4.11	1.53	22.61	2.63	0.63	5.53	8.04	2.21	63.4
Ocean											
(including tidal)		0	1.20	0	0.99	0	1.25	0	0.76	10.47	4.70
Geothermal											
(including non-electric uses)		0.03	2.65	0.92	4.02	3.14	42.02	0.2	0.17	5.96	143.0
Solar photovoltaic		1.08	0.86	0.04	38.12	6.74	10.08	3.02	0.5	0.60	131.5
Solar thermal		0	0	0	0	3.3	29.72	0	0.52	0	139.1
Total	61.9	1.25	8.82	2.49	65.74	15.81	83.70	8.55	9.54	19.24	524.0

uncertainties about technical, environmental and economic questions related to the technologies and the economic prospects in the countries themselves, but there are also great differences in opportunities, aspirations and cultural conditions between countries broadly classified as 'developing'. Many possess only the seed of an industrial and technical infrastructure and in many, the resource which these technologies might offer is known only in the broadest terms.

The future of alternative energy sources in these countries has recently been the subject of considerable debate at the UN conference on new and renewable energy sources in Nairobi (August 1981). This conference discussed not only those alternatives considered here, but also a number of sources such as biomass, fuelwood, draught animals and hydro power which are (and are likely to remain) of prime importance to many of these countries, but which are not within the compass of this book. For a fuller appreciation of the problems facing developing countries across a wide range of energy technology, the reader is referred to papers prepared for this conference, particularly the report of the so-called synthesis group [4] and the final conference report [5].

Some developing countries have been particularly affected by the fifteen-fold rise in oil prices during the last eight years, because more of their valuable foreign currency is now spent in buying the same amount of oil and less is available for developing other energy sources, raising the standard of living and developing their industrial base. The result is therefore a frustrating and crippling form of poverty trap. If industry cannot be developed or if higher prices cannot be obtained on world markets for the raw materials which the country possesses, then the position gets progressively worse with each rise in oil price. This, of course, only applies to the non-oil producing countries, those few who have oil of their own are in many cases enjoying rapid industrial growth. Energy availability has an important bearing not only on industrial expansion but also on agriculture, transport, health care, education and housing. It has been estimated that to achieve 'reasonable' progress in these areas over the last two decades of this century, developing countries will need to increase their energy consumption by some 300%, compared with an increase of only about 30% for the developed countries.

Alternative energy sources for the production of electricity may offer a number of advantages to developing countries where at present it is frequently largely derived from oil. Hydro power is already being developed wherever possible but geothermal, tidal and perhaps wind and solar power may also be important. There appears to be some reluctance on the part of many developing countries to welcome the introduction of these technologies not least because they believe that second-best technologies are being foisted on them while the developed nations continue to enjoy the benefits of the prime fuels and more advanced technology.

Furthermore, many alternative energy sources do not enjoy the same prestige as nuclear power stations and other very large power plants. On the other hand, in addition to removing dependence on imported fuel, alternative sources offer other real advantages for the developing world. Firstly, many of these countries enjoy geographical advantages in that there is a good resource of solar power and frequently of wind and geothermal sources as well. Secondly, being smaller in scale these sources are more able to meet the needs of rural communities. Thirdly, many are less demanding on expertise and on an industrial infrastructure than are conventional sources. Lastly, their generally modular nature brings the advantage that initial capital expenditures do not necessarily have to be enormous. It has also been argued that alternative energy sources for electricity generation in developing countries may be more compatible with the social and cultural conditions in many of their societies.

If these advantages are accepted and there is an increasing desire to develop these resources the question arises as to what must be done at national, regional and world levels, to introduce the technologies over a suitable transition period. A coherent policy must include provision for the following.

(1) Assessment of the resource by data collection and analysis for the various technologies in regions and individual countries.

(2) Collaboration between developing and developed countries to carry out the necessary research and development, particularly to adapt technologies to local needs.

(3) Assistance with planning and financial appraisal of possible schemes.

(4) Funding of demonstration projects to show compatibility with technical, industrial, environmental and cultural conditions and assistance with the evaluation.

(5) Assistance in providing suitable education and training programmes for the skilled personnel who will be required to build and operate plant, and provision of adequate information and spare parts when the project is underway.

(6) The transfer of technology to developing countries with the minimum of restrictive conditions.

(7) A continuing commitment to providing financial aid to establish programmes.

There would seem to be little doubt that if progress can be made along these lines over the next few years, the prospects for the development and use of alternative sources in many developing countries would appear to be good. This would almost certainly be to the benefit of both the rich and poor nations.

1.3 World potential of alternative energy sources

It is very difficult to assess the overall world potential for the various alternative sources, or for alternative energy as a whole. At the Nairobi conference, the Secretary-General produced an estimate of the present world use of various new and renewable energy sources and an estimate for comparable figures in the year 2000. Table 1.2 shows the figures for the electricity producing renewables and must probably be taken as the best available estimate, although in some cases it may be rather optimistic.

Table 1.2 Estimated use of renewable energy sources for electricity generation and projected use for the year 2000.

	Present use (TW h)	Use in 2000 (TW h)
Solar†	2–3 chiefly thermal	$2\text{–}5 \times 10^3$
Geothermal†	55 partly electrical	$1\text{–}5 \times 10^3$
Wind	2	$1\text{–}5 \times 10^3$
Tidal	0.4	30–60
Wave	0	10
OTEC	0	1×10^3

† No breakdown was given of electrical and non-electrical use.

Each of the technologies will be discussed in depth in the following chapters and the advantages and disadvantages of each highlighted. The brief concluding section of the book draws these various strands together in the form of a table summarising the pros and cons of each of the alternative sources considered.

References

1 1980 Research, development and demonstration of energy technologies for the UK *Energy Paper* 39 (London: HMSO)
2 Leach G *et al A Low Energy Strategy for the UK* (London: IEED Science Reviews)
3 United Nations Conference on New and Renewable Sources of Energy, Nairobi 1981 *UK National Paper* (United Nations Publication)

4 United Nations Conference on New and Renewable Sources of Energy, Nairobi 1981 *Report of Synthesis Group* A/Conf. 100/PC/41 (New York: United Nations)

5 United Nations Conference on New and Renewable Sources of Energy, Nairobi 1981 *Report* A/Conf. 100/11 (New York: United Nations)

2

Wind energy

2.1 Historical introduction

Man has used wind energy in one form or another for thousands of years. The earliest use was for the propulsion of ships but nearly two thousand years ago in Persia, China and Japan, windmills are believed to have been used widely. In the thirteenth and fourteenth centuries windmills of the traditional Dutch type came into widespread use for grinding corn and it is estimated that by the early nineteenth century 10 000 windmills of this type were in use in Britain alone. As Musgrove [1] has recently pointed out, it is little wonder that the windmill was so widely used. In a 7 m/s wind (force four), a windmill could develop 30 kW and average perhaps 10 kW through the year. It could therefore do the same mechanical work as 200 men. The first 'modern' windmills were developed in the late nineteenth century for pioneer agricultural use in Australia and the United States. These were originally developed for water pumping and were generally of the multi-bladed type still familiar in many parts of the world. Indeed, it is probable that the development of such windmills was an essential feature of the growth of cattle ranching in these countries. An excellent review of the early development of wind energy has been given by Minchinton [2].

Some small windmills were used for electricity production ahead of rural electrification programmes but the development of the internal combustion engine and steam cycle for electricity production led, in the early part of this century, to a rapid decline in the use of the wind as a source of power in most parts of the world. This occurred in spite of a growing appreciation of the aerodynamic principles of windmill construction, largely originating from the development of aircraft technology during the First World War. A notable exception to this general trend occurred in Denmark where La Cour developed windmills for electricity production. To overcome the variations in output, La Cour stored energy by electrolysing water to produce hydrogen which was then used to light his laboratories. At the turn of the century, a Royal Commission was set up in Denmark to develop windmills for agricultural use and several hundred such windmills were produced, notably by the Lykkegaard Company.

Likewise, in the USSR after the revolution, attempts were made to modernise agricultural production and prior to the Second World War a substantial number of small windmills were constructed. The largest of these generated 100 kW and was sited in the Crimea. From the limited information available, it appears to have performed quite·successfully.

The first really large scale modern windmill was built on a hill called Grandpa's Knob in Vermont, USA in 1941. This ambitious project was well ahead of its time and has unfairly been cited frequently as an example of the unreliability of very large machines. An excellent account of the building and performance of the machine is given in Putnam's book *Power from the Wind* [3]. It had a diameter of 55 m and gave an output of 1.25 MW in a windspeed of 13.5 m/s. The blades rotated at a constant speed of 29 rpm and it was an example of a variable pitch windmill, in which the blades can be feathered to maintain the output as the speed of the wind changes. In October 1941, the wind turbine was synchronised with the local utility grid and an extensive programme of measurements was carried out. At times, it generated up to 1500 kW and continued to operate in windspeeds of more than 30 m/s (70 mph) and survived winds of well over 50 m/s. After about eighteen months a main bearing failed and some breaking of the steel blade skin near the blade roots was observed. Because of the difficulties of providing new blades during the war, field modifications were carried out and, largely as a result of the difficulties encountered in carrying these out satisfactorily, in March 1945 one of the 8 ton blades broke off at the blade root and was thrown some 250 m from the machine. In 1945, electricity from conventional sources was about 40% cheaper than the projected cost from large windmills and it was decided to abandon the project. Much useful information and experience, however, was gained from this early attempt at operating a giant windmill with a power output nearly an order of magnitude greater than its predecessors.

After the Second World War, energy demand grew very rapidly in Europe and because of a shortage of fossil fuel, a number of European countries—notably the UK, Germany, Denmark and France—turned again to wind energy. In the UK, a national wind energy committee was set up in 1948 and a programme of wind surveys, site selection and prototype designs up to 100 kW was initiated. Much of this work is recounted in Golding's excellent book [4]. The Enfield Andreau machine was constructed in the early 1950s and had a rated output of 100 kW in a windspeed of 15 m/s. It was 25 m in diameter. The design was novel, having hollow blades open at the tips. The designers believed that high efficiency might be achieved by using an auxiliary air turbine close to ground level which was driven by the flow of air up the tower and out of the blade tips as a result of the centrifugal force on the air in the blades. The machine was not as efficient as had been hoped but the then British Electricity Authority attempted to obtain permission to site the machine on a high-windspeed site at Mynned Anelog in North Wales. Planning permission was, however, not obtained (the site turned out to be common land) and the windmill was sold to a supply authority in Algeria where it operated successfully for a number of years.

Two other machines were built in the UK in the fifties. A 100 kW, 15 m diameter prototype was erected by the John Brown Company on a very high windspeed site at Costa Head in the Orkneys but the difficulty of access to this prototype led to major problems in ensuring mechanical reliability. A more successful machine was operated on the Isle of Man by the Electrical Research Association funded by the Ministry of Fuel and Power. This was constructed in 1959. It had a diameter of 15 m and gave a rated output of 100 kW in a windspeed of just over 18 m/s. The machine used low-cost aluminium blades, drove an induction generator connected to the local grid and had a novel system of spring-loaded air brakes which were operated by excessive centrifugal forces to prevent overspeeding. The machine operated successfully for a number of years and it was claimed that it would have been economic at a number of very windy sites in the UK.

The Gedser windmill built in Denmark in 1957 was one of the best known and most successful modern machines. It was 25 m in diameter and achieved 200 kW in a windspeed of nearly 15 m/s. The machine was basically fixed pitch but the tips of the blades were pivoted so that they could be feathered at very high windspeeds. The Gedser windmill operated successfully for about eight years, although the energy it produced was not economic. It was recently refurbished for experimental purposes but has now been dismantled.

One of the most important machines in influencing current designs was the 100 kW rated, 34 m diameter machine built at Stötten in West Germany in 1957 under the direction of Hütter. The windmill had a number of features of particular interest. It was variable pitch with thin blades fabricated from glass-fibre-reinforced plastic rather than the steel or aluminium used for other windmills built at the same time and it also produced its rated power at a very low windspeed of 8 m/s. This is why the output is rather low for its diameter. The machine design has been the basis of some of the recent prototypes built and operated in the USA which will be discussed later in this chapter. The Stötten machine ended operation in 1964.

Finally, mention should be made of the French windpower programme of the late 1950s and early 1960s. Although France is one of the few northern European and North American countries which is not currently taking a particularly active interest in building large machines, its contribution to the post-war effort was considerable. Three machines were constructed between 1958 and 1963. A 132 kW, 21 m diameter machine was designed for a windy site and a 30 m diameter machine generated up to

800 kW at a lower windspeed location. Although minor technical problems were encountered, these machines generally worked well until the early sixties. A 35 m diameter machine rated at 1000 kW in a windspeed of 17 m/s was built on the same site as the 21 m prototype. Unfortunately this larger windmill did not operate for a long period before the French programme was discontinued, but for the limited time that it was tested it appeared to work quite well. One failure of a generator bearing was recorded. A good account of the development of wind power in this century has been given by Warne and Calnan [6].

This brief summary of the history of windpower and particularly of the major machines built during the first half of the present century provides the backcloth to the recent upsurge of interest in wind energy for power production. In the next two sections of this chapter, wind characteristics and an elementary theory of windmill performance will be presented. The ways in which present day technology might improve on the rather mixed success of earlier programmes and the way that this technology is being incorporated in machines designed in the seventies and eighties will then be discussed.

2.2 The wind and its characteristics

As will be shown later, the power in the wind is proportional to the cube of the windspeed. Thus detailed knowledge about the wind is essential if the design and economics of large windmills are to be properly understood and evaluated. Generally speaking, the highest windspeed sites are on exposed hilltops, offshore or on coastal sites. Because of the difference in terrain, wind characteristics may be widely different between these locations. Information is required with regard to various parameters, both in general terms for each type of site and in detail for specific sites. These include mean windspeed, the distribution about the mean windspeed, directional data, variations in windspeed in the short term (gusting), daily and annual or seasonal variations and the changes of windspeed and direction with height. Although records of windspeed have been kept at some places for very long periods of time and meteorological stations recording windspeed and direction are very widely distributed over an area like the United Kingdom, it is perhaps surprising that there are still many gaps in our knowledge about the detailed behaviour of the wind. This is particularly true for wind behaviour over the sea. One reason for this is that the wind is, in reality, a series of gusts and lulls. Weather stations record winds for synoptic purposes and are more interested in establishing the mean windspeed and direction averaged over periods of up to a few minutes at one standard height (usually 10 m). An excellent review of meteorological aspects of wind energy and the resource has been given by Lindley *et al* [7].

2.2.1 Source of the wind

The driving force for wind is the solar radiation incident on the earth's atmosphere and, in particular, the pattern of heat transfer between the surface and the atmosphere. Unequal heating of the surface establishes a pressure gradient and the air is thus subject to a force leading to acceleration in the direction of this gradient. The rotation of the earth also plays a major part in establishing global patterns of air flow. The Coriolis force resulting from these factors acts at right angles to the direction of air flow, and in the northern hemisphere this tends to establish clockwise rotation of the flow. The use of these forces to calculate the wind velocity distribution gives only part of the story, however. The air flow encounters a frictional force, due to the earth's surface, which acts approximately in the direction opposite to the wind and when the isobars are curved a further force—the cyclostrophic force—operates. If the pressure gradient is changing with time, there also exists a so-called isallobaric component [8]. The interaction between these forces and their final effect on air flow is beyond the scope of this book, but they are mentioned merely to demonstrate the complicated nature of the various forces driving the wind.

2.2.2 Variations with height

The combined effects of pressure gradient and the earth's rotation largely determine the character of the high altitude free stream or geostrophic wind. In the first few hundred metres above the earth's surface, however, there exists a turbulent layer known as the boundary or mixing layer. Within this boundary layer, windspeed and direction vary with height due to surface friction and temperature gradients. The variation depends on the type of terrain, atmospheric stability and the distance (fetch) over which the wind has already travelled. The latter parameter determines whether steady state conditions have been achieved.

Even the largest wind turbines operate within the boundary layer and it is therefore important to understand how the wind changes over the vertical scale of the machine and its supporting structure. The windspeed, in general, increases with height above the surface and empirically it has been found that a power law gives a good fit over a limited height range. The simplest relationship between the velocity V_h at some height h, and the measured velocity V_o at some reference height h_o, is given by the equation

$$V_h/V_o = (h/h_o)^\alpha$$

where the exponent α depends upon the roughness of the surface. For open land, α is frequently taken to be about 0.14, whereas for a calm sea, α may be as low as 0.10. A plot of windspeed with height for these values of α

is shown in figure 2.1. The nature of the surface also influences direction in the boundary layer. Over the land, the air stream tends to flow at the surface at about 25° to the direction of the isobars in the free stream, whilst at sea the equivalent figure might be as low as 10°. It is to be emphasised, however, that the real situation is much more complex. Unstable atmospheric conditions tend to reduce the directional difference compared to more stable conditions, and temperature inversions (due, for example, to surface radiation loss at night or cooler sea areas) can lead to compression of the stream lines at the top of the boundary layer forming what is known as a low-level jet. Surface topography, as distinct from surface roughness, is also important. Winds tend to align themselves with valleys, estuaries and tracts of water (such as the English Channel). Such variations may be particularly important close to the coast where coastal contours can influence local wind conditions.

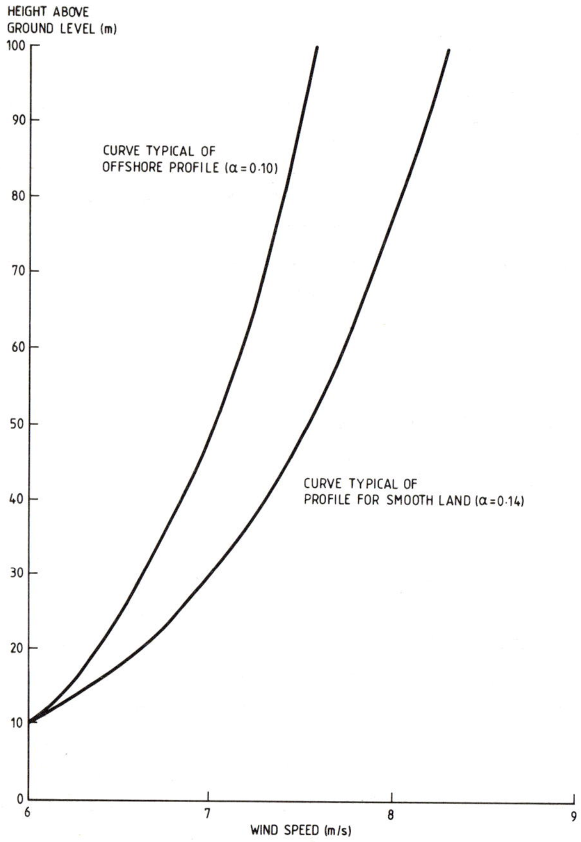

Figure 2.1 Typical variations of windspeed within the boundary layer.

It is a matter of common experience that the wind rarely flows with a constant velocity close to the surface. Because of the low viscosity of air, it tends to become turbulent at low velocities leading to the phenomena of gusts and lulls. These are associated with the vertical motion of eddies, bringing down faster moving air from above in the case of gusts. This vertical motion is, of course, essential in maintaining a surface wind against the tendency of the frictional force to reduce the air flow close to the surface. The occurrence and behaviour of gusts are important in two respects to the windmill designer. Firstly, they must be taken into account in evaluating the stresses to which the rotor and supporting structure will be subjected and secondly, they will influence the design of the logic used to determine machine orientation, pitch control and consequently power output. In the case of clusters or arrays of windmills (see below), the spatial correlation as well as the temporal variation of gusts is particularly important.

2.2.3 Variation with time

In areas like the United Kingdom, weather patterns tend to be dominated by the passage of fronts over a timescale of a day or two and these determine the coarse structure of the wind. Within these general patterns, variations occur over periods of tens of minutes and these are very significant in terms of windmill performance. Changes over this timescale influence the amount of conventional generating plant which may be needed to respond rapidly to changes in wind power output. Daily, seasonal and annual variations also occur and may have a significant impact on wind energy economics.

On a daily basis, the pattern of windspeeds, and thus the energy available from the winds, is not entirely random. Above land, atmospheric heat loss is greatest in early afternoon and this is particularly pronounced in the summer months. This is reflected in both higher mean windspeeds (see figure 2.2) and greater turbulence at this time. At sea, the position is much less clear; there is little diurnal variation in sea surface temperature and an inverse relationship between onshore and offshore atmospheric stability is to be expected. However, this will be strongly affected by land and sea breeze effects, distance offshore, wave height, seasonal variations and wind direction. These diurnal changes in wind are important because there exist well-defined variations of demand for electricity during the day and this affects the value of the wind energy. The fact, for example, that demand peaks in late afternoon in the UK, means that the statistically significant extra wind energy produced at this time by windmills sited on land may slightly increase its value to the electricity supply system (see §2.13).

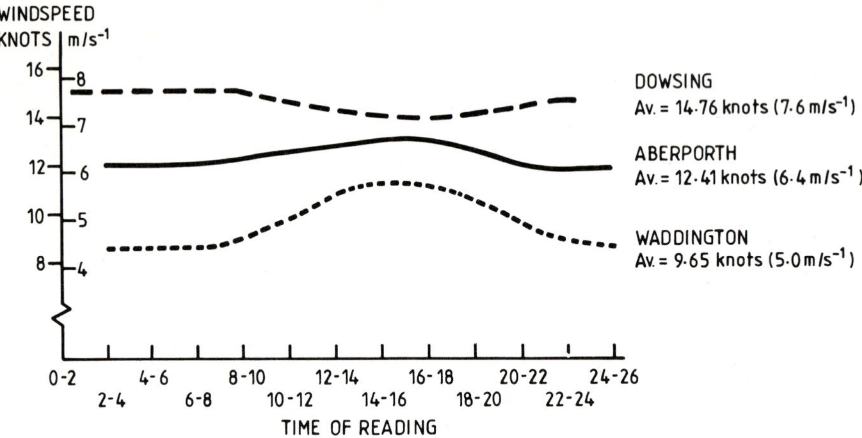

Figure 2.2 Diurnal variations in windspeed.

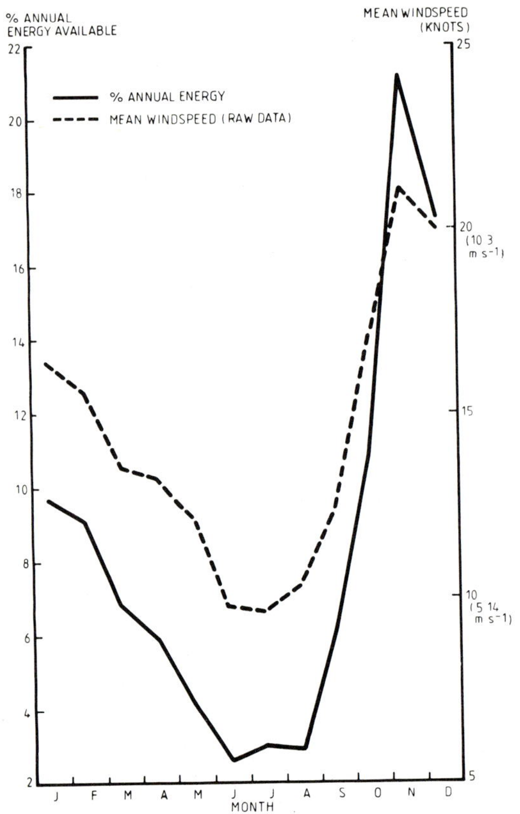

Figure 2.3 Dowsing—variability of wind velocity and percentage annual energy month by month.

A clearer correlation exists on a seasonal basis. In distinction to direct solar energy which peaks strongly in the summer months, wind energy (like wave energy) provides a large part of its annual output in the winter months. Typical data taken from the Ocean Weather Ship *Dowsing* in the southern North Sea is shown in figure 2.3. Almost 75% of the energy is obtained in the six winter months. This is very satisfactory as far as electricity production in the UK is concerned but the large-scale use of air conditioning in some parts of the world means that a peak in demand occurs in the summer. In this case, such a seasonal match might clearly be less satisfactory.

Finally, in calculating the output of energy from wind plant over the lifetime of the installation (which may be as long as thirty years), it is very important to base calculations on a typical year or a long term average. Figure 2.4 shows how windspeeds at several sites in the UK have varied over an extended period and it is clear that the statistical variation is very significant.

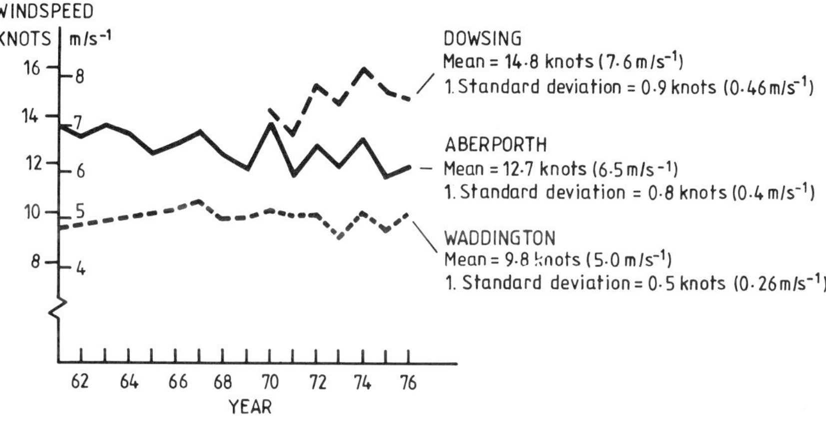

Figure 2.4 Yearly mean windspeed (historical).

2.2.4 Statistical representation of wind data

Knowledge of the mean windspeed for a site is of great importance and can be used for estimating the power in the wind (see next section) and the energy outputs from windmill systems. It is also important, for a number of purposes, to obtain statistical information about the variation of the windspeed about its mean. Firstly, windmills begin to generate power at a certain cut-in speed and are usually designed to cut-out when the windspeed exceeds a certain value. It is clearly important to assess the frequency and duration of calms and strong winds. Secondly, and most importantly, the designer needs to be able to assess the maximum stresses on blades and structures which are likely to occur over the lifetime of the installation.

Analysis of extreme conditions can be carried out by presenting data in the form of velocity duration or exceedance curves (i.e. curves showing the amount of time that various windspeeds are exceeded). A more convenient method, however, is to find the characteristic windspeed, c, which is exceeded for $1/e$ of the time (i.e. 3223 h in a year). Many sets of data then fit Weibull distributions of the form

$$t = \exp(-u^n)$$

where t is the proportion of time for which the wind exceeds any given speed u. The shape of the half-bell-shaped distribution is determined by the exponent n. Weibull distributions give straight lines when plotted on ln–ln versus ln or Weibull graph paper and n can be determined from the slope of the plot. It is extremely useful that a wide range of wind characteristics can then be summarised by specifying only two parameters—the characteristic windspeed, c, and shape factor, n. The use of Weibull distributions has been discussed by Swift-Hook [9]. A typical velocity duration curve and Weibull plot for the same data are shown in figure 2.5.

2.2.5 Power in the wind

The kinetic energy in the wind is $\frac{1}{2} V^2$ per unit mass or $\frac{1}{2}\rho v^2$ per unit volume and thus at a windspeed V, the energy flow rate or power per unit area is

$$G = \tfrac{1}{2}\rho V^3 \qquad (\text{W/m}^2).$$

Using a typical figure for the density of air at sea level, this is approximately given by

$$G = 0.6 \, V^3.$$

By analysing data in terms of the frequencies, $f, \ldots . f_n$, of winds within certain wind velocity ranges, and calculating the average power level corresponding to the velocity range considered, $G, \ldots . G_n$ over time intervals, Δt s, the energy, E, in the wind over that period can be estimated from:

$$E = \frac{\Delta t}{3600} \, \Sigma \, (f_1 \, G_1 + f_2 \, G_2 + \ldots f_n \, G_n) \qquad (\text{Wh/m}^2).$$

The energy resource available from the wind is frequently plotted in the form of annual energy isopleths in MWh/m². Details of such calculations are given in Appendix III of *UK Energy Paper* 21 by Allen and Bird [10].

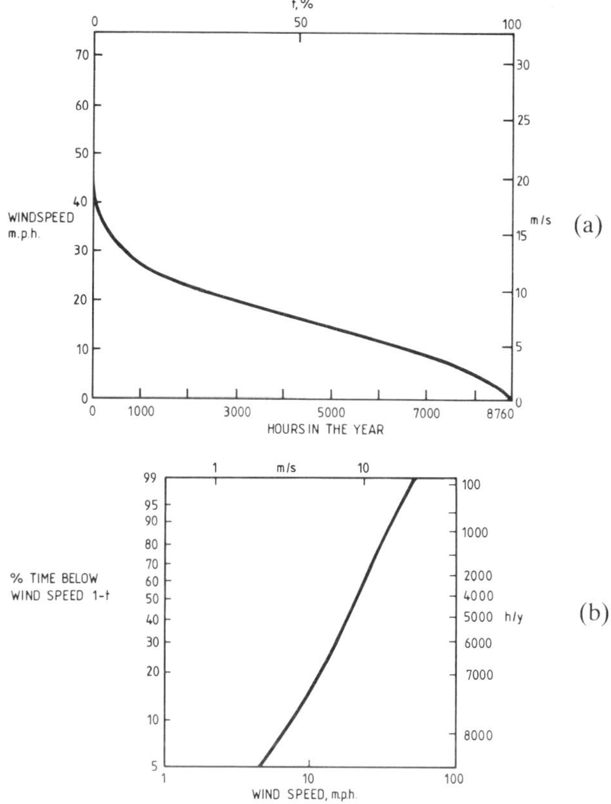

Figure 2.5 (a) A typical velocity duration curve (b) Weibull plot of data in part (a).

2.3 Simple aerodynamic theory of windmills

The idealised flow through a horizontal axis windmill can be represented as shown in figure 2.6. The main features are that because the windmill takes power from the wind, the windspeed behind the plane of the rotor (downstream) is less than the windspeed upstream. Furthermore, for the fairly low velocities encountered in normal conditions, the air flow can be regarded as incompressible and thus the streamlines must diverge as they pass through the rotor disc. In the simple momentum theory considered here it is assumed that the wind velocity outside the area affected by the rotor (i.e. outside the bounding streamlines) remains at the same value, V, and that the pressures immediately upstream of the blades, P_u, and downstream, P_d, are constant. The latter assumption effectively assumes that the two or three blades typical of a modern windmill have been replaced by an infinite number of thin blades, giving the same force on the approaching airflow as the average for the real rotor system.

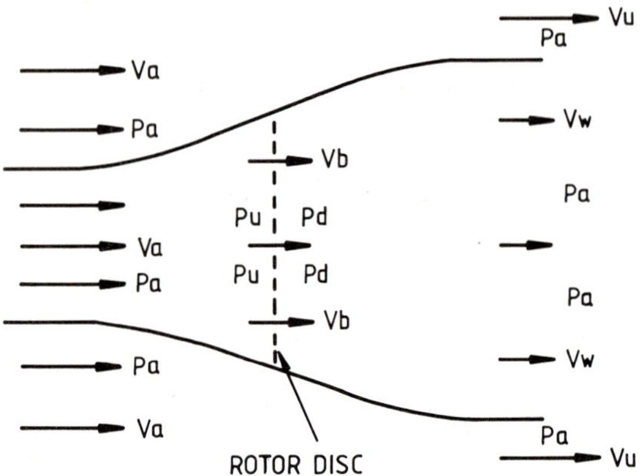

Figure 2.6 Schematic flow of air through a horizontal axis wind turbine blade, showing the symbols used in section 2.3.

If the air density is ϱ and V_b the air speed at the blades of area A, the mass flow rate, $\dot{m}$, of the area is given by

$$\dot{m} = \varrho V_b A. \tag{1}$$

The force on the disc of the rotor is given by

$$F = (P_u - P_d)A. \tag{2}$$

At some distance downstream of the blades, the streamlines will again become parallel. If the air flow velocity is V_w and the pressure P (which is the same wherever the streamlines are parallel), then the force can also be written

$$F = \dot{m}\,(V - V_w). \tag{3}$$

Applying Bernoulli's equation upstream and downstream, respectively, gives

$$P + \tfrac{1}{2}\varrho V^2 = P_u + \tfrac{1}{2}\varrho V^2_b \tag{4}$$

and

$$P_d + \tfrac{1}{2}\varrho V_b^2 = P + \tfrac{1}{2}\varrho V_w^2. \tag{5}$$

Subtracting

$$P_u - P_d = \tfrac{1}{2}\varrho(V^2 - V_w^2) \tag{6}$$

from (1–3) above

$$(P_u - P_d)A = \dot{m}\,(V - V_w) = \varrho V_b A(V - V_w). \tag{7}$$

From (6) and (7)

$$V_b = (V + V_w)/2. \tag{8}$$

The power output of the windmill can be written

$$\dot{m}\left(\frac{V^2}{2} - \frac{V_w^2}{2}\right) = \frac{\varrho V_b A}{2}(V^2 - V_w^2). \tag{9}$$

Substituting (8) in (9) gives

$$\text{power} = (\varrho A/4)\,(V + V_w)\,(V^2 - V_w^2) \tag{10}$$

and this is a maximum when $V_w = V/3$, thus the maximum power, P_{max}, which can be generated by a windmill according to this simple theory is

$$P_{max} = \frac{16}{27}\,\tfrac{1}{2}\varrho A V^3. \tag{11}$$

Since the power in the wind is given by $\tfrac{1}{2}\varrho A V^3$, the coefficient of performance (or power) of the windmill, conventionally labelled C_p, has a maximum value of 16/27 or 0.59. This theoretical upper bound for the efficiency of a windmill is known as the Betz limit. Modern designs of windmills for electricity generation operate at values of $C_p \sim 0.4$. The major losses in efficiency for a real wind turbine arise from the drag on the blades, the swirl imparted to the air flow by the rotor and the power losses in the transmission and electrical system. Milborrow [11] has recently discussed some of the shortcomings of the simple momentum theory.

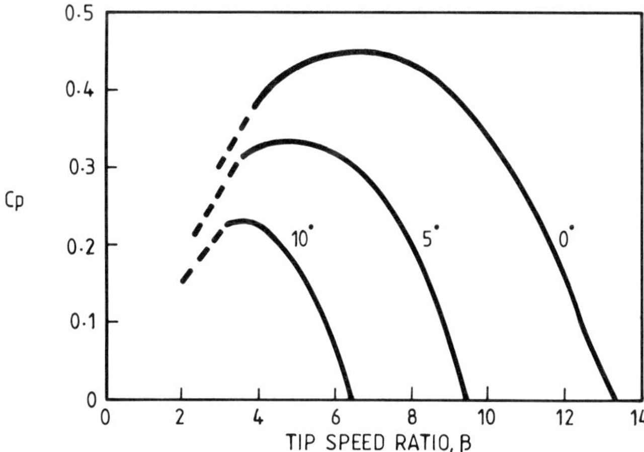

Figure 2.7 The general form of curves of performance (power) coefficient against tip speed ratio for different pitch settings.

Designers of windmills use aerodynamic theories which take into account the characteristics of the aerofoil section used for the blades. These theories supplement rather than supersede that of Betz. The results of these theories and of experimental studies are frequently summarised by plotting C_p against the ratio of the blade tip speed to the windspeed. This latter number is often denoted by β. Figure 2.7 shows typical plots of this type, one line for each angle of pitch of the blades. The dotted region indicates that in this range of values of β, the flow of air over the majority of the blade is stalled. A typical output-windspeed characteristic for a variable pitch machine is shown in figure 2.8 and compared with the characteristic for a fixed pitch machine without the flaps at the blade tips which are sometimes used on fixed pitch machines to limit output. For optimum performance a blade must be twisted with a large pitch angle near the roots which gradually decreases as the tip is approached. The reasons for this can be seen by reference to the vector diagram in figure 2.9. This shows a blade section of a horizontal axis windmill viewed from above. The resultant velocity of the air at the blade is the vector sum of the retarded windspeed, V_b and the tangential blade speed and it is clear that ϕ, the angle between this resultant and the plane of rotation, varies along the blade length because the velocity of the blade is greater towards the tip. Two forces are defined: a lift force, L, which is perpendicular to the resultant air flow and a drag force, D, parallel to the air flow. The resultant tangential force on an element of the blade in the direction of the blade motion will be of magnitude $L \sin \phi - D \cos \phi$. For the best performance, the lift to drag ratio L/D should be as high as possible. To achieve this, an angle of attack, a, of around 4–8° is required where $a = \phi - \delta$ and δ is the blade pitch. In some designs of windmill the value of δ is fixed; in others δ is a variable, controlled by a pitch change mechanism often located at the blade root. There is, however, a constraint on a. Although both drag and lift increase as a increases, above a certain value (between 11 and 15°) stall occurs and the lift decreases sharply whilst the drag increases. In normal operating conditions stall is to be avoided, although in some fixed pitch designs power is limited at very high windspeeds by allowing the blade to stall progressively from the tip. To keep the optimum angle of attack it is clear that the variation in ϕ along the blade must be compensated by a radial variation in the angle of pitch. That is, for best performance the blade must be twisted with a larger pitch angle at the root than at the tip. The actual value of L/D at which a windmill operates depends partly on the choice of aerofoil section, but also upon the Reynolds number (which is proportional to the air flow velocity over the blade), the blade width (chord) and air density. It is inversely proportional to the air viscosity. For large windmills, Reynolds numbers of several million are frequently achieved giving L/D values ranging up to or above 100. At peak efficiency, L/D at the high end of this range will be achieved. The lower Reynolds numbers achieved by small windmills (for which there are more limited data on aerofoils) is one of the reasons why modelling of the performance of large machines by small wind tunnel models is not always very reliable.

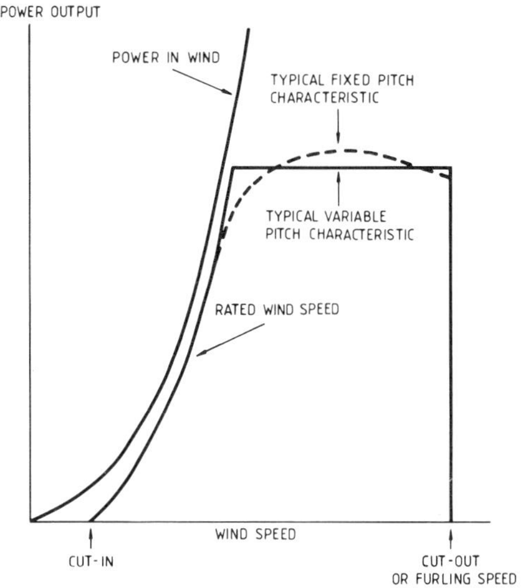

Figure 2.8 Schematic diagram showing typical fixed and variable pitch characteristics.

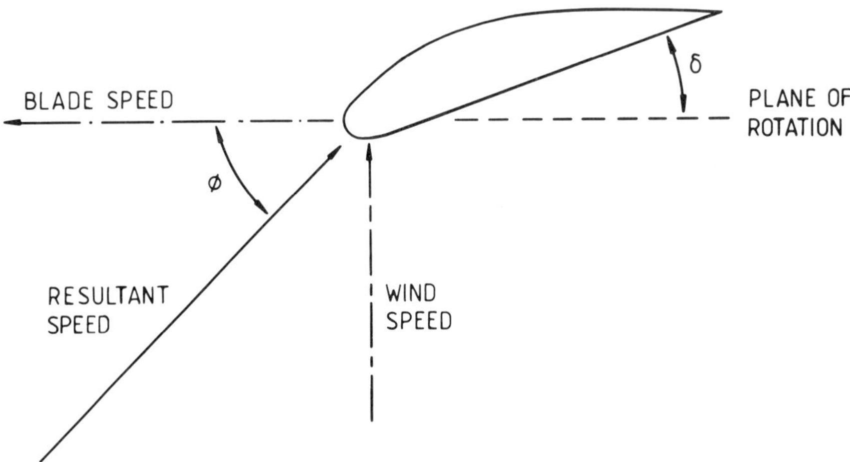

Figure 2.9 Vector diagram for a horizontal axis wind turbine blade.

Another important aerodynamic parameter influencing design, is the solidity, S. This is defined as the ratio of the total area of blade facing the wind to the swept area of the blades and uniquely defined for maximum C_p†. It can be shown using fairly simple theoretical arguments that for optimum performance $\beta^2 S$ should be constant and thus if for mechanical or other reasons the value of β is changed from optimum, S will also change. This implies that high speed rotors should have low solidity and vice versa. For large electricity-producing windmills, high rotational speeds reduce gearing costs and the weight of the gearbox because the maximum torque which the gears have to transmit is lower. High rotational speeds and low solidity are thus preferable from this point of view. On the other hand, low solidity means that what blade area there is carries a relatively high aerodynamic load. Once the optimum solidity has been calculated the windmill designer has to decide over what number of blades to spread the solidity, bearing in mind that the more blades there are the higher the fabrication cost, that one blade with a counter-balance may produce unbalanced forces leading to a shorter life and that three (or more) blades may be strutted in the plane of the rotor to give greater strength. The usual compromise for large electricity producing machines is either two or three blades. This will be discussed further in the next section.

2.4 Machine design options

2.4.1 Vertical axis or horizontal axis machines

Windmills which rotate about a horizontal axis are not the only type of wind turbine which are currently being considered. There are several designs of machine which rotate about a vertical axis. One such design, known as the Darrieus (or egg-whisk) windmill is illustrated in figure 2.10. This is receiving considerable attention in the United States and Canada and has blades with a symmetrical aerofoil section; the so-called troposkien shape of the blades being designed to eliminate bending stresses. Another type of vertical axis windmill currently being studied in the UK is illustrated in figure 2.11. This type of machine has evolved from a design by Dr Peter Musgrove of the University of Reading and is known as the variable geometry vertical axis wind turbine. The output in the original H-shaped configuration was controlled by the centrifugal forces on the hinged vertical portion of the blades which forced these blade sections to tilt outwards, giving a characteristic similar to that of a variable pitch horizontal axis machine. The arrowhead configuration is an improvement because the large vertical forces on the rotor with the blades inclined

† Sometimes, alternatively (and somewhat confusingly), the solidity at a specific radius on the blade is defined.

(reefed) are eliminated, the actuator mechanism for reefing is horizontal and blade-bending moments are reduced due to the presence of the support struts.

Figure 2.10 17-metre vertical axis wind turbine (VAWT) at Sandia National Laboratories. It can generate 60 kW of electrical power at a windspeed of 51 km/h. (Photograph courtesy of Sandia Laboratories.)

Figure 2.11 An artist's impression of the Musgrove vertical axis wind turbine being developed by a consortium led by Sir Robert McAlpines. It is hoped to build a 25 m diameter version of this machine at Carmarthen Bay power station. (Photography courtesy of Sir Robert McAlpine and Sons Ltd.)

It is still too early to say whether windmills of the future will be of the horizontal or vertical axis types. Horizontal axis machines have so far received greater attention and are developed to at least one design cycle ahead of vertical axis machines. It was once believed that horizontal axis machines could attain higher coefficients of performance than vertical axis windmills but recent experience seems to have shown that there is little difference in aerodynamic performance. Vertical axis machines have the advantage of being omnidirectional, so that no yaw orientation sub-system is required. For the Musgrove variable geometry design, the windmill blades are of a particularly simple shape, i.e. straight, symmetical aerofoil sections, without twist or taper. The Darrieus design does not offer advantages in this respect, since the complex troposkien blade shape requires considerable engineering to ensure satisfactory performance. It has also been argued that vertical axis machines will offer advantages if machines several times larger than existing designs are to be built. With increasing size aerodynamic loads increase, but because blade cross-sectional areas also increase the corresponding blade stresses do not change, to first order. Similarly, the stresses due to centrifugal forces are not increased at larger diameters. For very large horizontal axis windmills, however, it has been argued that gravitational loadings become significant because bending stresses due to the blade weight reverse every revolution as the blade moves from the three o'clock to the nine o'clock position. It is not clear at what size this constitutes a major design problem. Advocates of horizontal axis designs argue that this problem would only be significant at diameters of a few hundred metres and that stresses can, anyway, be kept at the same level by thickening the blade root. There is also the argument that, since for variable geometry machines the swept area is reduced by the blade attitude control, these machines can continue to operate safely at higher windspeed than could other designs of the same blade size. However, feathering of horizontal axis designs would seem to achieve the same end. Vertical axis machines allow the gearbox and generator to be housed at ground level thus allowing, in principle, a simpler nacelle and support structure.

The two major disadvantages of the variable geometry windmill compared with a horizontal axis fixed pitch machine, are likely to be possible additional complexity involved in hinges and hydraulic mechanisms for blade inclination control and, perhaps more seriously, that the aerodynamic loads on the blades change sign from radially inwards to radially outwards as the blades move from the upwind to the downwind side of the tower (this is also true for the Darrieus type). It is argued by some designers of vertical axis machines that this is less than the fluctuating gravity load on horizontal axis machines but the question remains open. It is worth noting that horizontal axis machines also experience cyclic aerodynamic loads because of the gradient of windspeed with height. Such loads are not generally large but may be more severe if the machine is yawed relative to the wind. Darrieus designs have the disadvantage that they require a tall tower to support the tall blade structure and are inherently fixed pitch. The main arguments for and against vertical axis designs are summarised in table 2.1.

Table 2.1 Possible advantages and disadvantages of vertical axis wind turbines.

Advantages	Disadvantages
1. Omnidirectional, so no yaw control required	1. In most designs, methods of feathering or pitch change are complex or impossible
2. Aerofoil can be of a simple shape in some designs	2. Aerodynamic loads on blades change sign from radially in to radially out giving rise to some cyclic power variation
3. Aerodynamic and gravitational stresses could be lower in very large machines	
4. Gearbox and generator can be housed at ground level	3. Some designs require a tall tower to the top of the blade and frequently need to be guyed
5. Avoids cyclic aerodynamic loads arising from wind-speed gradient with height	4. Dynamic analysis may be more complex than for horizontal axis machines

2.4.2 *Stiff or soft designs*

Most large aerogenerators have been built as rigidly as possible in order to avoid structural resonance problems. This has led to heavy, costly machines. Recently, however, in designs in the USA and Sweden, there has been a move to build machines which are lighter and thus less costly. This may produce resonances at a lower frequency than the rotational frequency of the blades and these can thus be excited on bringing the machine up to speed. However, the systems contain sufficient damping to ensure that stresses are kept within acceptable limits. Such designs have not yet been fully proven at full scale but if successful could lead to cheaper energy. Often found in association with 'soft' or 'compliant' towers are teetered hubs. In these designs the rotor is attached to a central hub which is allowed to see-saw as it rotates. This teetered hub allows stresses in the rotor to be relieved. This is particularly important when blades are situated downwind of the tower and pass through its shadow on rotation. If the rotor is free to yaw to follow the wind, the teetered hub reduces loads which would result from gyroscopic forces in the blades during yawing.

2.4.3 *Number of blades*

There is still much disagreement on the question of how many blades a machine should have. This was discussed to some extent in §2.3 but, in practice, the decision is a complex one. Three or more blades lead to a smoother output and they are marginally more efficient aerodynamically (perhaps by one or two percent). Furthermore, strutting may be incorporated in a multi-bladed machine to reduce stresses in the plane of rotation. On the other hand, three or more blades may be considerably more expensive both in terms of the extra blading and because of the greater complexity in the hub; gearboxes may need to be larger to accommodate the increased torque and the problems of mounting a multi-bladed ensemble onto a tower may be considerable.

2.4.4 *Upwind or downwind*

If blades are mounted downwind they can be coned outwards in a downwind direction to give adequate blade-tower clearance with a small and less costly nacelle. The price to be paid for this, however, is that the blades operate in the wind shadow from the tower and variations in output result from this (twice per revolution for a two-bladed machine). There is again no clear consensus, although there appears to be some tendency to upwind operation with the consequently larger nacelle. In the upwind configuration the gravitational moment of the blades about the tower is counterbalanced by the wind loading.

2.4.5 *Fixed or variable pitch*

Variable pitch blades have been favoured in German, American and Swedish designs and offer the advantage of being able to feather the blades when the rated power is achieved in order to maintain constant power output. But this mechanism is a source of extra cost and may reduce the reliability of the machine. The alternative is to maintain the blades at a constant pitch angle and to rely upon progressive stall in high windspeeds to limit the output and to reduce it at the highest speeds. Ultimate control on fixed pitch machines to avoid storm damage is sometimes achieved by braking or by stalling the blade by rotating its tip. Because the fixed pitch windmill usually generates more power above the rated output than its variable pitch equivalent (see figure 2.8), higher transmission and generator costs are often incurred.

2.4.6 *Generator*

Most large windmills drive either a synchronous or an induction generator, although other options such as hydraulics or AC/DC/AC conversions with thyristor control are sometimes employed. The synchronous generator is self starting but must run at a precise speed in order to remain in synchronism with the grid. Because pitch change mechanisms cannot always respond to rapid gusts, there is always the danger that such generators will pull out of synchronism. Some attempts have been made to minimise this effect by using a flexible transmission between the rotor and the generator, sometimes referred to as a quill shaft, and by other techniques. The advantage of the synchronous machine is that it can operate into weak grids (e.g. in some parts of the USA or on island sites) and because it can provide a greater degree of voltage stabilisation against changes of system load and power factor. In the case of distributed land-based machines, output can then be fed into lower voltage parts of the distribution network.

The major alternative, which is acceptable when energy is being fed into a 'strong' grid, is the use of the induction generator. Such generators draw their excitation from the grid and their main disadvantage is the reactive power demand that they impose on the system. Starting large numbers of generators at close intervals may lead to voltage variations resulting in some flicker. This may normally be minimised, however, by a suitable starting control logic. In addition to being simple and reliable, induction generators minimise the effect of gusting on the system and may have advantages when several machines are interconnected. Improvements in energy recovery over a standard induction generator may be achieved by slip energy recovery or by using a pole-amplitude-modulated induction generator which can accept two (or even three or more) different discrete rotor speeds determined by the ratio of the number of pole windings on the generator [12]. At some extra capital cost this in principle allows for more energy to be collected at lower windspeeds. Two-speed induction generators look as though they may be economically attractive for some wind regimes.

2.4.7 Size and rating

It is usually argued that the economics of wind turbines for utility use improve as a function of size. Whilst this is almost certainly true offshore (see below), curves of the projected cost of electricity from series-produced wind turbines sited on land against size show a fairly flat minimum above a diameter of 60 m [13]. Others argue that maintenance costs will be lower for smaller machines and that, for the foreseeable future, the benefits of series production would be more quickly gained for these sizes. From a siting point of view, the position is equally confusing. Large machines may be more intrusive in the landscape, but on the other hand fewer machines would be required. Currently most of the major manufacturers seem to favour the production of large units (2–8 MW, depending on site windspeed and up to 125 m in diameter).

The peak output achieved by a wind turbine is to some degree a design option. Reference to figure 2.8 shows that the peak output can be made very high if the machine is rated to achieve its full power only at the highest windspeeds. This has the consequence, however, that for much of the time the wind turbine is operating on the steep part of its characteristic leading to rapid changes in output. Furthermore, rated output is achieved for only a short time. Its load factor is thus reduced. The optimum rating for a wind turbine depends upon its design (e.g. fixed or variable pitch) and the wind characteristics of the site. It is usual to refer to the ratio of the windspeed at which peak power is produced (sometimes, confusingly, peak efficiency) to the site mean windspeed as the wind turbine rating.

For fixed pitch designs, a number of authors [10], [12] have suggested that maximum energy may be obtained when rated to achieve peak power at windspeeds of about 2 to 2.3 times the mean. For variable pitch machines, this optimum probably lies between 1.4 and 2 times the mean. The trade-off between energy capture and load factor and rating is illustrated for typical machines in figure 2.12.

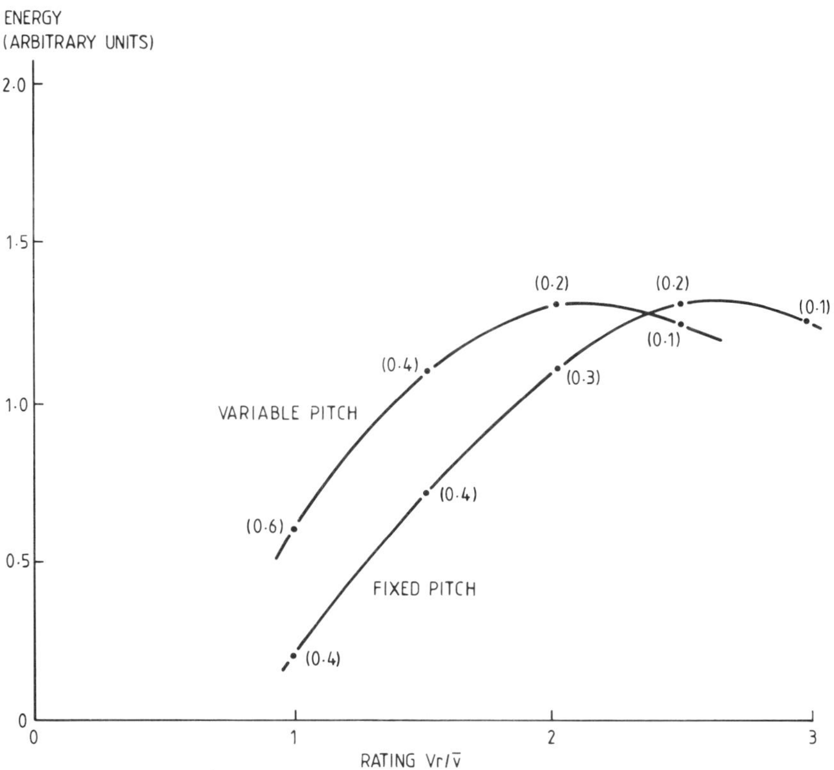

Figure 2.12 Energy output as a function of rating for a fixed and variable pitch machine. Approximate load factors are given in brackets next to each curve.

2.5 Materials for wind turbines

In determining the acceptability of various types of material for building wind turbines, as with all structures, it is essential to know as much as possible about the loads and stresses expected, the existence of effects such as resonances and instabilities and the environment to which the machine will be subjected. The general principles involved in analysing thse factors and the compatability of some materials will now be discussed. For a more detailed examination of the subject the reader is referred to references [14] and [15].

Most designs of large wind turbines are extensively analysed to calculate the static and dynamic loads which various components will experience. Although the stresses arising from the aerodynamic, gravitational and centrifugal loads on blades and gravitational loads on the tower are relatively small, many fatigue cycles can be accumulated in a short time. In a lifetime of 30 years, the blades of a large wind turbine will rotate more than 10^8 times. For most materials, data on this number of fatigue cycles (particularly in salt spray for offshore siting) is very sparse. In addition to calculating loads on the blades, towers and other components, machines need to be particularly carefully analysed to understand dynamic effects, and sophisticated computer codes have been developed to achieve this. When resonances are found to occur at critical frequencies, standard techniques involving changing the stiffness or mass of some part of the structure or the addition of damping can then be applied. The process is, however, an iterative one. The major components in which resonances are likely to occur are the blades, towers, cross arms on vertical axis machines and occasionally in struts or cables. Designs also require to be analysed for instabilities. These involve an interaction between degrees of freedom, reinforced by an external source of energy which builds the interaction up to very high values. A number of possible mechanisms which are reinforced by the wind can be identified in the blades, for example. Many of these are well known in helicopter technology. Rather than occurring at a specific rotational speed, as does a resonance, instabilities tend to occur at a particular windspeed. Instabilities in yaw, particularly for machines which yaw freely, have been the subject of considerable study.

It is not possible to review the choice of materials for all components of a wind turbine, but the blades and tower are particularly important. The choice of materials for towers is fairly narrow—chiefly being steel lattice, thin steel shaft or concrete. Stiffness in lattice towers can be varied considerably and this is a useful design variable in avoiding resonances. Particular attention is usually paid to weld quality and to the areas around the holes used to fasten lattice tower cross members. Aesthetic questions are also important in the choice of tower design and for large machines there is some evidence that thin towers are visually more acceptable than those of steel lattice construction.

Worthington in his contribution to [14] has reviewed the range of materials available for blades. This is summarised below.

(1) Aluminium and aluminium alloy. Problems with fatigue have been encountered in using these materials and it does not appear that they are suitable for long term use.

(2) Steel. This is widely used for blades and appears to perform well in dry conditions. Coatings must be used in marine environments. Steel blades tend to be heavy and have the disadvantage of giving rise to higher levels of television interference than composites.

(3) Titanium and titanium alloy. These have high specific strength, stiffness, good fatigue properties and good corrosion resistance. Titanium is a strategic material, however, and is expensive.

(4) Composites. These are gaining in favour because they have good specific strengths, fatigue properties and corrosion resistance. Stiffness can be varied with choice of fibre type and content. Although more expensive to fabricate than steel blades, costs are coming down and large blade production is now technically feasible.

(5) Wood. Wood and wood composites are among the most promising blade materials tested in the US programme. Moisture absorbtion leading to cracking may be a long term problem which would militate against their use offshore.

(6) Concrete. This has been considered because of its cheapness, but weight considerations alone will probably rule it out.

2.6 Advanced concepts

There are three routes by which radical improvements in performance of wind turbines might be achieved—variations on existing concepts, the use of devices to enhance the flow of air through the machine (augmentors) and unconventional approaches which use the wind in an entirely different way from a wind turbine.

Several methods of obtaining better performance by using variations on present designs have been proposed. One concept is to use a single blade counterbalanced by a weight. This has the dual advantage of allowing faster rotational speeds to be developed thus saving on gearing costs and, of course, saving the cost of the second blade. The disadvantages are that aerodynamic efficiency is reduced by a percent or so, the counterweight increases the drag, dynamic balancing is difficult and the chord of the single blade needs to be greater to achieve optimum solidity. There is some evidence that single-bladed machines may be noisier. Nonetheless, the idea is being actively pursued as part of the German wind energy programme (see below). An idea at the opposite extreme is to build multi-bladed machines, with thin blades rather like the spokes on a wheel. Mensforth [16] has advocated building a 200 m diameter 24-bladed machine generating up to 20 MW. Suggestions have also been made to improve the performance of vertical axis machines. These include setting two blades 90° apart (rather than 180°) or arranging for cyclical pitch control of blades. In the first case, the variations in torque which are obtained from conventional vertical axis designs would be evened out, because each blade would give maximum torque whilst the other was giving its minimum value. In the second, the blade angle of attack can be optimised throughout each rotation.

Most augmentation systems involve costly and large structures which, to be effective, must be yawed to face the wind. One concept is to produce discrete vortices of high power density by the interaction of the wind with a delta wing and to extract power from the vortices using a conventional rotor. In the author's view one of the most promising concepts is the tip vane augmentor which is being studied at the University of Delft in Holland and at Cranfield Institute of Technology in the UK. Tip vanes are small aerodynamic winglets mounted at the tip and at right angles to the main blade. In theory, circulation around this aerofoil will cause the streamlines to contract upstream of the rotor and expand downstream, thus augmenting the pressure difference across the rotor disc. The idea has produced good results in very simple wind tunnel tests on a rotating cylindrical rod, but it is not yet clear whether the use of lift producing blades will yield an appreciable effect. The extra drag of the tip vanes may counterbalance any gain in efficiency if this turns out to be small.

Several radical ideas for generating electricity from the wind have been suggested but none proven. These include the tornado generator, a machine for generating electricity from humid air, the electrofluid dynamic generator, the Madaras system using the so-called Magnus effect and wing flutter (oscillating) devices. These have been described in the reviews by Vas and South [17] and Ainslie [18].

2.7 Wakes and clusters

If electricity is to be generated from wind energy on a large scale, wind turbines will need to be built in clusters (arrays). Understanding how they interact with each other in various configurations and at various spacings is an interesting scientific and engineering problem. It has been reviewed extensively in an excellent paper by Milborrow [19].

There are three principal factors which must be considered. Firstly, the effect of wind turbines operating in the wake of those in front and the consequent reduction in output of the array. Secondly, the depletion of momentum from the boundary layer that a large array can produce (this means that for a considerable distance behind the array there will be a momentum deficit and the boundary layer windspeed–height equilibirium value will not be re-established). Thirdly, the effect of imparting turbulence to the air and thus changing the pattern of airflow over other machines.

These effects can be studied in a number of ways. Boundary layer theory can be used by treating the wind turbines as additional roughness elements increasing the drag on the boundary layer. Mathematical modelling using the assumed or measured wakes behind single machines can be used to calculate the development of subsequent interactions between the wakes. Wind tunnel tests can be performed using small model rotors (e.g. anemometers) to simulate a field of wind turbines, or direct measurements can be made on small clusters of machines in the field. Milborrow has summarised the capabilities, limitations and data requirements for several techniques in a table which is reproduced below (table 2.2).

Table 2.2 Various methods for modelling performance of wind turbine arrays.

Technique	Capabilities/ limitations	Data inputs and additional requirements
Mathematical modelling of wake interactions	Suitable for small arrays only (< 100 machines)	Wake data from field tests of large machines required
Detailed models of flow field	Suitable for detailed studies of interactions between few machines only	Accurate aerodynamic performance data from wind tunnel tests in arbitrary flows required
Wind tunnel model studies	Size of arrays limited only by tunnel size	Experimental data needed to confirm that simulations of wind turbines are accurate
Boundary layer methods, including time marching techniques	Most apply to large arrays only; not possible to cater for changes of array configuration nor define performance of individual machines	Boundary layer theory as verified by theoretical and experimental studies. Difficult to validate precisely; tests of an outdoor array needed

The results obtained from these methods appear to be in quite good agreement. For a large array (infinite) a spacing between machines of about fifteen diameters downwind and crosswind, appears to lead to an overall array efficiency of just over 80%. Efficiency is here defined as the ratio of the summed output of the cluster to the sum of the outputs of isolated machines for a particular incident windspeed. Closing the spacing either downwind or crosswind reduces the overall efficiency and to first order the product of the two distances must be kept constant to achieve the same efficiency. For large clusters with spacings of ten diameters, efficiency drops to around 65–70%. In the case of such large clusters, of course, the addition of further wind turbines gives a poor energy return but it is unlikely that clusters of such size (say 1000 machines or more) will be built for some time. For small clusters the reduction in efficiency is much less. Some of these conclusions are illustrated in figures 2.13 and 2.14.

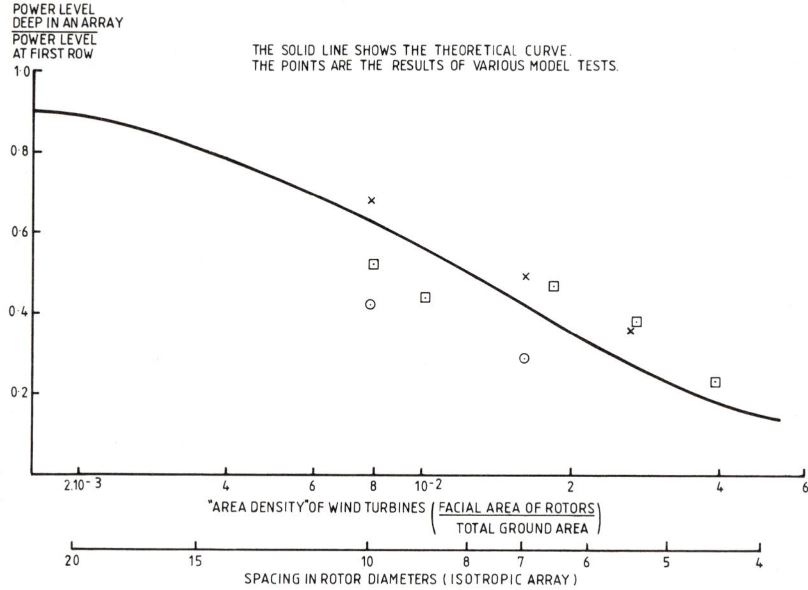

Figure 2.13 Power reduction in 'infinite' arrays of aerogenerators.

Incident wind direction and array patterns are also important, since for certain wind directions (along the diagonal in a square array for example) the spacing between machines will on average be greater, and higher array efficiency obtained. It is possible to design configurations of arrays in which machine interactions are minimised. Another possibility, where there is a strong prevailing wind direction, is to build thin rectangular patterns with the long face perpendicular to the prevailing wind. This gives higher efficiency and can minimise interconnection costs in some cases. In general rotor size, height, performance characteristics and incident turbulence levels are second-order factors affecting array efficiency.

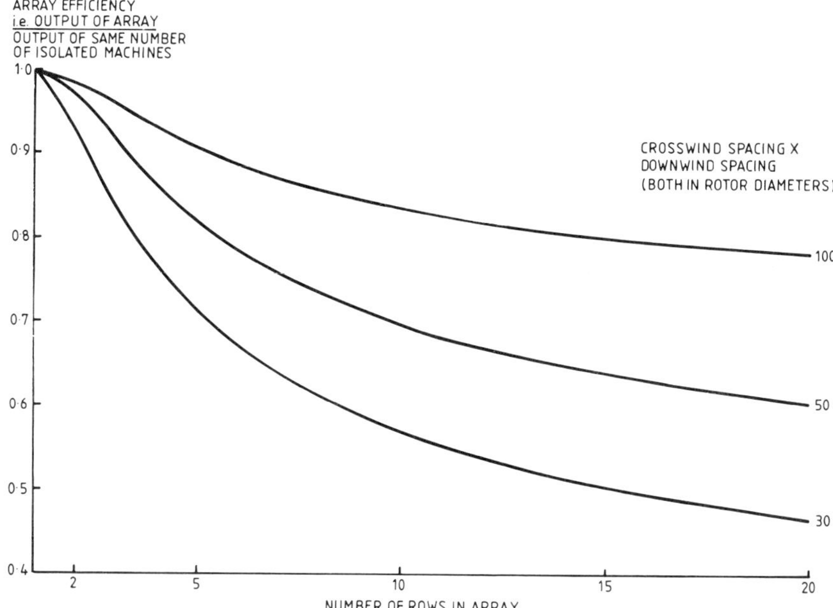

Figure 2.14 Efficiencies of 'finite' arrays of aerogenerators. (Published with permission of D J Milborrow.)

Important thought these reductions in efficiency undoubtedly are, in real arrays the development of high levels of turbulence may be of even greater significance, especially for close spacing. Preliminary evidence suggests that all wakes—whether they originate from horizontal axis or vertical axis machines—degenerate into similar wake patterns a few diameters downstream. Further information is required on turbulence levels, however, because these could be of considerable importance to wind turbine designers. Field measurements are very difficult to obtain because both spatial and temporal resolution needs to be achieved. Various techniques such as laser anemometry may be capable of being developed and used to measure such effects. A field measurement programme around small and intermediate machines in the UK has been in progress since 1979 and the techniques developed to study wakes on this scale will, it is hoped, be used shortly to study wakes from much larger machines. The UK collaborative wakes programme has been described by Swift-Hook [20].

In practice the positions of machines within clusters and their extent will probably be decided, at least on land, more by reference to amenity and environmental considerations than to those considerations mentioned above, and these will now be reviewed.

2.8 Environmental effects and siting

It is clear from the illustrations of modern wind turbine generators (figures 2.10 and 2.11) that they bear little relationship to the old fashioned pastoral windmills used for grinding corn and irrigating farmland. A modern wind turbine is a very large piece of rotating machinery built on an imposing tower to allow clearance of the blade tips from the ground and make use of any increase in windspeed with height. The problem of the visual acceptability of these machines is not one that we can be sure about yet because few modern machines have been constructed. Their visual impact is generally made greater by the fact that to gain the highest available wind, and thus benefit fully from the cube law relationship between wind velocity and power output, the best sites on land are usually on hilltops or other exposed sites. In the UK it seems unlikely that a very large number of wind turbines could be sited on hilltops because many of these are in areas of outstanding natural beauty. In other parts of the world, these sites may be more easily used. The UK Central Electricity Generating Board (CEGB) believes that it is possible that some flat lowland sites might be suitable for developing clusters of wind turbines and to this end has announced a step-by-step programme to examine the prospects for doing this (see § 2.12). Whatever sites are chosen onshore, however, wind turbines will have to prove their environmental acceptability with regard to a number of potentially sensitive issues which will be examined below.

(1) Rotor and tower safety. Making the pessimistic assumptions that a blade could fail while rotating at high speed and simultaneously be cleanly detached from the rotor hub with the optimum take-off angle, a naive projectile calculation for a large machine shows that it could be thrown a thousand metres or so and a fragment from the tip even further. This makes no allowances for air resistance which would be appreciable, and the probability of failure at both top speed and at 45° the optimum angle is very low. More detailed computations made in the USA by NASA for the 40 m diameter MOD 0A machines (see below) suggest that the maximum blade throw would be about 165 m. In the USA and Sweden an exclusion zone for habitation of about 200 m has been taken even for very large machines. Modern machines are being designed in accordance with codes and standards to withstand extreme wind loads and cyclic stresses so that the chance of failure is likely to be very low indeed [21]. In addition, many machines have monitoring systems to detect crack propagation in sensitive

areas. As a further safety measure, machines are automatically shut down when sensors detect excessive vibrations or dynamic imbalance. It is claimed that a combination of good design and early detection can reduce the likelihood of catastrophic blade failure to negligible proportions.

A problem may also arise from blade icing. The Clayton MOD 0A machine has, for example, lost availability due to icing. The operating philosophy in the USA seems to be to incorporate ice detectors on the rotor and close machines down when significant icing occurs.

(2) Television and other electromagnetic interference. Electromagnetic interference, particularly to television reception, seems to be very strongly site dependent. Of the large machines currently in operation only one or two have given rise to any problems and these can, in principle, be overcome albeit at some cost, by the installation of cable television or by improving the quality of the signal.

The phenomenon of interference occurs because the plane of the rotor disc acts like a mirror, reflecting signals from the transmitter. At certain orientations, the reflected signal interferes with the direct signal to the receiver. The strength of any interference is determined by the size of the rotor, blade material, the distance of the wind turbine from the receiver, signal frequency, the geometry of the receiver with respect to the wind turbine and transmitter and the strength of signal and quality of the receiving set and antenna. The worst conditions exist for high frequencies and weak signals with a large wind turbine with metallic blades on the line of sight between a poor quality receiver and the transmitter.

Experience in the USA [22] suggests that with large rotors, reception difficulties could, in principle be experienced up to a quarter of a mile away from the receiver for VHF but up to three miles away for UHF. These distances are, however, likely to be much reduced by the use of fibreglass, wood or other non-conducting blade materials. It has been estimated that fibreglass blades would reflect only about 40% as much of the transmitted signal as metallic blades. Addition of a carbon layer and the use of a honeycomb core structure could lead to a reduction to about 20% of that of the metallic case [23]. It is not clear to what extent these reductions in the level of interference would be offset by the need to incorporate metallic material in the blades to provide lightning protection.

There is little published information on the effect of wind turbines on microwave transmission but it does not appear to be great when the wind turbine is off the direct line of sight.

(3) Noise. Audible noise has frequently been cited as one environmentally unacceptable consequence of wind energy. Modern designs, however, can be rather quiet. Noise is related to a number of factors including whether the blades are mounted upwind or downwind of the tower, speed of rotation and the clearance between the blade tips and the tower.

Noise at low frequencies (infrasound) may be more important. There have been complaints from local residents about vibrations causing windows to rattle in the locality of the MOD 1 60 m diameter machine at Boone, North Carolina. This appears to have been overcome by reducing the rotational speed of the machine although with a consequent loss of power. It is believed to have arisen because vortices shed from the lattice tower interacted with the downwind blades for some wind directions and at certain rotational speeds. It should not be difficult to remove the effect in later designs.

(4) Hazard to birds. A number of studies have now been carried out into mortality of birds [24] and it it appears that, except in particularly adverse conditions, birds avoid wind turbines and there is no evidence of frequent impacts.

The conclusion of most environmental studies into wind turbines seems to be that if properly designed and carefully sited, they should present little environmental disturbance. The major unknown is the public acceptability of the visual effects of wind turbines, particularly of clusters.

2.9 Offshore siting of wind turbines

If there are severe constraints on siting wind turbines on land in areas like northern Europe where there is a high population density, another option is to construct arrays in shallow offshore waters. This is now being taken increasingly seriously and has thus been given a separate section in this book. The technology would be more costly than for similar constructions on land, but the windspeed is generally higher and siting offshore might be expected to minimise amenity and environmental objections. The resource for the UK alone, even making rough allowance for fishing areas and shipping lanes [25], appears to be vast (see §2.10). Offshore siting has recently been studied in detail by a UK consortium (Taylor Woodrow, CEGB, Electrical Research Association (ERA)) for the Department of Energy, and by government funded teams in Sweden, Holland and the USA. Some of the main results of the UK study will now be presented.

(1) Siting. Whilst no preferred sites for future development were chosen, three areas were selected for more detailed study in order to give a range of seabed and other engineering conditions. These were Burnham Flats near the Wash, Morecambe Bay and Carmarthen Bay. Some of the shallower offshore areas where machines could, in principle, be sited are shown in figure 2.15.

(2) Meteorology. The mean speed, speed probability distribution, variation with height and extreme speeds were assessed for the above locations. Because of the smoothness of the sea, the windspeed is considerably higher than at the coast and is probably only exceeded by the speed on exposed hilltops. For the three sites, annual mean windspeeds of

7 to 7.5 m/s at a height of 10 m were predicted although there are only limited data available to confirm this. The smoothness, however, tends to reduce the rate of increase of windspeed with height ($\alpha \sim 0.1$), so hub height windspeeds were estimated to be about 9 m/s.

(3) Cluster configuration. Several possible configurations were examined in detail. Clusters contained either 196 or 64 machines arranged in square or rectangular patterns with efficiencies ranging from 60–90%. Typical values of about 70–85% were assumed. In some locations, rectangular arrays were found to have a distinct advantage over square arrays.

(4) Engineering geology. Desk studies were carried out of seabed geology at the sites in order to allow structure types to be chosen.

Figure 2.15 An isovent map of the UK showing some offshore areas of interest for wind power. Estimates of annual average mean windspeed (m/s) are given at 10 m above ground in open country, based on data for 1965–9.

(5) Support structures. The most cost effective and durable structures for water of less than 20 m in depth were found to be piled steel, low-level rafts and gravity spread footing. The cost of providing the support structure for the machines was found to be about eight times greater than onshore.

(6) Materials and marinisation. Various modifications to the wind turbines were found to be required in order to ensure reliability. Inspection and maintenance costs were assessed, and the costs of an onshore support base was included in these estimates. Concrete structures were preferred as they give the highest durability.

(7) Electrical interconnection and control. Methods of starting and control of the cluster were analysed and an interconnection scheme, both between machines and to the shore, was costed. For 196 machines, the costs (all at 1979 prices) were estimated to be about £180 per kilowatt.

(8) Environmental effects. Several possible constraints to siting were identified, commercial fishing probably being the most important. Figure 2.16 shows an artist's impression of the view from a boat inside a cluster of 80 m diameter wind turbines separated by ten diameters upwind and across wind.

(9) Availability. The loss of single machines in a large cluster does not have much impact on availability but loss of the submarine link to the shore or at critical interconnections in the array clearly does. It was estimated that overall availability approaching 90% might be achieved.

(10) Costs and performance. It was assumed as the basis of the study that a design of UK multi-megawatt machine which has subsequently been superceded was to be sited offshore. The baseline machine was thus only 60 m in diameter and was not optimised for offshore conditions. Sensitivity studies to various parameters including machine diameter, assumed mean windspeed, rating, water depth, availability and rate of return were carried out and the results are shown in figure 2.17. For production of 196 machines of 80 m diameter in a water depth of 15 m, 7.5 km from shore, the cost of generation was predicted to be about 5 pence per kilowatt hour (p/kWh) assuming a required rate of return of 5% and a machine lifetime of 20 years. For larger machines and more favourable wind conditions, it was thought that costs might reduce to 3–4 p/kWh. The study was, however, conservative, using non-optimised components and spreading costs over a relatively small number of machines. It did not consider technology which might be available in future, such as innovative support structures or compliant machines. The results of this study are described in greater detail in references [26] and [27].

2.9.1 Other studies

Some of the other national studies have been more optimistic. The Swedish work [26] suggested that costs for a large programme (1 TW p.a. over 10

Figure 2.16 Artist's impression of an offshore array of horizontal wind turbines.

years) of 5 MW machines in water of 20 m depth and in similar wind regimes to those considered in the UK studies, could be 0.17 Swedish kronor (about 2 p) per kilowatt hour at 1978 prices. This is now regarded as optimistic. The USA study concentrated on deeper water and assumed generally lower mean windspeeds. For 500 MW installations comprising fifty 10 MW, 92 m diameter machines costs in the range 7–17 cents per kilowatt hour (c/kWh) were estimated using US accounting methods. The Dutch study costed only the construction and installation of support structures. A detailed intercomparison and a critical review of the various studies has been given by Dixon [28] who has highlighted some of the widely different assumptions which have led, perhaps fortuitously, to broadly similar cost estimates.

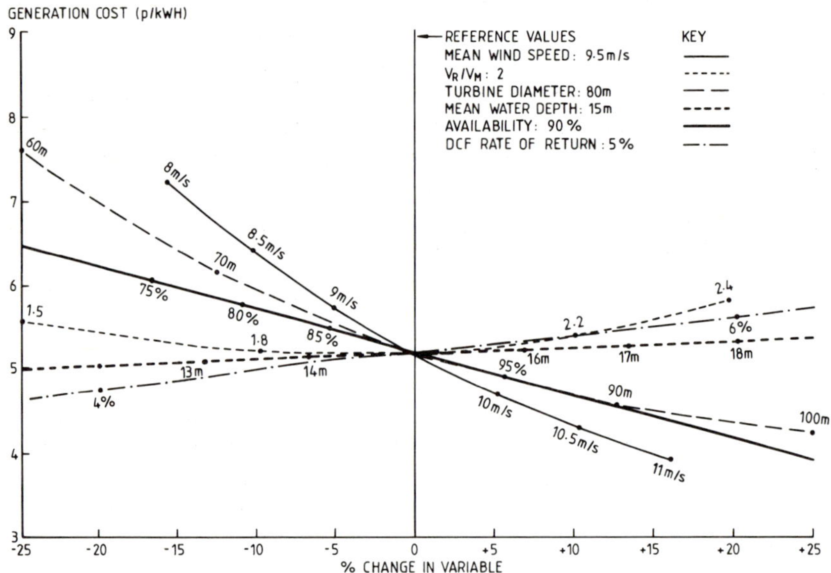

Figure 2.17 Graph of generation cost (1979 prices) variation for offshore wind energy with various parameters for a 49×4 machine array with $10_D \times 10_D$ spacing. Mean wind speed is measured at hub height.

2.9.2 *Further work*

Offshore wind power appears to be technically feasible and is an interesting longer term option particularly if onshore siting proves difficult. More studies need to be carried out to see if costs can be reduced and these are now proceeding in several countries including the UK. There are three main areas in which these cost reductions might be achieved. Innovative structures such as sand islands, the use of advanced compliant designs (if these are compatible with the wave forces and frequencies to which they will be subjected) and building much larger diameter machines (100 m and greater) to utilise to the full the investment in the offshore support structure.

It is hoped that a demonstration offshore wind turbine will be constructed in the next few years, perhaps as part of an international collaborative programme. The CEGB has expressed interest in hosting such a machine.

2.10 Resource

Relatively large variations in windspeed occur over quite small distances because of the effects of terrain. Unlike the cases of solar, OTEC or wave energy, it is difficult to make broad generalisations about areas of high windspeed. In the UK, for example, windspeeds vary from annual averages (measured 10 m above ground) of up to 10–12 m/s on hilltops in Scotland, to about 7 m/s offshore, 5–6 m/s near the coast and down to 3–4 m/s in some inland parts of England. Generally speaking, sites with mean windspeeds in excess of 6 m/s annual average are considered to be excellent and those with annual averages in excess of 5 m/s or so are considered good. One of the difficulties in exploiting wind energy is that it is very important to obtain accurate wind data from specific sites before building commercial machines. A general idea of the windier regions of the world is shown in figure 2.18.

The world-wide size of the resource which might be exploited commercially depends upon several other factors. These include the ability to integrate the fluctuating wind output into existing utility grids—especially small networks on islands, the proximity of the windy area to a source of demand and in particular the development of advanced machines to exploit lower windspeed sites. The soft or compliant designs being built in the USA and Sweden are designed for sites with mean windspeeds of just over 6 m/s and may be economic in some cases on sites with mean windspeeds of 5 m/s or less. The degree to which the resource can be exploited thus depends critically on progress in developing these machines.

An isovent map of the UK is shown in figure 2.15. Even for a country for which wind measurements have been taken for many years at a number of locations, this must be treated as conjectural as regards detail. It takes no account of terrain effects or different variations with height. In general, three types of area can be identified which could offer a significant resource in the UK.

(1) Hilltops. Studies carried out by the ERA have been reported in reference [10] and suggest that there might be some 1500 hilltops with very high windspeeds. These would provide sites for over 3000 machines in the UK and generate up to 30 TWh p.a. This, however, assumes that consent

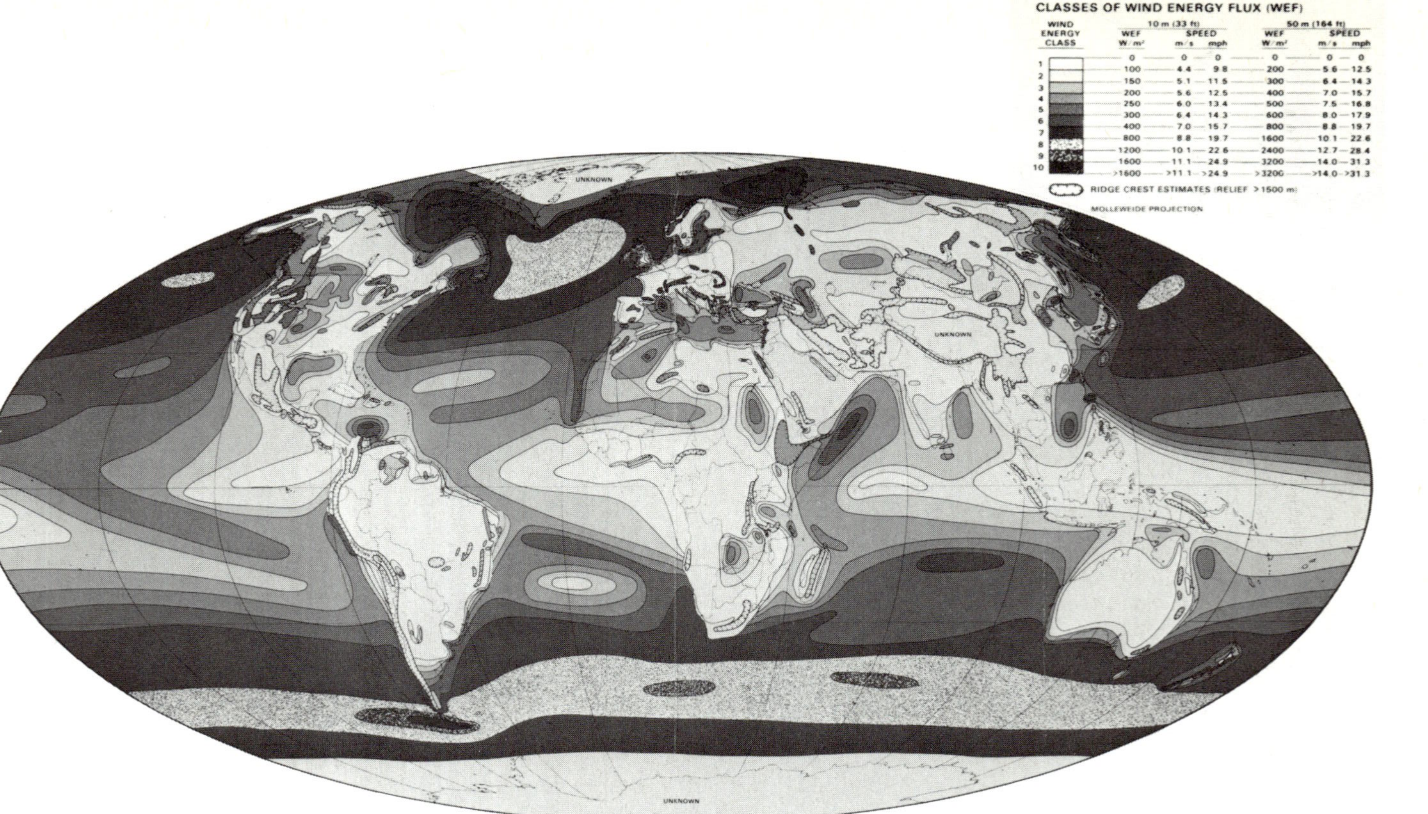

Figure 2.18 World-wide wind energy resource distribution estimates. The map is a preliminary estimate of the annual mean wind energy available at typical well-exposed locations throughout the world. The average energy in the wind flowing in the layer near the ground is expressed as a wind energy class. The greater the average wind energy, the higher the wind energy class (see key).

The wind energy class is defined in relation to the mean wind energy flux (WEF) at 50 m above ground level. The WEF is the rate of flow of wind energy through a unit vertical cross sectional area perpendicular to the wind direction. At 10 m the WEF estimate represents large areas that are relatively free from obstructions. Local terrain features can cause the mean wind energy to vary considerably over short distances, especially in coastal, hilly and mountainous areas. (Courtesy World Meteorological Office and Pacific Northwest could be obtained to build wind turbines on these sites. Since most of them are in areas of outstanding natural beauty this seems rather unlikely. It is for this reason that the electricity utilities in the UK have not actively pursued the possibility of siting wind turbines on hilltops except on islands such as Orkney.

(2) Coastal areas. Whilst siting large numbers of windmills on the coast may not be acceptable for aesthetic reasons, figure 2.15 suggests that there are quite large lowland areas of the UK within 50 km or so of the coast (particularly in the east of the country) where average windspeeds of 5–6 m/s might be found. Until recently, such windspeeds would not have provided a resource which could be exploited economically, but if the advanced designs currently being developed can be proven and reach their target costs (see below), there is some prospect that wind energy could provide electricity from such areas on a fairly large scale. The size of the actual resource depends critically on whether clusters of large wind turbines will prove to be acceptable to the public and it is in an effort to begin to evaluate public response that the CEGB intends to work towards building up a cluster of about 10 proven machines by the end of the 1980s.

(3) Offshore. The UK offshore resource is likely to be very large indeed. One estimate [25] has suggested, on conservative assumptions with regard to windspeeds and cluster efficiencies, that the gross resource in areas of less than 30 m water depth and more than 5 km from the coast would in principle be capable of generating up to 240 TWh p.a. (cf. the current UK demand of about 220 TWh p.a.). Even making rough allowance for fishing, shipping and areas where the seabed might be unsuitable, the resource might still be greater than 100 TWh p.a. It may be possible to exploit greater depths without greatly increasing costs, and studies are now underway to calculate the resource at various water depths and ranges of cost. By going to water depths of up to 50 m, very recent studies have shown that the resource off the UK coast which falls within areas regarded as probably capable of development is comparable to present annual electricity generation although the possible reduction in this figure due to fishing is very difficult to assess. Figure 2.15 also shows the areas of shallow water which have been considered in order to arrive at these estimates.

A more detailed review of the UK resource has recently been published by Lindley *et al* [7].

2.11 Economics

Reference to the likely cost of electricity from offshore sited wind turbines has already been made. At very high windspeed sites (greater than 8 m/s or so) such as good hilltops and some islands, wind energy is probably already economic (particularly if electricity is being produced from diesel generators) even using intermediate size or large wind turbines of rigid, 'conventional' design. If the advanced machines prove capable of operating over a long life in very high windspeed locations, the economics of wind energy from these sites will be much improved.

Table 2.3 Breakdown of the costs of a typical modern large wind turbine such as MOD 2.

	% Cost
Blades/hub/pitch charge etc	25
Gearbox/generator/shaft etc	20
Tower	10
Nacelle and yaw mechanism	10
Foundations/site work	10
Assembly etc	10
Other (e.g. operation and maintenance, spares, transport etc)	15

Advanced machines such as the Boeing MOD 2 (see below) cost about $5–10M at present, and in series production it has been estimated that the cost will drop to about $2–3M (at 1980 prices) per machine [29]. An approximate breakdown of component costs for such a machine is given in table 2.3. For a site with a mean windspeed of just over 6 m/s (for which the MOD 2 has been designed), advanced wind turbines such as the MOD 2 might be expected to generate 9–10 GWh p.a. at a load factor of about 40%. In the USA, the cost of energy is calculated by multiplying the capital cost by an annual charge rate which is currently set at 18% (this allows for depreciation, taxes, return to investors, insurance and assumes a lifetime of 20 years) and dividing by the annual energy produced. On this basis, and allowing an extra 2% p.a. for operation and maintenance, in the USA the cost of the energy produced is (at 1981 prices) 5–6 c/kWh. Estimated cost trends (at 1977 prices) are shown in figure 2.19. Musgrove [30] has recently converted the 1981 figures to the corresponding cost of energy under UK conditions. He estimates that for a 10 m high site with a mean windspeed of 5.4 m/s, MOD 2 would give an output of 7150 MWh. Assuming an exchange rate of $2 = £1, using the UK requirement for utilities to achieve a 5% real rate of return (equivalent to an annual charge rate of 8%) and adding 2% for operation and maintenance to give a charge rate of 10%, he shows that the cost of electricity from this machine would be 1.82 p/KWh. Not all factors have been taken into account in this analysis but even so if the above target costs are achieved, the economics of large wind turbines sited in intermediate windspeed sites thus look attractive compared with some other forms of generation.

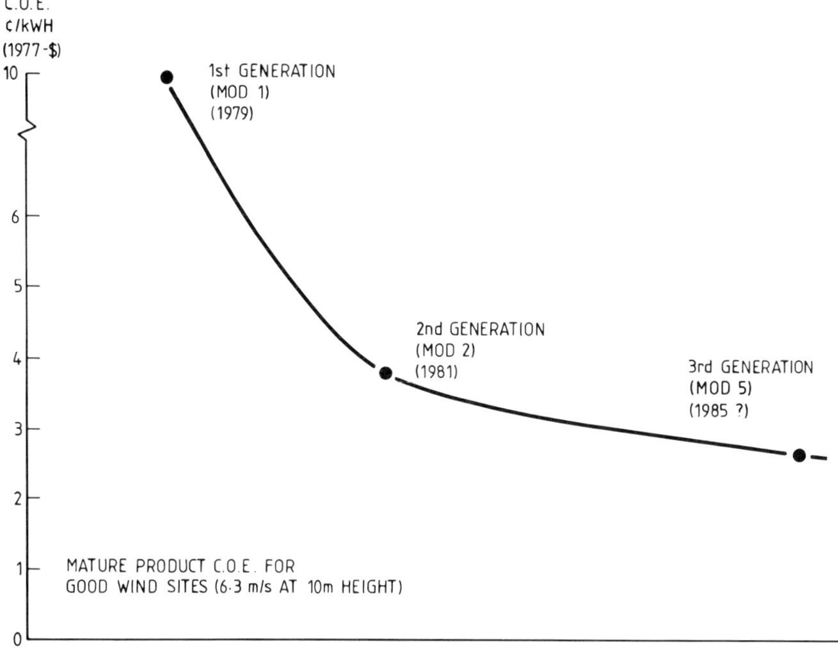

Figure 2.19 Cost trends of large wind systems.

2.12 World R, D and D in wind energy

In this section the progress towards developing reliable, economic, wind turbines within the various national wind power programmes will be described. By far the largest effort has been mounted in the United States, but recently Sweden, Germany, Denmark and the UK have started to invest substantial resources towards the development of wind turbines. Details of the various large machines which have been constructed or are planned are given in table 2.4.

2.12.1 United States

The US wind energy programme has until recently been growing very rapidly. In 1979, federal funding for R, D and D was $60M and in 1980 $80M was set aside. The programme has been structured around several laboratories: NASA Lewis Research Center in Cleveland, Ohio is responsible for megawatt and intermediate size wind energy systems, the Solar Energy Research Institute (SERI) administers the innovative systems programme, the Rocky Flats test centre at Golden Colorado monitors

Table 2.4 A survey of large horizontal axis wind turbines (about 40 m diameter or more).

Contractor	Bendix (US)	Westinghouse (US)	General Electric (US)	Boeing (US)	Hamilton Standard (US)	General Electric (US)	Boeing (US)	Elsam (Danish)	Elsam (Danish)
Wind turbine characteristics	Bendix design	MOD 0/0A	MOD 1	MOD 2	UTC WTS 4	MOD 5A	MOD 5B	NIBE 'B'	NIBE 'A'
Number of blades	3	2	2	2	2	2	2	3	3
Rotor diameter (m)	80	38	60	91	78	122	127	40	40
Rated windspeed (m/s)	15.8	7.7 (10 m) (MOD 0A)	11.5 (10 m)	8.9 (10 m)	13			13 (hub)	13 (hub)
Hub type	Rigid	Rigid	Rigid	Teetered	Teetered	Teetered	Teetered	Rigid	Rigid
Orientation	Upwind	Downwind	Downwind	Upwind	Downwind	Upwind	Upwind	Upwind	Upwind
Machine rating (MW)	4.5	0.1/0.2	2.0	2.5	4.0	7.2	7.2	0.6	0.6
Rotor speed control (rpm)	Fixed (25)	Fixed	Fixed (23)	Fixed (17.5)	Fixed (30)	Fixed two-speed	Variable	Fixed (33)	Fixed (33)
Electrical generation equipment	Induction	Synchronous	Synchronous	Synchronous	Synchronous	PAM Induction	Advanced	Induction	Induction
Pitch control technique	Full span	Full span	Full span	Partial span	Full span	Partial span	Partial span	Full span	Partial span
Tower type	Cylindrical Soft	Truss Rigid	Truss Rigid	Cylindrical Soft	Cylindrical Soft	Cylindrical Soft	Cylindrical Soft	Cylindrical Rigid	Cylindrical Rigid
Tower material	Steel	Steel	Steel	Steel	Steel	Steel	Steel	Concrete	Concrete
Blade material	Laminated wood or fibreglass	Laminated woods, Al or fibreglass	Steel, fibreglass?	Steel	Fibreglass	Laminated wood	Steel–laminated wood tips	Fibreglass and steel	Fibreglass and steel
Status (No. complete)	Design	Operational since 1975 (5)	1979 (1)	1981 (5)	1982 (1)	Design	Design	1979 (1)	1979 (1)

Table 2.4 Continued.

Contractor	Swedyard/ Hamilton Standard (Swedish)	VOITM (German)	MAN (German)	MBB† (German)	Private (Danish)	Bendix/ Schachle‡	WEG (UK)	KMW– ERNO (Sweden)
Wind turbine characteristics	WTS 3	—	Growian I	Growian II	Tvind	SWT 3	WEG 3 MW	WTS 2
Number of blades	2	2	2	1	3	3	2	2
Rotor diameter (m)	78	52	100.4	145	54	51	60	75
Rated windspeed (m/s)	13 (hub)	—	12.2 (hub)	11.3 (hub)	12 (hub)		22 (hub)	13 (hub)
Hub type	Teetered	Teetered	Teetered	Teetered	Rigid	Rigid	Teetered	Rigid
Orientation	Downwind	Downwind	Downwind	Downwind	Downwind	Upwind	Upwind	Upwind
Machine rating (MW)	3.0	0.26	3.0	5.0	2.0	3.0	3.0	2.0
Rotor speed control (rpm)	Fixed (25)	Fixed (37)	Variable (18.5 ± 15%)	Variable	Variable	Variable	Fixed	Fixed (25)
Elecrical generation equipment	Synchronous	Synchronous	Advanced	—	Asynchronous AC	Synchronous	Synchronous	Induction
Pitch control technique	Full span	Full span	Full span	Full span	Full span	Full span	Partial span	Full span
Tower type	Cylindrical Soft	Soft	Cylindrical Soft	Cylindrical Rigid	Cylindrical Rigid	Rigid Truss	Cylindrical Rigid	Cylindrical Rigid
Tower material	Steel	Steel Guyed Mast	Steel Guyed	Concrete with cables	Concrete	Steel	Concrete	Concrete
Blade material	Fibreglass	Carbon fibre	Steel spar —fibreglass shell	Fibreglass	Fibreglass	Wood and fabric	Steel	Steel box spar, GRP leading and trailing edges
Status (No. complete)	1982 (1)	1981 (1)	1982 (1)	Design	1978 (1)	1980 (1)	Design	1982 (1)

† A 1/3-scale model of this machine (Monopteros) has recently been completed.
‡ This machine has recently been extensively modified from the original Schachle design by the Bendix Corp.

commercial small machines, Battelle Pacific Northwest Laboratory (PNL) is responsible for research into wind characteristics and siting and the Sandia Laboratories in Albuquerque, New Mexico are developing and testing Darrieus machines.

At the time of writing, (mid-1982), the new administration in the USA appears to have substantially cut the annual budget for wind energy. Perhaps more significantly, it does not seem likely that a major bill to promote the commercialisation of wind energy will be voted funds. The Carter administration, towards the end of its period in office, passed a bill (The Wind Energy Systems Act) which would have led to nearly $1000M being spent in 'commercialising' large and small wind turbines. One object of this bill was to provide a federal subsidy to companies building large wind turbines in order to reduce costs to an economic level. As mass production, resulting from the orders stimulated by the policy, got underway, the subsidy would have been progressively reduced. The target was to achieve a goal of 700 MW of installed large wind turbines by 1987. Other legislation passed by the Carter administration offers tax incentives for the development and use of renewable sources. The legislation also requires utilities to accept electricity supplies from these sources on to their grids (subject to certain conditions) and to pay a 'fair' price for the electricity. This legislation is still having a major impact. These laws are leading to a fairly rapid commitment to wind energy by some utilities in the USA because various entrepreneurial groups are proposing to put up risk capital to construct farms of wind turbines and sell the electricity generated to utilities at a rate tied to the cost of fuel oil. One such company, Windfarms Inc. recently attempted to negotiate a contract to build 80 MW of wind turbines, the electricity from which would have been sold to a utility in Hawaii. This has apparently not been successful because the initial capital could not be raised, but is typical of the type of project now being considered. Other major projects involving utilities which are largely dependent on oil-fired generation include negotiations for 200 MW of wind turbines for Southern California Edison (a letter of intent has been signed for the first 20 MW, and a longer term goal of 350 MW has been set) and 350 MW to be built by 1990 for Pacific Gas and Electric. Thus despite the cut-backs in government subsidies, the prospects for wind energy utilisation on a large scale in the USA still look promising although there may be an adverse affect on timescales.

Five federally funded designs of large wind turbine have now been built in the USA [31] which will be briefly described below.

MOD 0. This was the first of the large machines built by Westinghouse for the Department of Energy programme. The 125 foot (38 m) diameter, 100 kW, variable pitch, horizontal axis machine was built at Plumbrook, Ohio in 1975. It has been used as a test bed for machine development and has proved an invaluable tool in the design of later machines. In particular, compliant design options and blade materials have been extensively studied.

MOD 0A programme. Four MOD 0A wind turbines have been developed by Westinghouse from the original MOD 0 design, primarily to gain experience with machines connected to utility grids. Each 125 foot (38 m) diameter machine generates a peak power of 200 kW. In 1979, power generated by MOD 0A machines was fed into utility grids from sites at Clayton, New Mexico, Culebra, Puerto Rico and Block Island, Rhode Island. A fourth machine which has achieved 80% of its theoretical energy output has been operating in Hawaii in the last year. The operation of the machines has been quite successful (although some problems with cracking in the aluminium blades at Clayton have been experienced). It is claimed that public acceptance has been good and the environmental effects minimal.

MOD 1. This was the first of the megawatt size machines planned for the US programme. The 200 foot (61 m) diameter, 2 MW (at 25 mph or 11 m/s), downwind variable pitch, horizontal axis machine has now been synchronised with the local grid at Boone in North Carolina. It has operated for short periods close to specification but control and mechanical problems have to some extent limited its operation. As discussed in § 2.8, one of the most serious problems encountered with the machine has been the generation of low frequency sound (infrasound) which has brought complaints from a small number of people living nearby. This has been diagnosed as being due to interference between the steel lattice tower and the downwind blades amplified by local topography. The problem has been overcome by reducing the rotational speed with a consequent loss of power. It is claimed that some television interference has also been generated by MOD 1.

There appears to be a feeling that the machine will not contribute as much to the US programme as had previously been anticipated, since recent MOD series designs are significantly different. The machine has been more costly than expected and produces electricity at about 20 c/ kWh. It is believed that if a production run of 100 such machines were ordered, the hundredth unit would supply electricity at about 12 c/kWh. The machine has been placed at a poor site from a meteorological point of view but has provided useful data on system integration and visual acceptability. It is not known whether MOD 1 will continue to be operated, following a mechanical fault in 1981 and the US budget cuts.

MOD 2. Three 300 foot (91 m) diameter machines generating 2.5 MW each in a 25 mph (11 m/s) wind were completed by Boeing at Goldendale in the Columbia River Gorge, Washington State in late 1980 and 1981 [32]. The machines are of the same design with two upwind steel blades on a teetered

Figure 2.20 The Boeing MOD 2 wind turbine. One of three at Goldendale in Washington State generating a total of 7.5 MW. (Photograph courtesy Boeing Engineering and Construction Ltd.)

hub and a 'soft' tower. They employ tip control rather than full blade pitch control. These advanced machines should produce electricity for the Bonneville Power Administration at a cost of about 10 c/kWh. A fourth machine is being built in Wyoming and a fifth near San Francisco for

Pacific Gas and Electric Inc. As discussed above, for series produced versions of this design, electricity costs should be as low as 3–4 c/kWh (at 1977 prices and using US accounting criteria). If recent control problems can be overcome, and if Boeing decide to proceed with series production in the absence of federal financial incentives, MOD 2 may itself be commercially attractive to utilities. Figure 2.20 shows one of the machines.

MOD 5 and MOD 6. General Electric and Boeing were awarded contracts to design an advanced wind turbine which might achieve first machine electricity costs of 6–8 c/kWh and 3 c/kWh in production (again at 1977 prices). Both designs (5A and 5B) envisage machines of about 125 m in diameter generating 5–7 MW and employ advanced concepts such as a soft tower, teetered hub, quill shaft, epicyclic gearbox and the ability to operate at more than one speed. Due to budget cuts it now seems unlikely that full federal funding will be available for the machines, and their future is uncertain.

MOD 6 was to be designed by Rockwell and was an intermediate size machine for remote sites. A series target cost of 6 c/kWh was set for this advanced design but it has now been cancelled as a result of the US budget cuts. In addition to the federally funded programme, several other large machines have been built or are under construction in the USA. In particular, Hamilton Standard have invested substantial sums to develop an advanced design of machine, one of which (WTS 4, 4 MW) has been built at Medicine Bow in Wyoming and the other (WTS 3, 3 MW) in Sweden. Swedyards, the Swedish state shipyard, are collaborating with Hamilton Standard [33]. The machines will be built on soft towers with two downwind glass-reinforced plastic blades on a teetered hub, employing free yaw and full blade pitch control. A cut-away of the nacelle layout of the Swedish machine is shown in figure 2.21.

Southern California Edision have purchased a 3 MW, 49 m diameter, three-bladed, horizontal axis machine from Bendix which uses a hydraulic power take-off. This prototype has recently been extensively redesigned. Bendix have now designed an advanced multi-megawatt machine. Southern California Edison also purchased a 500 kW Darrieus machine built by AlCoA but this recently failed catastrophically due to a control fault.

2.12.2 Major European programmes

Sweden currently has one of the largest wind energy programmes in Europe [34]. An 18 m diameter, 63 kW machine was commissioned in 1977 and this has been used as a test bed to assess design and performance data and to test reliability and operating costs. Two very large wind turbine generators have now been ordered, each costing about £5M in total. Swedyards and Hamilton Standard are designing and building the 3 MW machine described above (WTS 3) and this is expected to be operational in

late 1982. KMW and ERNO are constructing a further machine of about
the same size [34]. This will also be a two-bladed, variable pitch design but
will be upwind and mounted on a rigid tower, driving an induction
generator. It is due to operate in early 1983.

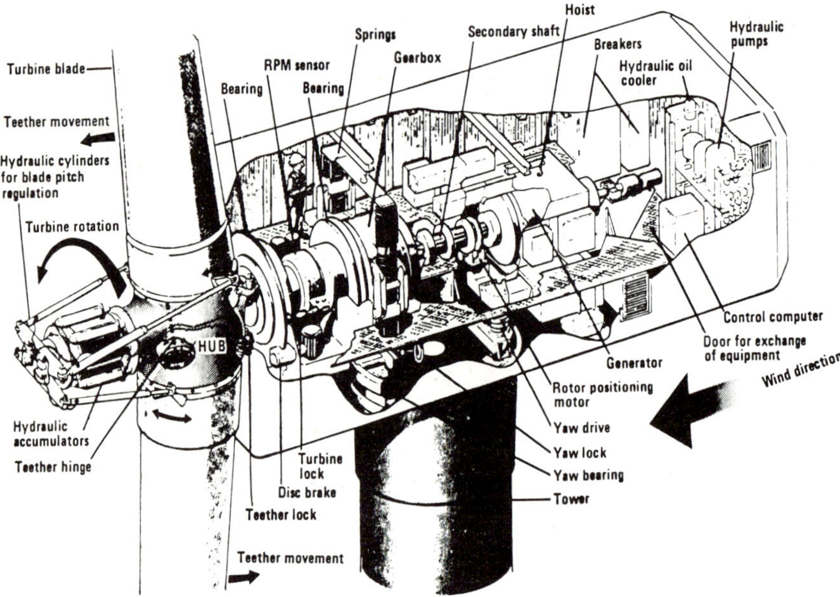

Figure 2.21 The layout of the nacelle in the Hamilton Standard/Swedyards 3 MW
wind turbine. (Courtesy Swedyards Corporation.)

The other large European national programme—which until recently
has enjoyed rapid growth—is that of West Germany. As with Sweden,
much work is being carried out on wind measurement and integration into
grids but the programme is dominated by the two large machines planned
as part of the Growian (Grosse Wind Anlage) programme. Growian I is an
advanced design, downwind, teetered hub, variable pitch machine with
two 100 m diameter blades constructed with steel spars and glass fibre
composite. Its peak output is designed to be 3 MW. It is presently under
construction. Growian II is one of the most ambitious projects in the
world. The design is still not complete but it is likely to have one blade, be
some 145 m in diameter and generate 5 MW. A third-scale version of this
machine, named *Monopteros* is now operating. A 52 m diameter prototype
machine (the VOITM machine) has recently been completed for studies of
automatic operation on a grid. Its peak output of 265 kW is rather low for
the diameter because it is intended for low windspeed locations. It has two
slender carbon fibre blades.

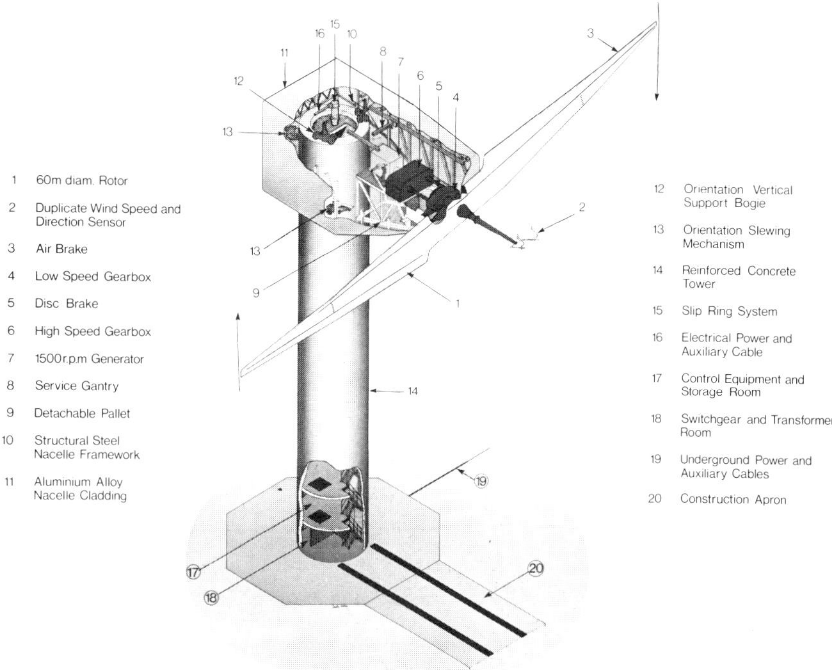

Figure 2.22 60 m diameter horizontal axis wind turbine generator—Taywood 3.7 MW machine (1980 design). (Picture courtesy Taylor Woodrow Ltd.) The most recent design of this machine is illustrated on p. 59.

The Dutch and Danish windmill programmes have, as described in §2.1, been of great historical significance. They are both making significant contributions to modern technology. Denmark has a fairly sizeable wind power R, D and D programme. Refurbishment of the Gedser machine has been followed by the completion in 1980 of two machines on a common site at Nibe Bredning. Both are 40 m in diameter and give a peak output of 630 kW. Construction only 220 m apart will give useful information on machine interactions and an intercomparison of machine performances. Both machines are rigid, upwind, three-bladed designs employing variable pitch control. Also in Denmark is the remarkable Tvind windmill, built by a cooperative effort at a college. This 54 m diameter, 2 MW machine has had a number of teething troubles but shows what can be achieved in the field of windmill design by a team of determined amateurs. It has now been operational for more than three years, although it has not apparently operated at full output.

Holland has a thriving programme of generic studies, particularly of interactions between machines and offshore siting. A 25 m diameter,

200 kW horizontal axis machine employing many design options has been completed at Petten as a test bed for future MW size machines. Proposals to build a cluster of 1 MW machines are being examined. Further medium- and large-size wind turbines are planned, including a project for 40 machines each of a few hundred kilowatts.

In the UK, design work for a 60 m diameter multi-megawatt wind turbine has been carried out by Taylor Woodrow, British Aerospace, GEC and others since 1977. A decision has now been taken by the UK Department of Energy to construct a 3 MW version of the design [35] at Burgar Hill, Orkney, which has an exceptionally high annual mean windspeed (10–12 m/s). Figure 2.22 shows a cut-away illustrating the main features of this machine before it was recently redesigned to include partial span pitch control and a teetering hub and figure 2.23 shows an artists impression of this latest design. A 20 m machine similar to the large one will also be built [36] to prove the design, and for studies of integrating wind energy into small grids which employ diesel generation. This in- termediate size machine should be completed by late 1982 and the large machine by 1985. There appears to be potential for some tens of megawatts of wind power on Scottish island sites and the 3 MW machine may be a prototype for further similar machines.

A 25 m diameter (about 130 kW) vertical axis wind turbine (of the Musgrove arrowhead type) is being designed by a consortium led by Sir Robert McAlpine [37]. It is hoped that this will be constructed in Wales in 1983, and if successful may be the prototype for a 100 m diameter machine of similar design. A model of the 25 m machine is shown in figure 2.11 and the method of controlling output is shown in figure 2.24.

The prospects for the large-scale generation of electricity from wind turbines in the UK has been given a further stimulus by the announcement in 1980 by the CEGB that they intend to purchase a proven multi- megawatt advanced machine for siting on a low land site in England in about 1985 [38]. If progress continues to be encouraging, and if machines at such locations prove to be environmentally acceptable, the CEGB intends to build up a cluster of about ten proven machines by the late 1980s. A 200 kW, 25 m diameter horizontal axis machine purchased from James Howden Ltd is nearing completion. The machine design has operated successfully in the USA, where it is manufactured and sold by WTG Energy Systems Inc. The machine will be sited at Carmarthen Bay Power Station in South Wales.

The programme of generic studies in the UK has until recently been its major contribution to the international programme. This has concentrated on wake studies using models in wind tunnels and small machines in the field. The UK study of offshore wind energy has arguably been the deepest analysis of this concept. The UK programme has recently been reviewed by Pooley [39] and by Bedford *et al* [40].

Figure 2.23 Artist's impression of the Wind Energy Group 3 MW wind turbine on Orkney. (Courtesy Taylor Woodrow Ltd.)

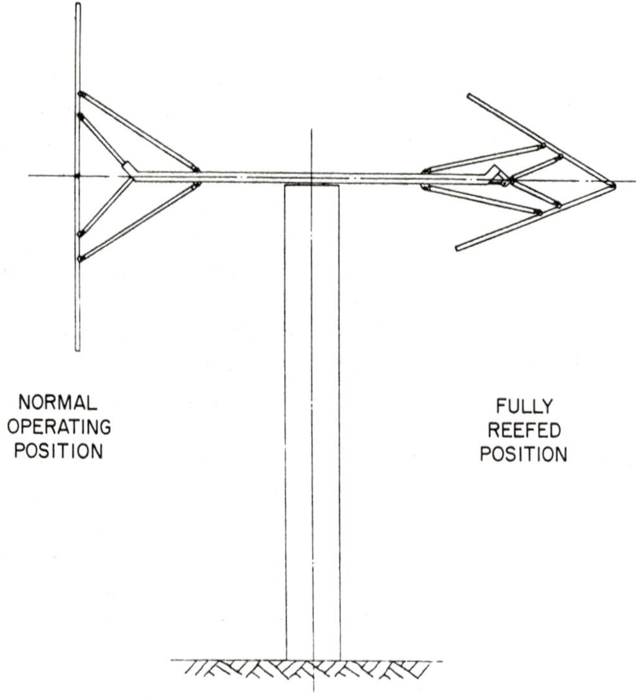

Figure 2.24 Vertical axis wind generator of the Musgrove/McAlpine design. (Courtesy Sir Robert McAlpine and Sons Ltd.)

2.12.3 Other countries

Canada probably has the most significant programme other than those mentioned above. A number of small and intermediate sized Darrieus machines have been constructed and recently plans have been approved to build a megawatt size Darrieus machine.

Little is known about work in the USSR, although there have been reports of large clusters of wind turbines being planned in the Kola Peninsula near the Barents Sea. Norway, Italy, Spain and Greece have small windpower programmes. The hybrid wind–solar system built at Manzanares in Spain will be described in Chapter 6. Although they currently have no national R and D programme, the electricity authorities in France have constructed an intermediate sized machine on an island site.

There has recently been a resurgence of interest in wind energy in Australia and New Zealand but there are, as yet, no plans to build major machines.

2.13 System integration

The problems of how a variable source of power such as that produced by the wind, waves or sun can be integrated into a conventional utility plant mix is of great importance. They have recently been reviewed [41]. Because the output from a cluster of wind turbine generators, for example, can rise or fall over a fairly short period of time, it is necessary to ensure that other generating plant on the system is capable of having its output increased or decreased over a similar timescale in order to minimise any mismatch between electricity supply and demand. Modelling of such systems has been carried out [42]. In operating the system, the output from a renewable source such as wind, wave or solar energy would tend to be used whenever available since no fuel costs are incurred in operation. This integration problem, common to each of these sources, will now be discussed. The considerations affecting the integration of tidal power are very similar but some additional points will be discussed in Chapter 4.

In fact, wind or wave energy conversion systems may be valuable to utilities both because of the fuel savings that they may achieve and because of the capital investment in conventional plant that can be delayed or foregone if they are used in the system. This latter value is frequently referred to as the capacity credit and has been discussed with reference to wind energy by Rockingham [43].

The proportion of these two components (fuel saving and reduced capital investment) which determines the total value of these variable sources to a system will depend on a number of factors. In the case, for example, of wind power these would include

(1) The wind conditions; particularly strength, predictability and perseverance.

(2) The type of wind turbine generator (WTG) and the shape of the output characteristic curve.

(3) The rating of the WTG and its cut-in, rated and cut-out windspeeds.

(4) The future peak electricity demands and the daily load shapes—their susceptibility to load management and the required reliability of the system.

(5) The future plant mix that the utility will use to serve demand.

(6) The future cost of fuel.

(7) The future cost of plant.

An upper bound to the energy saving value of these variable sources to a system may be obtained by relating it to the running costs of the least efficient plant on the system at the time that the electricity from the wind, waves or sun is produced (ignoring the operating problems concerned with the variability and unpredictability). At the opposite extreme, if forecasts of wind or wave conditions are so unreliable that they have no operational value, other plant on the system would need to be scheduled on the assumption that no wind or wave energy will be available for the period of time being considered. Under these conditions, in which the back-up conventional plant would rarely be shut down but would require to be run part-loaded, ready to respond to changes in wind turbine output, the value of the wind energy would be reduced by the costs incurred in operating the back-up plant in this fashion.

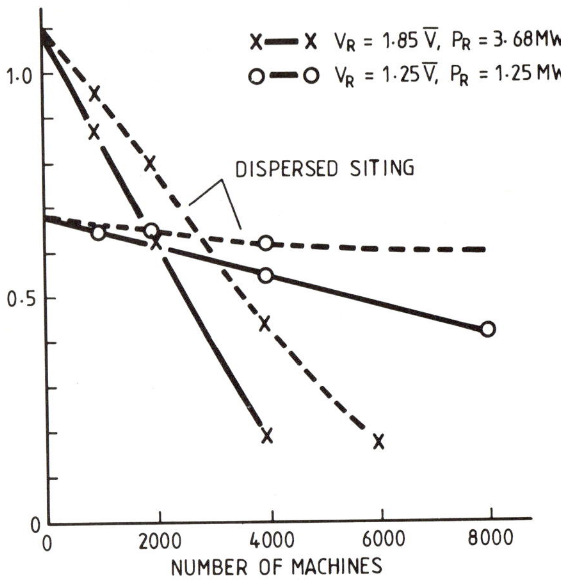

Figure 2.25 Reduction of capacity credit as a function of the amount of wind energy on the system. Curves for dispersed siting and for siting at one location are shown.

The value also depends on the amount of such plant incorporated in the system. The first wind energy plant would clearly displace the least efficient steam plant, but as more renewable sources are added to the system, they would increasingly displace lower fuel cost plant. In addition to this decrease in value, it would be more and more difficult to integrate the renewable source into the system for operational reasons. Large, generally more efficient, steam plant is less capable of responding to changes in output. The actual limits of penetration of wind, wave or solar power into a system might thus be set by either economic or operational criteria (although these are frequently difficult to separate).

Capacity credit is usually derived on a statistical basis by comparing records of output with periods of peak demand. In this way, it can usually be shown that the contribution of wind, wave or solar plant to overall system reliability may allow a utility to reduce its reserve margin of conventional plant on the system and thus, in principle, forego or delay the construction of such plant. The value depends on the amount integrated (see figure 2.25) and on the correlation of the output from the sites from which the power is being generated. A number of analyses have been attempted for various utilities and the results are very variable. They have been reviewed in reference [41]. The theoretical capacity credit for a given utility clearly depends upon many factors including details of plant mix and seasonal and daily load curves. The results of one utility compared with another are difficult to correlate. Figure 2.26 shows the estimated proportion of the total value ascribed to fuel displacement and capacity credit for the CEGB.

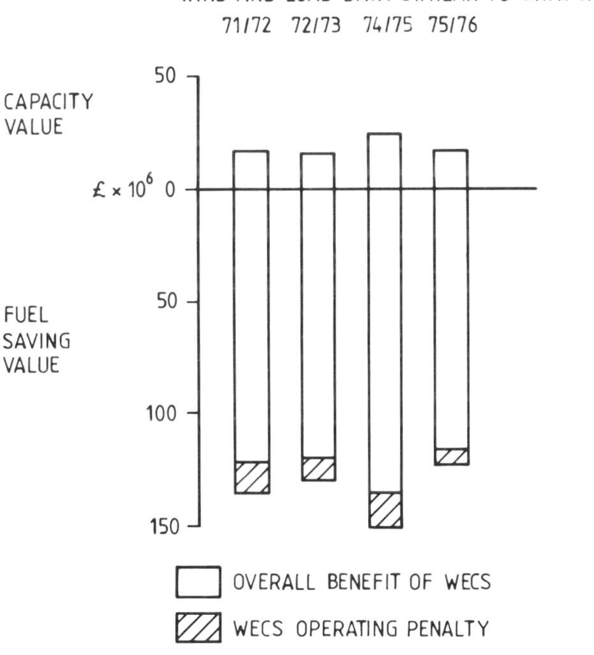

Figure 2.26 Proportion of the value which can be absorbed to fuel saving and capacity. Operating penalties are also shown. For 1000, 90 m diameter machines at a moderately windy location ($\bar{V}_{10} = 5.8$ m/s, rated wind speed at 10 m is 7.3 m/s, machine rating is 1.25 MW).

In common with other studies carried out in this area, the need for operational experience before such capacity values can be credited to wind, wave or solar systems is stressed.

It is frequently argued that what is required to increase the value of wind, wave or solar energy is a form of storage to smooth the output. The availability of medium term storage may eventually be necessary from an operational viewpoint, but the optimal use of such storage requires that it should be available to the whole system and not dedicated to any one type of generation. It is true that large amounts of such sources in a system may make storage more attractive for system use, but such a conclusion is reached only from a detailed *systems* analysis. (A discussion of methods of centralised storage is given in reference [45].) It has also been argued that wind energy, for example, should be associated with local storage in the consumer's premises and that this could radically decrease the need for other plant to meet the seasonal heating load. However, the counter argument that long cold and windless periods not infrequently occur in winter, would suggest that conventional plant would still need to be available to meet the demand during these periods.

Detailed aspects of the technical problems of system integration have been briefly mentioned above, including the options of synchronous, single- and multi-speed induction generators, reactive compensation and the problems of control. These are not believed to be insuperable but more R and D is required, preferably from the field testing of machines and groups of machines, before all the options can be evaluated.

One particularly important question which is now receiving considerable attention is how wind turbines or other devices can be incorporated into small grids using diesel generators [46]. These work very inefficiently at part load and thus the strategies designed for large systems may not be appropriate. Furthermore in some cases (e.g. a summer night), the wind turbine may be the lead generator in the system. Each system must be examined separately, but experience with several machine designs in a number of small grids will shortly be available.

It is for applications in remote communities or developing countries that renewable sources are likely to prove economic in the near future and thus verification of the ability to integrate into such systems is of immediate concern.

References

For an excellent detailed account of the technology and recent developments in wind energy the reader is recommended to refer to the British Wind Energy Association book *Wind Energy for the Eighties* published by Peter Perigrinus (1982).

1 Musgrove P M 1981 Economic costs of wind energy *Paper to Symposium on Economic Costs of Alternative Energy Sources, Caxton Hall, London*

2 Minchinton W 1980 *History Today* March issue 31–6

3 Putnam P C 1948 *Power from the Wind* (New York: Van Nostrand)

4 Golding E W 1955 *The Generation of Electricity by Wind Power* (London: E & F Spon)

5 England G 1981 *Speech to British Wind Energy Association Annual Workshop* (London: CEGB)

6 Warne D F and Calnan P G 1977 *Proc. IEE* **124** 963–85

7 Lindley D *et al* 1982 Resource *Wind Energy for the Eighties* (Stevenage: Peter Perigrinus) ch 1

8 Moore D J 1979 Offshore wind data *Proc. 1st BWEA Wind Energy Workshop, Cranfield* (London: Multiscience) pp199–208

9 Swift-Hook D T 1979 *Wind. Eng.* **3** 167–86

10 Allen J and Bird R 1977 *Energy Paper* 21 (London: HMSO)

11 Milborrow D J 1981 Measurement and interpretation of wind turbine wake data *Proc. 3rd BWEA Wind Energy Workshop, Cranfield* (Bedford: BHRA) pp153–64

12 Taylor R H, Leicester R J, Gardner W E and Franklin P 1979 Integration of wind power onto an electricity supply system *Proc. 1st BWEA Wind Energy Workshop, Cranfield* (London: Multiscience) pp134–6

13 Lowe J E and Engle W W 1979 Proceedings of the workshop on economic and operational requirements and status of large scale wind systems, Monterey *EPRI Report* ER–1110–SR

14 Swift-Hook D T *et al* 1982 *Wind Energy for the Eighties* (Stevenage: Peter Perigrinus) ch 2

15 Mortimer A R 1981 *Wind Eng.* **5** 207–18

16 Mensforth T 1979 Windpower generation on a large scale *Proc. IEE Future Energy Concepts Conf.* (London: IEE) pp268–72

17 Vas I E and South P 1980 *Proc. 3rd Int. Symp. on Wind Energy Systems, Copenhagen* (Bedford: BHRA)

18 Ainslie J F 1982 Innovative concepts in wind energy conversion systems *Wind Energy for the Eighties* (Stevenage: Peter Perigrinus) ch 8

19 Milborrow D J 1980 *J. Ind. Aerodyn.* **5** 403–30

20 Swift-Hook D T 1979 A collaborative programme of field measurements on wind turbines *Proc. 1st BWEA Wind Energy Workshop, Cranfield* (London: Multiscience) pp5–12

21 Reilly D H 1979 *DOE/NASA Report* 20305–79/3

22 Sengupta D C and Senior T B A 1978 *US DOE Report* TID-28828

23 ERDA 1977 Solar program assessment: environmental factors *ERDA Report* No. ERDA 77–47/6

24 Rogers F *et al* 1976 An evaluation of the potential environmental effects of wind energy system development *Report* (Battelle Columbus Laboratories)

25 Rockingham A P, Taylor R H and Walker J F 1981 Offshore wind and wave power—A preliminary estimate of the resources *Proc. 3rd BWEA Wind Energy Workshop, Cranfield* (Bedford: BHRA) pp63–9

26 Taylor R H and Simpson P B to be published A review of progress on offshore wind energy *Proc. Int. Conf. on Management of Oceanic Resources, 1981* (to be published by the Engineering Committee on Offshore Resources, Institution of Civil Engineers)

27 Simpson P B, Lindley D and Hardy W E 1981 An assessment of offshore siting of wind turbine generators in the United Kingdom *Proc. IEE Future Energy Concepts Conf.* (London: IEE) pp264–76

28 Dixon J C 1981 Large offshore wind turbines: system design and economics *Proc. 3rd BWEA Wind Energy Workshop, Cranfield* (Bedford: BHRA) pp80–7 and *Wind Energy for the Eighties* (Stevenage: Peter Perigrinus) ch 6

29 Thomas R L and Robbins W H 1979 Large wind turbine projects *Proc. 4th Biennial Conf. and Workshop on Wind Energy Conversion Systems* (US DOE Conf. 791097) pp75–98

30 Musgrove P M 1981 Wind energy: some comments on the economics *Proc. BWEA Int. Colloq. on Wind Energy, Brighton* (Bedford: BHRA) pp119–27

31 Divone L 1981 Wind energy development in North America *Proc. BWEA Int. Colloq. on Wind Energy, Brighton* (Bedford: BHRA) pp68–90

32 Axell R A and Woody H B to be published Test status and experience with the 7.5 MW MOD 2 wind turbine cluster *Proc. 5th Biennial Conf. and Workshop on Wind Energy Conversion Systems, 1981* (US DOE)

33 Brannstrom A 1981 Hamilton Standard–Swedyards large wind turbine *Proc. 3rd BWEA Wind Energy Workshop, Cranfield* (Bedford: BHRA) pp48–53

34 Hugosson S 1981 The European windpower scene *Proc. 3rd BWEA Wind Energy Workshop, Cranfield* (Bedford: BHRA) pp7–15

35 Lindley D and Stevenson W 1981 The horizontal axis wind turbine project on Orkney *Proc. 3rd BWEA Wind Energy Workshop, Cranfield* (Bedford: BHRA) pp16–32

36 Armstrong J R C, Ketley G B and Cooper B J 1981 The 20 m diameter wind turbine for Orkney *Proc. 3rd BWEA Wind Energy Workshop, Cranfield* (Bedford: BHRA) pp54–62

37 Clare R and Allan J 1981 Progress with the design of a 25 m diameter vertical axis wind turbine *Proc. 3rd BWEA Wind Energy Workshop, Cranfield* (Bedford: BHRA) pp33–40

38 CEGB 1980 *Press Release* No. PR693

39 Pooley D 1981 The wind energy research and development programme of the UK Department of Energy *Proc. BWEA Int. Colloq. on Wind Energy, Brighton* (Bedford: BHRA) pp2–11

40 Bedford L A, Lindley D, Stevenson W G and Swift-Hook D T to be published *Proc. 5th Biennial Conf. and Workshop on Wind Energy Conversion Systems, 1981* (US DOE)

41 Taylor R H *et al Wind Energy for the Eighties* (Stevenage: Peter Perigrinus) ch 5

42 Whittle G E 1981 Effects of wind power and pumped storage in an electricity generating system *Proc. 3rd BWEA Wind Energy Workshop, Cranfield* (Bedford: BHRA) pp88–97

43 Rockingham A P 1980 System economic theory for WECS *Proc. 2nd BWEA Wind Energy Workshop, Cranfield* (London: Multiscience) pp109–17

44 Rockingham A P and Taylor R H 1981 The value of wind turbines for large electricity utilities *Proc. IEE Future Energy Concepts Conf.* (London: IEE) pp348–53

45 Wright J K 1980 *Elect. and Power* February issue 153–7

46 Ballard L J 1980 *Proc. Conf. on Energy for Rural and Island Communities, Inverness* (Oxford: Pergamon) pp159–66

3

Energy from the sea

This chapter will concentrate on the major UK technology of extracting power from the sea—wave energy. Elsewhere in the world, considerable interest has been expressed in generating power from ocean thermal gradients (OTEC) and this will also be described. An outline will also be given of ideas which have begun to be evaluated for generating power from tidal streams at sea and from salinity gradients.

3.1 Wave energy

3.1.1 Introduction

The patent literature contains details of many devices which, it is claimed, will enable energy to be extracted from waves. It is only, however, since the mid-1970s that a major research and development effort has been mounted into this problem. Much of this work has been carried out in the UK, with significant contributions from Japan and Norway and lesser contributions from the USA, Canada and Ireland.

In early 1974, the DEn placed a contract with the National Engineering Laboratory (NEL) to carry out the first stage of a full technical and economic appraisal of ocean wave power. Independently the CEGB had also carried out a study of wave power as part of an overall review of the potential of renewable energy sources and had reached the conclusion that wave energy appeared to be particularly promising [1]. The early UK programme thus comprised the NEL study, a modest research programme within the CEGB, and the contributions of several university groups and interested individuals. Since that time, the UK programme has grown year by year and until recently has been quite substantial (£3–4M p.a. from government).

Early work was concentrated on four devices all of which will be described below: the Salter duck, Cockerell raft, Russell (HRS) rectifier and NEL oscillating water column (OWC). These devices achieved very high efficiency in tank tests and considerable innovation went into devising suitable seaworthy devices and developing power take-off systems. At the same time, in Japan, a device already used for powering navigation buoys (an owc) was being developed by Masuda for possible larger scale applications. The progress and publicity given to these early successes led to wide interest and many more university and industrial groups became involved, developing their own favoured systems.

In 1978, an independent study of the costs of a number of the better engineering devices was carried out by the consulting engineers Rendel, Palmer and Tritton for the DEn. The estimated costs (20–50 p/kWh) were found to be depressingly high. However, within a year the need for cost-cutting led to the development of a new generation of the previous designs which, it was assessed, might achieve about 5–15 p/kWh in mass production.

In the last two years, the UK programme has set out to attempt to tackle the centres of cost in these estimates and to attempt to identify a lead device which can be developed to full scale. Two further devices—the Bristol cylinder and Lancaster bag have shown considerable promise and have been funded. The 'clam' device has also shown promise. These devices are also described below.

Early estimates of the resource size in the UK were very encouraging. Indeed, using data from one of the few wave rider buoys in the Atlantic, a gross resource of 120 GW was estimated for the total mean power in the sea which could be obtained from the areas shown in figure 3.1. The western approaches to the UK are certainly among the best areas in the world for wave resource being both windy and enjoying a long fetch across the North Atlantic. However, later estimates based on more realistic wave data and device characteristics have reduced the early estimates to about 30 GW [2], [3]. This is clearly still a substantial resource.

This section will begin by describing wave behaviour, will then outline the types of device which have been proposed and then report on progress in a number of generic areas common to all devices, such as hydrodynamics, materials, moorings reliability and maintenance and environmental questions. Finally the most recently published figures (1981) for economics and resource assessments will be reviewed.

3.1.2 Wave characteristics

We begin here with a summary of simple wave theory which will be followed by a description of real seas and the methods commonly used to describe them. More detailed discussions are to be found in references [4] and [5].

Simple wave theory. Much can be learnt about wave power from simple

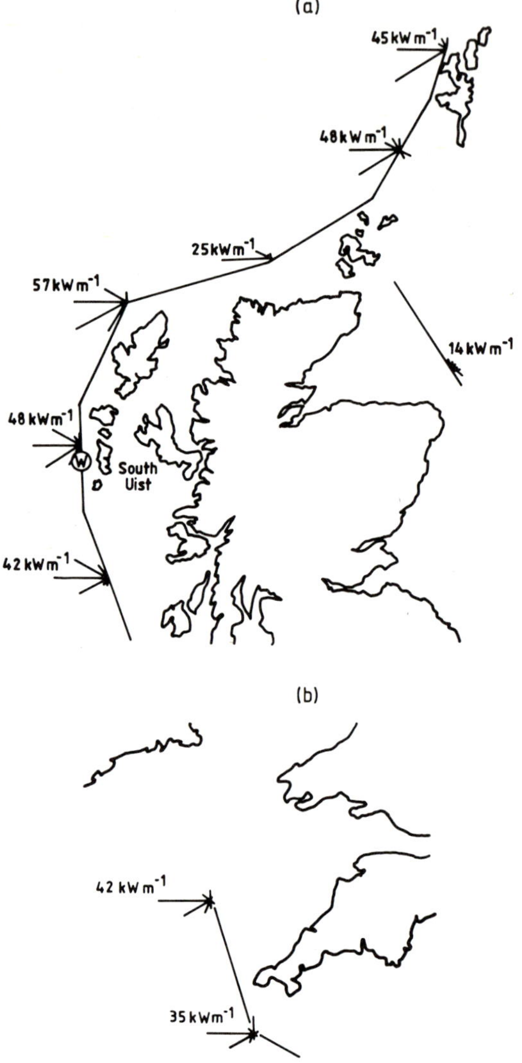

Figure 3.1 (a) Wave power at various grid points around the Scottish coast. The figures give the average non-directional power density over the two years for which data were taken and the 'crows feet' indicate the directional breakdown. (b) Wave power around the southwest coast of England. (Reproduced courtesy Energy Technology Support Unit.)

theory [6] which starts from the assumption that sea waves can be described as deep water (depth greater than half a wavelength), small amplitude, sinusoidal waves of a single wavelength (monchromatic). A progressive wave of this type, of amplitude a, wavelength λ and period t, will have a surface profile of the form

$$h_\mathrm{o} = a \sin (kx - \omega t)$$

where k is the wave number $(2\pi/\lambda)$, and ω the rotational frequency $(2\pi/t)$.

It may be shown that with the above assumptions, the speed at which the wave propagates, the phase velocity C, may be obtained from

$$C = [(g/k) \tanh kh]^{1/2} = \omega/k$$

where h is the water depth. This result shows that water waves are dispersive with different frequency components propagating at different speeds.

For deep water, $\tanh kh \to 1$, so

$$C \to (g/k)^{1/2} = (g\lambda/2\pi)^{1/2}.$$

The speed is thus proportional to the square root of the wavelength. Since C is defined as λ/t, rearranging:

$$\lambda = gt^2/2\pi$$

and thus a wave of period $10\,\mathrm{s}$ will have a wavelength of $150\,\mathrm{m}$.

In shallow water, where kh is small

$$C \to (gh)^{1/2}$$

and speed is proportional to the square root of the depth. Shallows can thus induce diffraction effects and this, for example, accounts for the fact that waves inshore tend to swing round to be parallel to the coast.

In deep water, each water particle at the surface moves in a roughly circular orbit of amplitude a, and angular frequency ω. With increasing depth, the fluid particles move with an amplitude decreased by a factor $\exp(-kh)$. This means that the radius of the circular motion decreases very rapidly with depth. At half a wavelength below the surface, the radius is less than 5% of that at the surface. The total energy E in the wave per unit surface area is made up equally of the kinetic energy of the water particles and the variation in potential energy above and below the mean water level. This is given by

$$E = \tfrac{1}{2} \varrho g\, a^2$$

where ϱ is the density of sea water.

This energy is transferred at the group velocity, which in deep water is one half of the phase velocity, v; thus the power, P, associated with a monochromatic wave is

$$P = Ev = \frac{\varrho g^2\, a^2\, t}{8\pi} \sim 3.8a^2t \qquad \text{(kW)}$$

per metre of wave front, where a is measured in metres. A 1 m amplitude wave of period 10 s will give a power of 40 kW per metre of wave front. The square term in a, however, shows that a 15 m wave of 15 s period will contain about 13 MW per metre.

The 'real' sea. The waves in the sea are, of course, very different from the ideal description given above. The sea surface is in some respects like a huge windmill, gathering wind energy over a wide area. Ocean waves transfer energy with very high efficiency and the attenuation distances are very large. Waves arriving at a point may either originate from local winds (the so-called wind sea) or from storms a very great distance away (the so-called swell sea). In the latter case, the effects of dispersion will lead to the long wavelength, long period waves spreading from their source more rapidly than shorter period waves. At some remote point, the swell sea will frequently appear as an almost monochromatic wave train which, however, becomes progressively shorter in wavelength. Superimposed on such a swell sea will be the wind sea. This may, in principle, be Fourier analysed into a set of 'ideal' waves of the type discussed above, with different wavelengths, phases and directions. It is usual to assume that the principal direction for these waves, σ, is the direction of the wind. The wind sea can then be represented as the product of a one-dimensional spectral density function $E(f)$ and a spreading function $g(\sigma)$.

The most commonly used spectral density function for a fully developed sea, is that of Pierson and Moskowitz, which is of the form

$$E(f) = Af^{-5} \exp\left(-Bf^{-4}\right)$$

where A and B are parameters related to fetch and wind speed. For convenience they can be related through probability theory by H_s and t_z which are measures of wave height and wave period. The significant wave height, H_s, is defined as $4\,h_{rms}$, where h_{rms} is the root mean square wave height and t_z, the zero crossing period, is defined as the number of times the water level moves through its mean position over a period of one second.

The actual relationships between A, B and H_s, t_z are

$$A = H_s^2/4\pi\, t_z^4$$

and

$$B = 1/\pi\, t_z^4.$$

The power, P, in kW per metre of wave front is then given approximately by [7]

$$P = 0.58\, H_s^2\, t_z \qquad (\text{kW m}^{-1} \text{ at some point}).$$

Typical Pierson–Moskowitz spectra are shown in figure 3.2. These may be generated from measurements of t_z and H_{rms} for a real sea which for convenience are often represented on a scatter diagram such as that shown in figure 3.3. Such frequency spectra can then be used for the modelling of the behaviour and energy productivity of devices. In practice the measured seas are not always fully developed and have swell and angular components. A representative set of some 46 sea states is used for modelling device productivity in the UK wave energy programme.

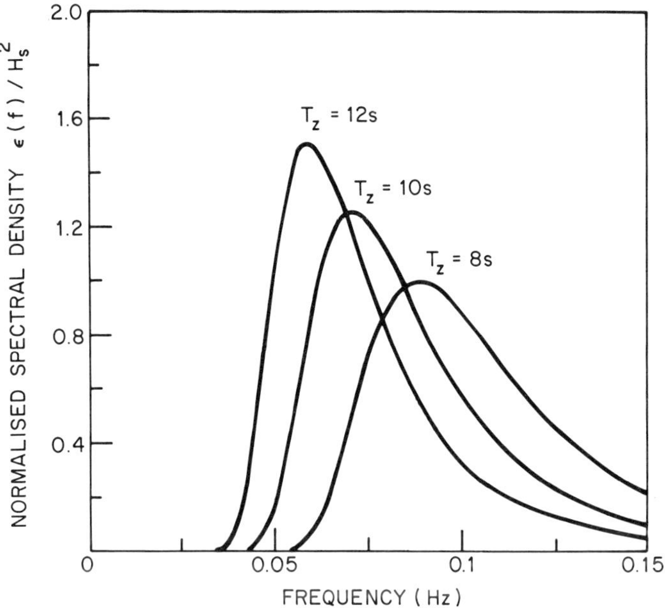

Figure 3.2 Typical Pierson—Moskowitz spectra.

Data on directionality are rather sparse but such information can be a very important consideration both in assessing the design of devices and the energy that they can extract from the sea. Results from measurements by the National Physical Laboratory (NPL) in the North Atlantic suggest that about 45% of the wind seas originate from a westerly direction, 24% from the south, 19% from the north and 12% from the east, but the energy content of these seas has not been fully established. This is particularly important to know, since some devices are essentially unidirectional and may be unable to fully utilise the energy from seas other than in (say) a westerly direction. Furthermore, some of the sites which might be used for wave power devices are quite close to the coast and, because of coastal geography, will be sheltered from some directions. The spreading function $g(\sigma)$ used to allow for directionality, mentioned above, is often taken to be of the form

$$g(\sigma) = |\cos \tfrac{1}{2}(\sigma-\sigma_0)|^{2s}$$

where σ_0 is the principal wave direction and the index $2s$ is often taken to be about two for locally generated seas. For swell seas in particular the value may be very much higher, reflecting the fact that the waves are effectively plane.

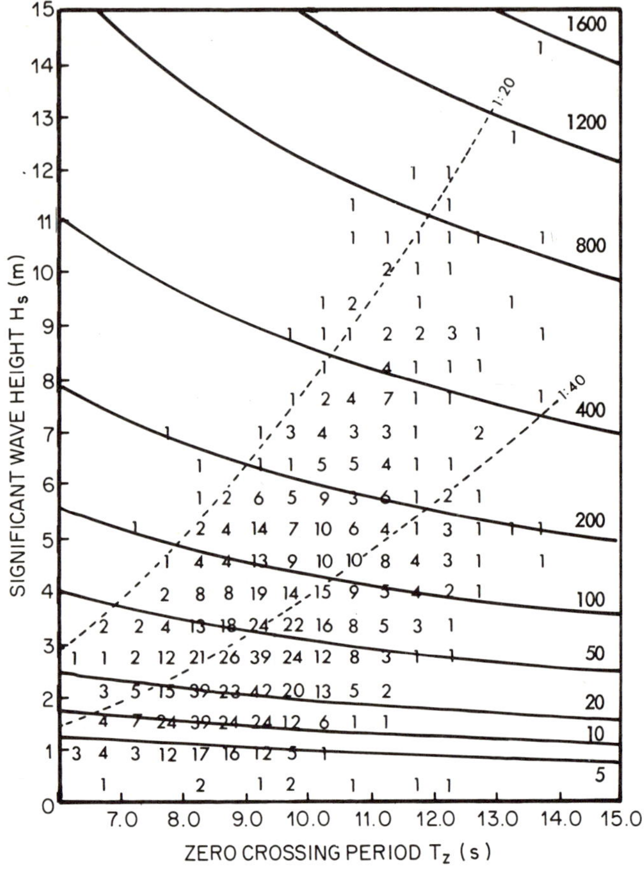

Figure 3.3 Scatter diagram showing the annual frequency/1000 of H_s, t_z combinations at Weather Ship *India*. The numbers indicate the frequency with which a given zero-crossing period and significant wave height was encountered in 1000 measurements. Significant wave height and zero-crossing period are defined in the text. – – – – lines of constant wave slope; ——— lines of constant wave power (kW/m).

The spread of directions from which waves approach a point, particularly away from the shore or in a predominantly wind sea, means that wave crests are not even and parallel but form a 'peaky' surface, characteristic of the mix of directions present. The distance between wave crests in such a sea can be an important design parameter, particularly for devices incorporating a spine (see below) on which individual devices are attached. If, for example, waves are approaching a point from two predominant directions, the ends of the spine may in some circumstances rest on crests, whilst the middle may be centred on a trough. Such a condition would impose severe loadings on the spine structure which would need to be allowed for in the design. Breaking waves, although relatively infrequent in deep water, are of great concern since these may cause very great damage. Further reference to breaking waves will be made in §3.1.8 on moorings.

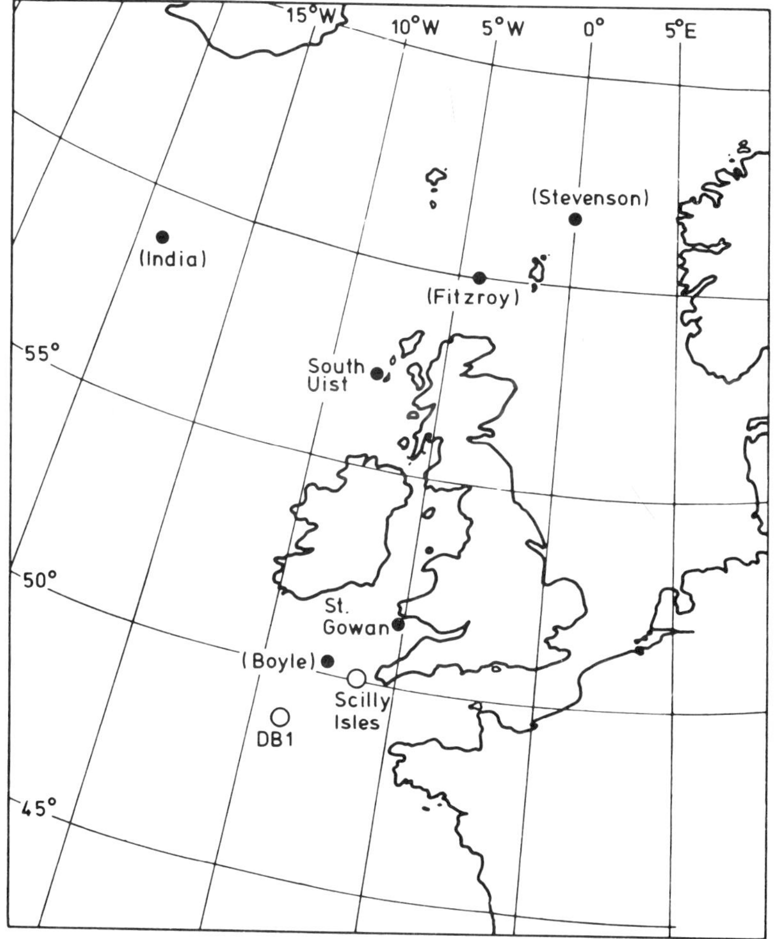

Figure 3.4 Wave-data collecting stations of particular relevance to the UK wave energy programme. ( ) denotes no longer operating; ○ denotes new buoys. (Reproduced from *Energy Paper* 42, courtesy HMSO.)

Availability of data. Apart from visual records, the amount of wave data even in the vicinity of a maritime nation such as the UK is very sparse. Some weather ships and lighthouses record wave data and those contributing useful data to the UK programme are indicated in figure 3.4. Ocean Weather Ship *India* and the data buoy at South Uist have been particularly important in this respect. Wave rider buoys, which convert their acceleration into a wavelength versus time record and send data back to shore by radio link are now providing an important supplement to these data, particularly from the areas closer to shore where wave energy converters might be stationed. Acquisition of such data has been an important aspect of the UK programme on wave power.

Analysis of OWS *India* data shows that power levels of 30–40 kW/m are most frequent with mean power levels of 80 kW/m. These data were the basis for the claims, mentioned above, that over the 1500 km of Atlantic coast, 120 GW of wave power would on average be available [8]. More recent data (covering only two years of recording time) from the wave rider buoy anchored off South Uist (more typical of areas which might be exploited) have recorded mean power levels closer to 50 kW/m over all directions, corresponding to perhaps 30–40 kW/m for a directional device parallel to the prevailing wave direction. It is not yet clear how representative this is of the general area off northwest Scotland but there is no reason to believe that the values are anomalous. The power levels are likely to be a little higher than those to be obtained off the southwestern approaches of the UK and much higher than those obtained in the northeast of the country.

3.1.3 Variability of wave power

Like wind power, the output from the waves is likely to be highly variable on a number of timescales. Again there is a good seasonal matching between wave power output and demand (see figure 3.5). There is also a great range of incident power levels. For about 1% of the time, wave heights in excess of 10 m and with periods of more than 11 s are likely to be experienced, giving incident power levels of 1 MW/m. The 100 y wave (the height of wave that could be expected once every 100 years on average) is estimated over 30 m in parts of the North Atlantic and this illustrates the extreme power levels (in excess of 20 MW/m) which might be encountered. At the other extreme, for about 1% of the time, no significant waves have been recorded.

As with wind energy, variations from year to year are strong. What is perhaps more surprising are the large changes in output that can occur over periods of a few hours. This is easily understood in the case of wind power, but is less obvious for wave power where the large collecting surface and attenuation distances would appear likely to lead to a longer time constant for output fluctuations. It has been reported that data from OWS *India* has shown power output variations of as much as 200 kW/m over a 3 h period. It is common experience that individual waves vary markedly in their height and a typical trace of surface level against time is illustrated in figure 3.6.

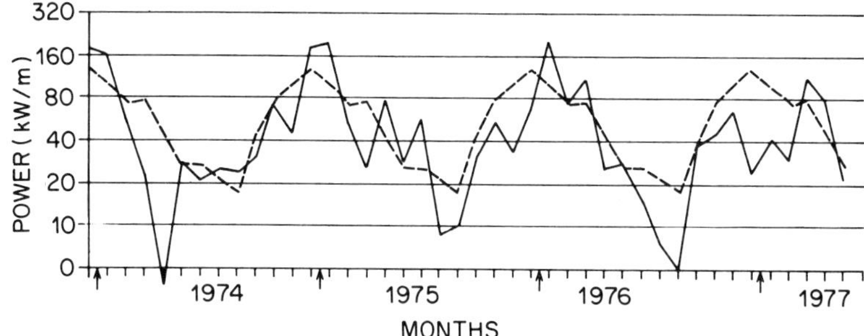

Figure 3.5 Monthly average power levels at South Uist, as measured by wave rider buoy, and fitted from wind data (full curve). The broken curve represents the twelve-year averages for each month. It should be noted that this figure shows a preliminary fit based only on the first of the two years of wave data from this site.

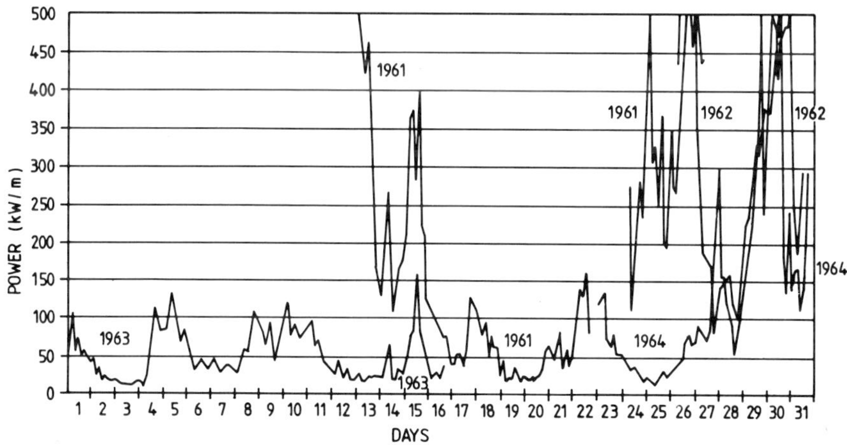

Figure 3.6 Wave power measured at Weather Ship *India* for the month of January. (Reproduced from *Proc. Heathrow Wave Energy Conf.* (1978), courtesy ETSU.)

These variations are important both in the design of devices and in ensuring that they give an acceptable electrical output over short time intervals, as well as in the analysis of the problems of system integration. It is clear that the comments made in Chapter 2 on the costs of integrating wind power into the supply system, capacity credits and the effects of dispersed siting and wind gusts are likely to be equally relevant to wave energy.

3.1.4 Desirable features of wave energy devices

The basic objective of any wave energy conversion device is to convert the motion of real seas into a form of usable energy. As will be discussed below, electrical energy appears more attractive than other energy vectors. Given the financial constraints which must be used to analyse the economics of any energy source, e.g. the appropriate discount rate, the major factors influencing its cost effectiveness will be capital cost, costs for operation and maintenance (O and M), lifetime and energy productivity.

Capital cost will be determined by the amount of material which is used in construction, by the complexity of the device, its ease of construction and particularly by its ability to be mass produced. These characteristics are perhaps well exemplified by the technology which was developed to build the liberty ships of the Second World War or modern large oil tankers.

Energy capture will clearly be determined largely by the ability of a device to take the wide range of frequencies and amplitudes present in a real sea and convert these efficiently to mechanical motion in the device, and then convert this mechanical motion to electrical output. Energy capture will also be influenced strongly by the rating of the devices in the same way that wind turbines are affected by this design optimisation as discussed in Chapter 2. Availability and the ability to accept waves from a wide range of directions are also important.

O and M costs will almost certainly be high for an offshore installation. It is important that designs should be simple, reliable (or contain provision for redundancy without a high capital cost penalty) and be easily main-tained. Careful provision must be made to reduce corrosion, marine fouling and to ensure adequate mooring. These considerations will also be important in making assumptions about the lifetimes which might be anticipated.

3.1.5 Types of device

Before describing individually many of the devices which have been studied in the UK and elsewhere in the last few years, a summary will be given of a number of broad classifications which have been used to describe the operating principles on which many of the devices work.

Glendenning and Count [9] have classified devices into ramps, floats, flaps, air bells and wave pumps. A ramp is a passive device which allows water to ascend a sloping sea-wall and uses the hydraulic head developed to drive a low head turbine. The best example of such a device is the Mauritius ramp (§3.1.6). Float systems are designs which heave in response to the surface motion of the water and in so doing drive a pump. Flaps operate on the horizontal component of the water motion which decays exponentially with depth. In principle, it is possible to optimise a flap operating from a hinge below the surface such that it operates between the free undamped condition, where it absorbs a wave and generates a new wave behind, and the overdamped condition where the wave is largely reflected. Air bells, such as Kaimei (§3.1.6) operate on the principle that a vessel with an open bottom held below the surface will translate the motion of a passing wave into a cyclic pressure variation in air trapped in the top of the bell. Wave pumps exploit pressure variations beneath the surface, pumping a hydraulic fluid or air around a suitable system containing a turbine. The pressure variations of the hydrostatic head due to the passing waves die away exponentially with depth, so such devices need to be close to the surface.

Another classification divides devices into rectifiers, tuned oscillators and untuned oscillators. This was used in *Energy Paper* 42 [4]. Rectifiers, such as the HRS (Russell) device referred to below, generally convert the motion of the waves into a head of water and the potential energy is thus exploited—rather like a conventional hydro generation scheme—in driving a water turbine. Tuned oscillators, which are represented by several devices in the UK programme, operate by responding to a narrow range of wave frequencies with a fall-off in response at frequencies away from optimum. As in their electrical analogues, bandwidth is an important parameter. Untuned dampers such as the flexible bag have, in principle, the ability to absorb a wide range of wave frequencies at high efficiency. In fact, even a device such as the flexible bag is almost certainly tuned to some extent and there is probably no system which is truly untuned.

Perhaps the two most useful types of classification are the simplest. The first of these is the distinction between dynamically active devices and those which are essentially passive [9]. In the first case, a structural element of the device is able to respond to the wave, and power is extracted from the relative motion of various components of the device. Passive arrangements merely capture the maximum available energy by presenting, in one form or another, a large immovable structure to the waves. The final classification considered here, and now the most popular, is that of terminators, attenuators and point absorbers [5]. Terminators are aligned parallel to wave crests and form a barrier to the oncoming waves. Attenuators are aligned perpendicular to the wave front and absorb energy progressively along the length of the device. Point absorbers, unlike attenuators and terminators, are able to absorb energy from waves approaching at all angles. They are rather like the wave energy equivalent of a vertical axis windmill. In the same way that wind turbines have to be carefully spaced to optimise their array efficiency, wave power point absorbers would need to be built in an array.

3.1.6 *The major devices*

In this section, fourteen devices will be described in varying detail. Some, such as the first four described (the Salter duck, Cockerell raft, HRS rectifier and owc), have been studied in depth in the UK programme. Work on two of these devices, the HRS rectifier and the Cockerell raft has recently been cut back within the UK programme, partly because they appear to show little potential of achieving further major cost reduction and partly because they appear to have now been developed fairly fully on a small scale. Some of their features are very important and have been incorporated in new devices. Other systems such as the clam, flexible bag and Bristol cylinder have recently received increasing support because of their early promise [4]. Many of the other types of device described are interesting because of their potential for specific applications (e.g. the Mauritius ramp) or because they contain novel features which suggest routes for further development (variants on the owc for example).

It is too early to predict whether any of the devices described below will be capable of commercial development. It is quite possible that entirely new concepts will appear which would replace the existing devices. The most likely result, however, is that if utilisation of wave energy is to succeed, it will do so gradually by merging successful features of various devices. This is why clear classification into types is useful and why space will be devoted in this chapter to describing most of the devices developed by industrial or by university groups, even if some of these are not currently fashionable. It should be noted that designs and costings are constantly changing, and the descriptions and figures given here reflect published versions of the favoured systems in 1981. The main details and estimated costs for a number of the devices are given in table 3.1 [3].

The Salter duck. This device is by far the best known of the wave power devices and was one of the first to be developed in the current UK wave energy programme [4]. Indeed, Stephen Salter of Edinburgh University started to design the duck in 1973. The shape of device (illustrated in figure 3.7), evolved from a series of experiments on flaps and floats. Its main property is that the beak rolls or nods about the centre of the duck in such a way as to extract the maximum energy from the incident waves over the widest possible bandwidth. Large numbers of ducks rotate on the same central shaft called the spine. In the classification schemes referred to in §3.1.5, the duck is variously a flap, tuned oscillator, active device and

Table 3.1 Main features of the leading wave energy devices (1979/1980 designs). Figures in brackets show estimates for 1982 designs as reported in *Offshore Engineer*, August 1982, p51.

Device	Salter duck	Lancaster bag	Bristol cylinder	NEL owc fixed	Vickers twin owc	Vickers chamber	Cockerell raft (HP)	HRS rectifier	Clam	Belfast buoy
Width/ length	26 m	>200 m	>50 m	60 m	30 m	120 m	100 m	—	300 m	25 m dia
Material	Concrete	Concrete	Concrete	Concrete	Concrete	Steel	Concrete	Concrete	Concrete	Steel and concrete
Mode of operation	Terminator	Attenuator	Attenuator	Terminator	Terminator	Attenuator	Terminator	Terminator	Terminator	Point absorber
Rating	2 MW	5 MW	2.5 MW	—	1 MW	1 MW		—	10 MW	1.5 MW
No. required to give 2 GW	1000 (~900)	450(~350)	800(444)	(589)	2000	2000	900	—	(~330)	1200–1500 (1500–1900)
Water depth	100 m (100)	50 m (75)	40 m (~45)	20 m (21)		60 m	50–100 m	15 m	50–100 m (80–90)	50 m (30)
Length of 2 GW station	40 km	65 km	~60 km	60 km	—	—	90 km	60 km	80 km	90 km
Device team cost (p/kWh)	5 (3.3)	4 (9.1)	~7 (8.6)	~10 (6.3)	Target of 6	Target of 6	9	—	6 (4.4)	8 (7.3)
Consultants cost (p/kWh) (1979)	10 (5.6)	6 (10.4)	up to 14 (10.2)	13 (9.3)	—	—	12	40	(6.7)	12 (15.5)
Reference designed	Yes	Yes	In progress	Yes	No	No	Yes	Yes	In progress	Yes

terminator. The efficiency of the basic duck device is very high. In monochromatic waves matched to the resonant frequency of the duck, efficiencies can reach 90%. In real seas, it is estimated that the efficiency of converting the wave motion to the mechanical motion of the device is likely to be about 50%.

The power take-off system presents a difficult problem with many wave power devices and in the case of the duck, development of an acceptable system in terms of cost, lifetime and reliability has been a major problem. The motion of the duck entails a relatively small angular movement at a frequency of about 0.1 Hz. If a few megawatts are to be transferred from duck to spine, this implies huge torques. Early duck designs employed mechanical systems such as gears, tapes or toothed belts driving radial hydraulic pistons or rams housed inside the spine. These, in turn, drove hydraulic swash-plate motors which regulated an electrical generator. Recent designs which have been produced in collaboration with John Laing's and the Scottish Offshore Partnership, are elegant but much more complex [11]. The principal innovation is the use of gyroscopes. The new concept avoids operating the power take-off system in a salt water environment. Four gyroscopes, with pairs spinning in opposite directions, each weighing 17 tonnes and rotating at about 1000 rpm, are required for each 24 m wide duck module and operate within a sealed unit maintained at a pressure of about 0.01 bar. As the duck nods, the gyroscopes precess in a plane mutually at right angles to the planes of nod and spin. The precession can be used to do work if it is opposed by a torque. Each gyroscope carries a pair of cam rings as illustrated in figure 3.8. These drive conventional piston pumps via cam rollers and specially designed links. It is important to match the torque resisting the precession of the gyro and this is achieved by controlling the number of pistons allowed to pump. The decisions about which pistons are to work are made by a computer which controls poppet valves on each cylinder. In recent designs there is a total of 384 cam pumps on each duck. Oil pumped from the cams, drives a swash-plate motor at each end of the gyro shaft and these are in parallel with another motor driving an electrical generator. If there is a large wave incident on the duck, the angular deflection of the swash-plates increases and the extra power developed is used to speed up the gyro. When the input from the waves is less than average, there will be less oil from the ring cam pump than is required to drive the generator, and the swash-plates will swing over so that they pump and draw the energy deficit from the gyro. Each duck stores about half a megawatt hour of energy and Salter has argued that ducks may have a value as spinning reserve in a power system. Because the inertia of the swash-plate is very small and the forces very large, the generator can respond to power changes in milliseconds. A sketch of the hydraulic circuit and a short section of power record is shown in figure 3.9.

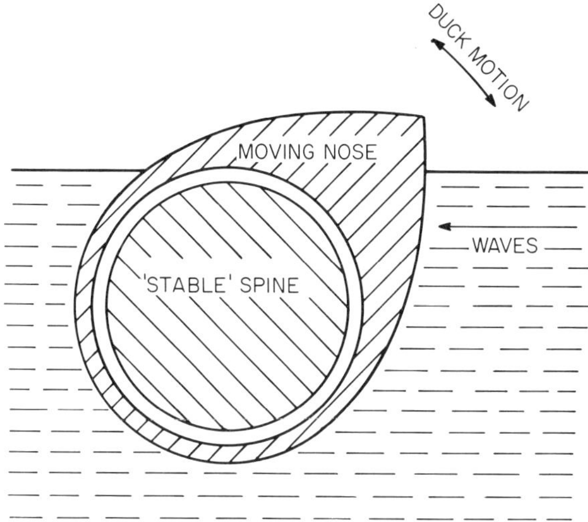

Figure 3.7 The Salter duck.

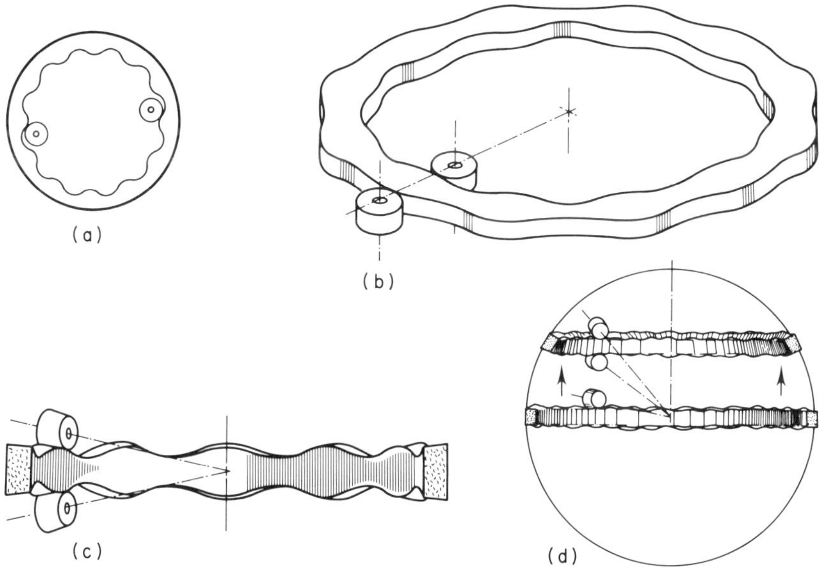

Figure 3.8 (a) A conventional ring cam. If these are constructed with large diameters the force from each roller has to travel further round the ring before it can be balanced by forces on the other side. Salter therefore proposes the use of pairs of rollers (b) working against each other through the thickness of the ring. To utilise space efficiently the ring is twisted (c) and is moved from the 'equator' to the 'tropics' of each gyro (d) so that two can be fitted into each gyro. (Courtesy Mr S Salter, University of Edinburgh.)

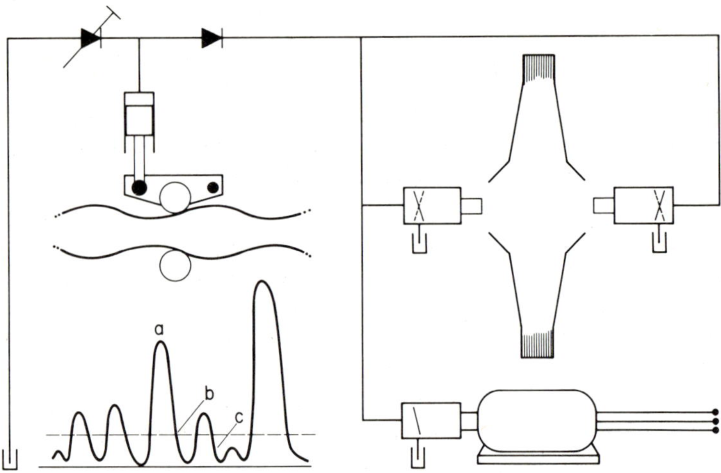

Figure 3.9 Hydraulic circuit and a short section of the power record for the Salter duck. At point *a* power is being fed into the gyros. At point *b* the incoming power is exactly right and the swash plates of the gyro drive motors go to their middle position. At point *c* energy is flowing out of the gyros to maintain the output at a steady value during the lull. (Courtesy Mr S Salter, University of Edinburgh.)

This power take-off scheme makes heavy demands on components but Salter takes the view that reliability of the entire system can be achieved by ensuring that duplication, intelligent control and regard for layout should overcome the unreliability of individual components. Others have argued that the complexity inherent in the Salter device should be avoided and that true reliability can only be achieved through simplicity. Several components such as the laminated steel fly wheel and the advanced swash-plate pump design (the Clerk motor) will require substantial validation before it is clear whether full-scale devices would be satisfactory.

The power take-off assembly is housed in the beak of the duck. Another important part of the design is the bearing between the duck and the backbone or spine. This is again novel. Here, long life with a minimum of maintenance and low frictional losses is essential. Whatever is chosen, whilst not having to handle high loads, must accept reversing velocities and operate in sea water. One solution that Salter has suggested is an elastomeric material called slubber which when subjected to external pressure, exudes a liquid. The operating surface of the bearing is made from concrete covered with a skin of cupro-nickel and on top of this rides a circular pad of this appropriately named material. Rather like a hovercraft, this floats on the surface, but in this case the floating effect is achieved by feeding the pad with water from a second chamber at a higher pressure. Recently a device called a magnetic squeeze film bearing has been suggested.

The spine is one of the important parts of the duck which has now been incorporated in a number of other wave power devices. Its main function is to provide a reference frame which will cancel the forces due to waves of different phase arriving at the device. If long enough, the spine will remain stationary. The concept has undergone extensive testing both in wave tanks at Edinburgh and by a team at Lanchester Polytechnic (Sea Energy Associates) working at Loch Ness [12]. The experiments have shown that for full-scale rigid spines of the sizes envisaged (10–15 m diameter, 100–500 m in length) there may be problems in designing them to give the required stiffness and at the same time withstanding predicted bending moments. Salter and his team have attempted to get round this problem by designing compliant spines. The present design envisages the use of pairs of precast concrete ducks mounted on 60 m lengths of spine formed from longitudinally prestressed concrete sections and jointed together by Hooke's joints. It has also been suggested that further power could be generated from the relative motion of the sections of spine. The motion of the joints is controlled by double-acting hydraulic rams which drive further swash-plate motors, themselves driving generators in phase with the main generators. Salter believes that intelligent control of the joints can aid device survival. In the most recent designs (1981) the spine is formed from 45 m lengths, adjacent lengths being connected by alternately rigid and flexible joints.

Much space has been devoted to this description of the duck system not because it is necessarily the leading wave power system, but for two other good reasons. Firstly because it has probably received more development effort than other devices and secondly because the designers have chosen the route of complexity rather than relative simplicity. Although the chances of many of the individual components being successfully developed would appear to be quite high, a question mark remains on whether a successful system can be engineered and mass produced. Of the fact that it is ingenious there can be no doubt. An artist's impression of a power station built from Salter ducks is shown in figure 3.10.

The Cockerell raft. Sir Christopher Cockerell began to develop ideas for possible wave power devices in 1972 and the raft emerged as an early leader among the designs considered. The basic idea of raft designs is again simple. If a series of floating pontoons is connected together by hinges, a raft can be formed which will adapt to the contour of the waves. Power can be generated from the relative motion between the pontoons.

Originally, the raft was conceived as an attenuator but recent designs have been adapted to form a terminator [13], [14]. It was found that so long

Figure 3.10 An artist's impression of the Salter duck in operation. (Reproduced from *Energy Paper* 42, courtesy HMSO.)

Figure 3.11 Cockerell rafts on test in the Solent. (Courtesy Mr J Platts, Wave-power Ltd.)

as the rear section is held stable, a doublet of pontoons will generate as much power as a string but, of course, with a saving in materials costs for the additional structure. Like the Salter duck, peak efficiencies of over 90% can be achieved at 1/50th scale with waves of about 1 Hz. A heavily instrumented 1/10th scale model (figure 3.11) has been tested in the Solent near Southampton and this has provided useful data on performance in real seas and on mooring forces.

As with the Salter duck, one of the major problems with the raft is that of power take-off. The relatively small and slow oscillations at the hinges again means that very large torques are developed. To optimise performance, the torques applied to the hinges need to vary roughly proportionally with the wave height and to be independent from one hinge line to the next. This led to the consideration of fluid transmission for power take-off. Oil hydraulics were rejected because it was felt that they could not be sufficiently reliable in a marine environment where inspection and replacement would be infrequent, and rejection of excess power in high amplitude sea states would be difficult in a closed hydraulic system. The most attractive system for power take-off appears to be a simple open circuit design using sea water as the working medium. In the 1978 design, displacement pumps at the hinges are fed into a common reservoir (see figure 3.12) and then into a turbine coupled to a generator, the water finally being released back into the sea. The reservoir ensures a fairly even output and can be used to spill power in very rough seas. The torque at the hinges can be adjusted by varying the water flow through the turbine thus maintaining the optimum pressure differential.

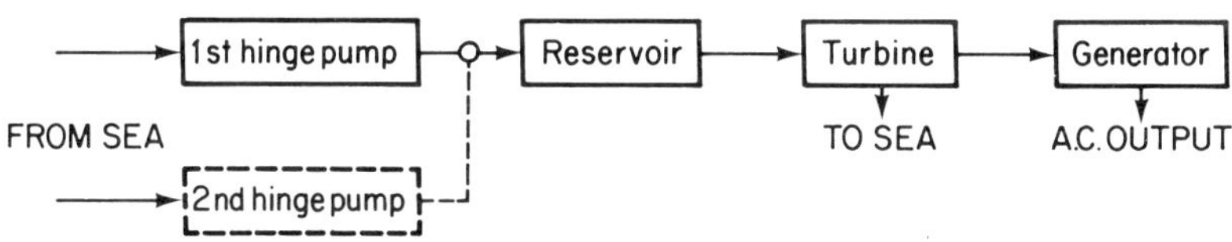

Figure 3.12 Open circuit seawater system.

The actual design adopted is shown in figure 3.13. This employs a single hinge line with two independent raft sections mounted on a common back section. The bearing is designed from bonded laminated rubber. It is possible that a number of units could be mounted on a spine. The pump at the hinge is in the form of a double-acting oscillating vane pump extending across the device and the pumped sea water flows via a transverse main into the central reservoir driving the vertically mounted axial flow Kaplan-type turbine.

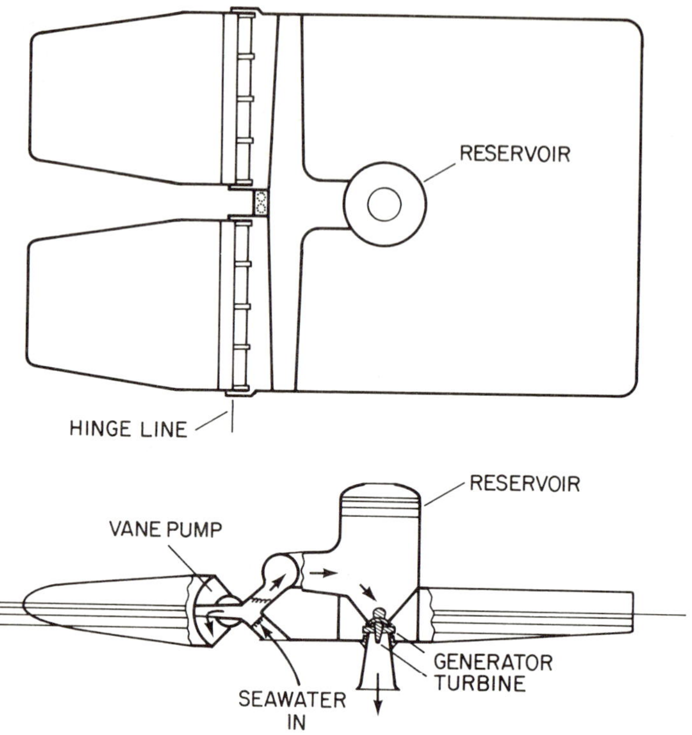

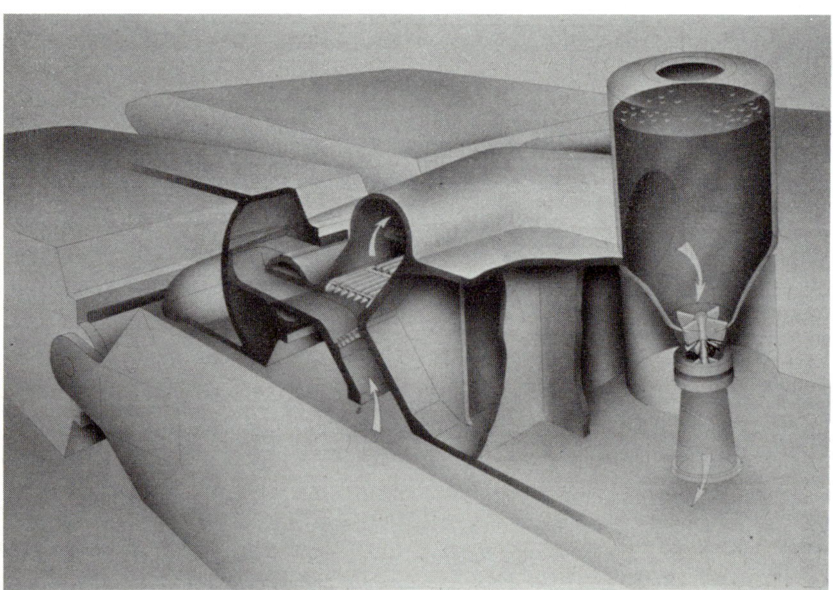

Figure 3.13 The actual design of one form of power take-off for a Cockerell raft. (Courtesy Mr J Platts, Wavepower Ltd.)

Subsequent work has also led to the design of a high pressure (about 4800 N/m^2) sea water system using plunger pumps which deliver water to a common sea water main. This runs along the seabed and is connected to a central turbine generator station serving about 40 rafts. This scheme offers some advantages in centralising the generation but the maintenance costs of the bottom mounted system may be high and the hydraulic piping would need to withstand the constant movement of the rafts. Although sea water systems avoid cooling, pollution and some maintenance problems likely to be encountered with a high pressure closed-loop oil system, they will also present several problems which would need to be examined in detail, such as corrosion and marine growth on moving parts and the ability to provide good quality, long life seals. Secondly air entrained in the water may lead to cavitation erosion and low efficiency in compression.

One possibly novel and important idea which has grown out of the work by Wavepower Ltd on the raft is that of the tube pump rode. It is believed to be desirable to limit mooring loads by using compliant moorings. The raft team have developed a system which could, in principle, be used to extract power as well as offer such a mooring system. As the mooring is stretched, layers of fibre windings straighten, thus reducing the volume of the line and compressing a core of rubber containing water. The device is not yet proven and it is possible that the large rotational and transverse shears (particularly in the rubber) may prove too great to give an acceptable lifetime but tests are proceeding.

Designing adequate mooring and power take-off systems remain two of the major problems with the raft concept. The design of a suitable hinge may also be technically difficult. One advantage of the design is its ability to be mass produced by ship building technology. Designs have been assessed using both steel and concrete. Steel construction would involve techniques used in the construction of super-tankers. Each raft assembly (mean power rating about 1 MW(e)) would be 50 m wide, 100 m long and 8 m deep and 12 000 tonnes of ballast water stored in bulkheads would be required to bring the 3300 tonne steel hull to draft of 3 m. The major disadvantage of steel is likely to be corrosion and much attention would need to be paid to providing adequate protection. Concrete, on the other hand has high corrosion resistance and durability but is rather heavy. Wavepower Ltd and others have carried out fairly detailed studies of the production facilities required and the feasible rates of production. They have argued that the ability to mass produce their structures in modular form could significantly reduce costs. Despite this, however, costs for the raft look significantly higher than for some other devices and interest in the device has recently waned.

The Russell (HRS) rectifier. The rectifer designed by H R Russell of the Hydraulic Research Station in 1975, is a deceptively simple device. It has some features in common with the Mauritius ramp; early versions of the rectifier were, in fact, designed with the island of Mauritius in mind.

Each module [4], [15] is in the form of a caisson divided into an upper and a lower reservoir with the seaward side containing a large number of one-way flap valves mounted on vertically aligned hinges. These are arranged alternately to let water flow in on the upper level and out at the lower level. Low head Kaplan turbines are mounted between the upper and lower reservoirs on each device as shown in figure 3.14. The head is maintained by operating the inlet valves such that water is collected in the upper reservoir at a wave crest and the outlet valves expel water from the lower reservoir at each trough.

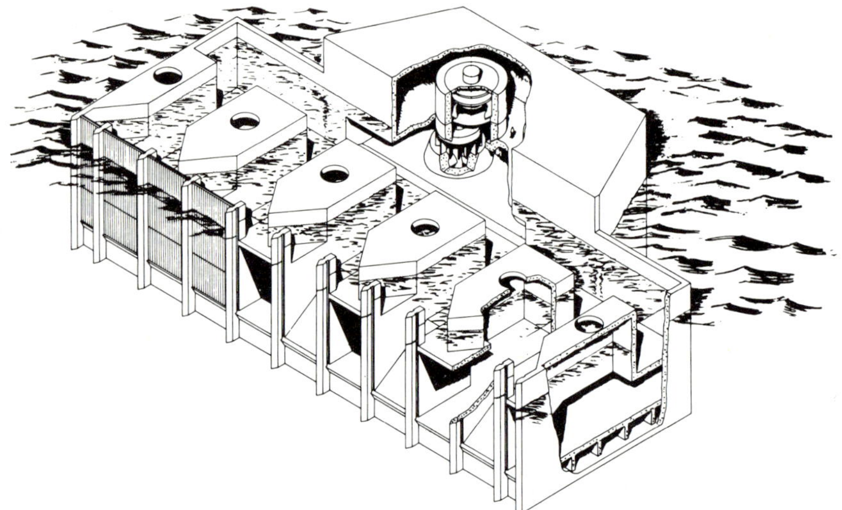

Figure 3.14 Cut-away diagram of the HRS rectifier. (Courtesy ETSU.)

Detailed design of the system of gates presents a number of difficult problems. Firstly, the horizontal movement of the water beneath the wave must be large compared with the gate width or else the gate will not open sufficiently. Secondly, the dynamic pressure of the waves decreases with depth (see §3.1.2) and this is opposed by a uniform hydrostatic head across the front of the gate; thus the gates tend to open differentially with depth. Long, narrow, torsionally flexible gates are therefore required for each compartment. The overall size of each compartment must also be designed taking into account the wavelength of the incoming water. Theoretical studies have shown that each inlet channel should be less than about 1/5 of a wavelength in width and the back to front distance should be such that minimal destructive interference can occur between incident waves and reflected waves from the back of the chamber. These features have been successfully modelled at 1/30 scale [15].

The civil engineering of the device would involve constructing caissons in dockyards, floating them out and sinking them onto a specially prepared area of seabed in about 15 m of water. Each device would be about 100 m long, with a peak output of about 3 MW. Each caisson would be built from about 50 000 to 60 000 tonnes of concrete. The large amount of material and the relatively low energy available in the shallow water appears to make the device look uneconomic. The technical problems to be overcome in designing suitable flap valves and avoiding fouling from marine debris have not been fully overcome. For these reasons, it is probably fair to say that the rectifier now looks one of the least attractive of the devices which have been pursued in some detail in the UK wave energy programme.

The oscillating water column. Several types of owc have been studied in the UK including the Vickers devices and the Belfast device. The Japanese Kaimei is also of the same general type. These will be described below, but in this section the floating and bottom mounted devices designed by the NEL will be discussed [4], [16]. An artist's impression of such a device is shown in figure 3.15.

Figure 3.15 Artist's impression of the NEL oscillating water column. (Reproduced from *Energy Paper* 42, courtesy HMSO.)

Once again the concept is basically simple. In its very simplest form it is no more than a box held partly below the water surface with its bottom open and with a closed top containing an orifice. The motion of the waves then causes the water in the box to move up and down and this induces an oscillating flow of air through the orifice. The principle is illustrated in figure 3.16. The device may thus be classified as a tuned resonator and can be either a terminator or point absorber. The major design problems are to produce a structure which is economic in material and which gives highly efficient flow and hydrodynamic stability, and to develop a method of rectifying the air flow or a turbine which will accept the oscillating flow of air and efficiently convert it to electricity. If this latter objective can be achieved, owc devices can offer one of the simplest power take-off methods for wave energy devices.

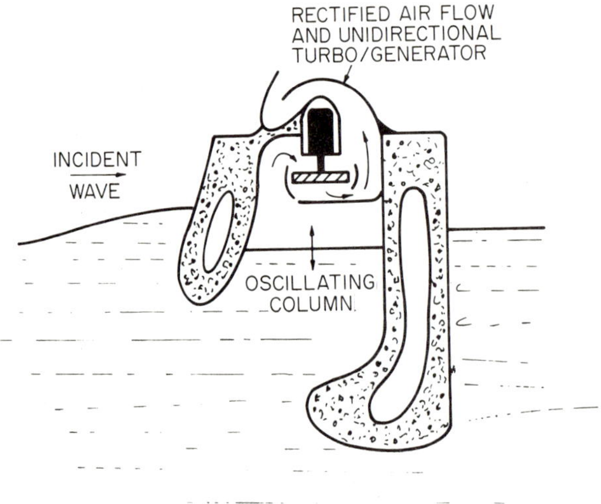

Figure 3.16 Advanced air bell of the type proposed by NEL.

Very considerable theoretical and small-scale experimental effort has gone into designing the optimum size and shape of device. It was found necessary to make the rear wall of the device deeper than the front in order to reduce the size of the transmitted wave. It was also realised that the best performance of a floating device would be achieved when the natural frequency of the structure in heave corresponded to the wave excitation frequency. Performance also depends on the damping of the column of air introduced by the orifice, i.e. on orifice size. Bearing these factors in mind, the NEL team investigated a range of shapes of device and found that the basic shape illustrated in figure 3.17(*e*) provided the best overall perform-ance. If it was made from reinforced concrete the shape shown in figure

3.17(*c*) would probably be preferred for production reasons. With the optimum shape and aperture size, efficiencies of about 90% can be achieved for monochromatic waves in tank tests and the bandwidth can be made wider than for many other devices. In fact, three-dimensional devices can, in principle, be designed to achieve capture ratios in excess of unity (where capture ratio is defined as the energy extracted compared with the energy incident in a length of wave equal to the diameter of the device). Experiments have confirmed that very high capture ratios can be achieved in the point absorber mode, but this is at the expense of bandwidth.

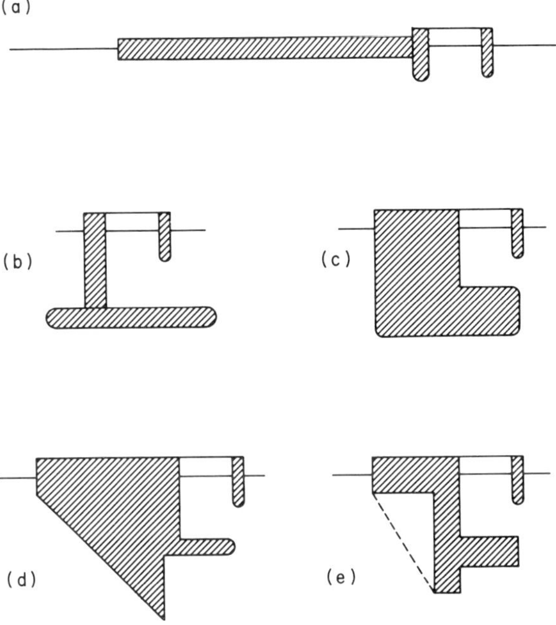

Figure 3.17 Some of the shapes of oscillating water columns which have been tested. (Courtesy NEL and ETSU.)

Having determined an optimum shape, it is necessary to engineer a device which can withstand the cyclic forces imposed by the incident waves. Three approaches are possible: seabed mounting, the use of spines and resisting the forces locally. The NEL team have examined each of these and have produced seabed mounted and floating versions of the device on prestressed concrete spines. Producing seabed mounted devices is clearly a practical way of reacting to wave forces, but will be very expensive in deep water. Its economics therefore depend upon whether sufficient wave energy will be available in moderate depths of water. Early results suggested that the reduction in energy with depth off northwest Scotland was very appreciable and bottom mounted devices appeared to offer little hope of ever being economic. Recently, however, there have been new measurements which suggest that these measured power densities were anomalously low. If this is the case, bottom mounted structures (break waters) may re-emerge as candidates for further development. This possibility is reinforced by the fact (see §3.1.8) that mooring costs for floating devices are likely to be high. Most effort is now being concentrated on the breakwater design. The NEL 1980 reference design for a bottom mounted device is illustrated in the cut-away diagram (figure 3.18).

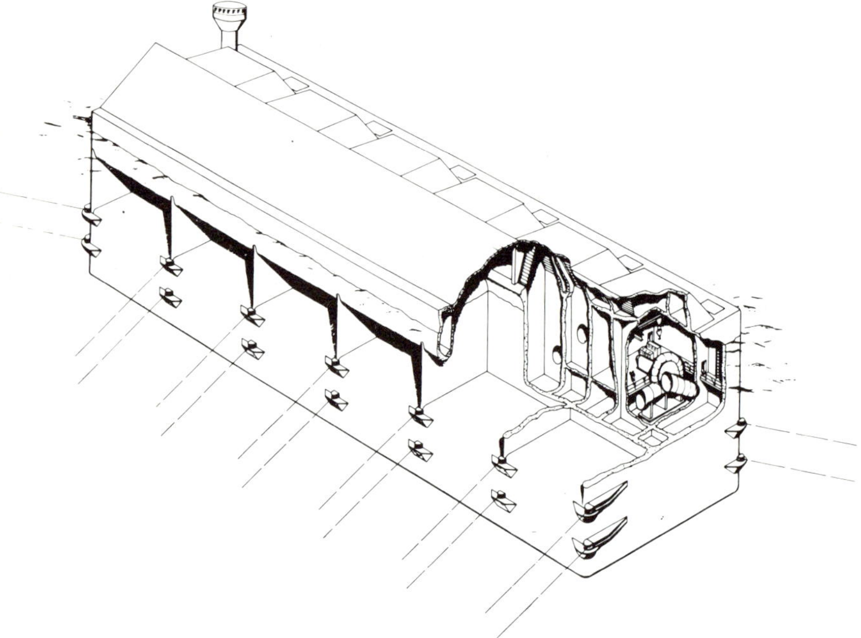

Figure 3.18 The NEL breakwater-type wave piston (1980 design). (Courtesy ETSU.)

There are at least two options for power take-off. Ideally the turbine system requires achievement of high efficiency over a wide power range, combined with good reliability, an ability to throttle back to dissipate excess power and a tolerance (or avoidance) of water in the air stream. Much work has been carried out by NEL on valve systems in the form of louvres (as illustrated schematically in figure 3.19) to rectify the airflow. As the water rises, the air column passes through the turbine from left to right and continues to do so even when the column is falling. With this rectified airflow, a Francis turbine is probably the favoured method of generating electricity, with one turbine for each of the units on the spine. The possibility of manifolding outputs is being studied.

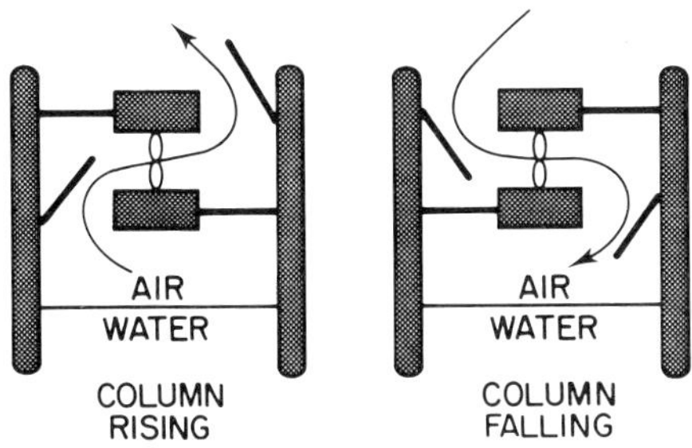

Figure 3.19 Diagram showing the principle of four-valve rectification. (Courtesy NEL and ETSU.)

Another possibility is to use a turbine developed by Dr A Wells of Queen's University, Belfast [17]. The Wells turbine rotates in the same direction and produces power whatever the direction of the airflow. The device has been investigated in some detail at the CEGB's Marchwood Engineering Laboratory (MEL) [18]. In its simplest form this self-rectifying turbine consists of a single row of rotor blades with guide vanes either side, mounted in an annular duct. Figure 3.20 shows such an arrangement. Simple aerodynamic arguments show that the blades move in the same direction regardless of the direction from which the air approaches. Theory and experimental tests both show high efficiencies (up to 70%) for a fairly wide range of flow rates. The conclusion of the MEL work is that the Wells turbine appears to be a practical solution to power take-off on owc devices and can lead to a much simplified design. The Wells turbine was tested on the Kaimei device. However, it can be argued that a central turbine serving a series of owc modules will be economically more attractive.

Reference was made above to the use of these devices as three-dimensional point absorbers. Falnes and Budal [19], working in Norway, have suggested that the fairly strong peak in the amplitude response at a particular frequency of the devices tested so far, could be overcome by an appropriate control philosophy. They have pointed out that all resonant systems exhibit a phaseshift in their response to an excitation as the exciting frequency moves away from the natural frequency. This power output depends not only on force and velocity, but also on the phase angle between them. Falnes and Budal suggest that if the device can be restrained in its motion until the incident wave is at a peak or trough, the force and velocity may be kept in phase over a wider bandwidth thus much improving efficiency. This form of intelligent control, or latching as it is known, may also be applicable to other devices and is an example of how new ideas in wave energy could, in principle, have a radical effect on its prospects. The ideas of Falnes and Budal are now being pursued in the UK as well as in Norway [20].

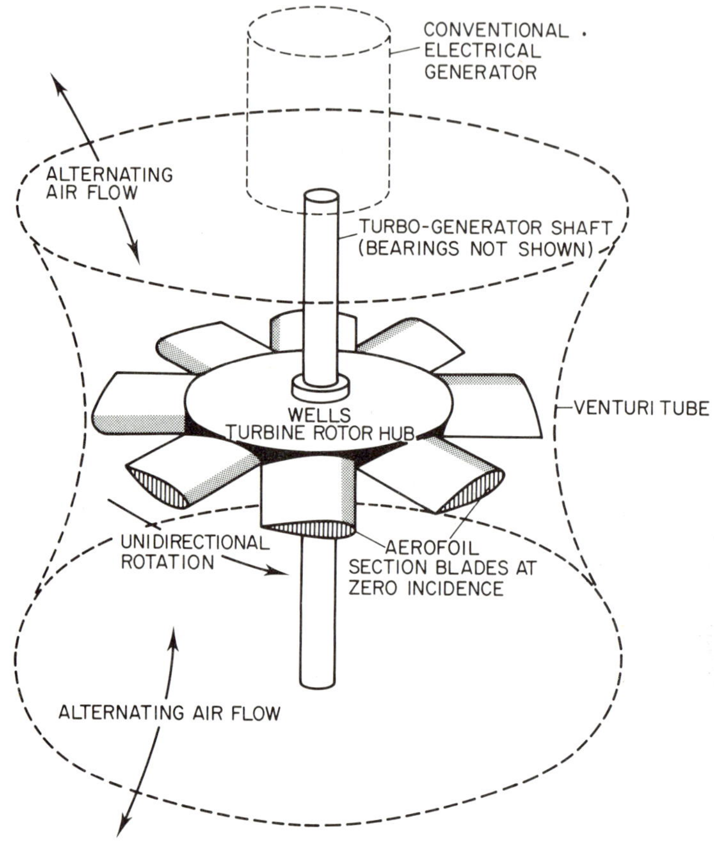

Figure 3.20 The Wells self-rectifying turbine. (By permission of The Queen's University of Belfast, Department of Civil Engineering.)

The Bristol cylinder. This device emerged from theoretical studies carried out by Dr D Evans of the University of Bristol in 1977 [21]. Starting from the fact that if the axis of a submerged cylinder is rotated orbitally, waves are produced downstream, Evans showed that the reverse was also true. Thus the motion of a wave over the submerged cylinder causes the axis of the cylinder to orbit (not necessarily with circular motion) about its static position. If a suitable power take-off system can be designed, such a device would clearly offer a potentially simple means of extracting wave energy.

Early experimental studies carried out by Evans and Shaw at Bristol showed that in tank tests, the device could convert wave energy with efficiencies of over 90% for monochromatic waves. By reducing the mass of the cylinder, high efficiencies could be maintained over a fairly broad bandwidth.

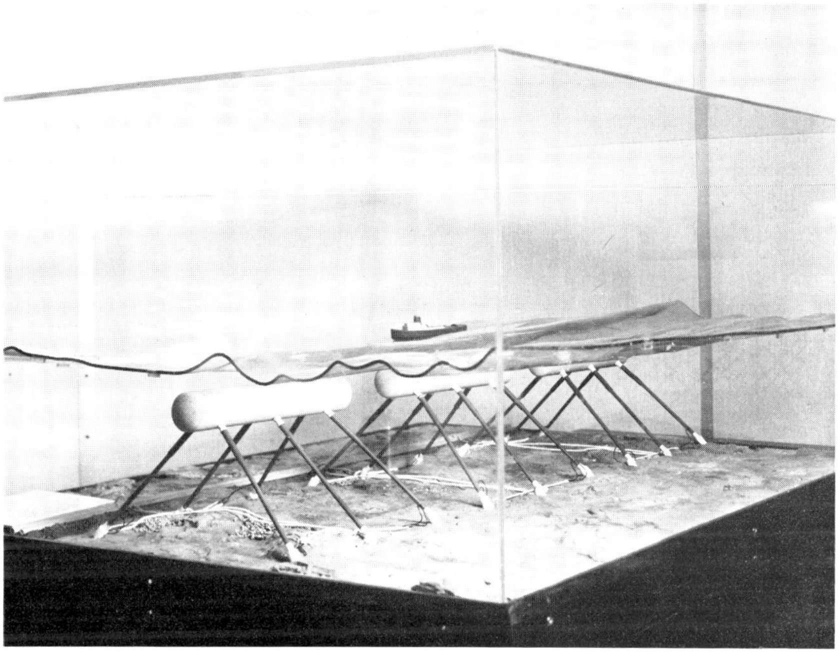

Figure 3.21 Tank model of the Bristol cylinder device. (Courtesy Dr T Shaw, Sir Robert McAlpine and Sons Ltd.)

In 1979, a more detailed engineering design was initiated under the leadership of Sir Robert McAlpine and Sons. A model of such a device is shown in figure 3.21. Although still at a relatively early stage, significant progress has been made in producing a reference design [22]. In its 1981 form, the power take-off, which also acts as a spring forming a resonant system with the cylinder, consists of a pair of double-acting pumps with four hydraulic rams and nitrogen accumulators at each of the mooring lines. These are positioned at each corner of the cylinder. The pumps operate on the outward stroke, whilst on the inward stroke, power is provided from the rams and accumulators. High pressure water pumped through this system is carried via a collector main to a platform mounted turbine which is operated by a hundred or more pump systems. The mooring system would probably employ chains but it is conceivable that tube pumps could be used. The cylinders would be fairly straightforward in design. Each would be about 12 m in diameter and a minimum of 50 m long and would be held with its axis about 12 m below mean sea level. The necessary density (about 0.6) could be achieved using concrete construction.

Although it is still too early to know whether the cylinder will be sufficiently cheap in a final optimised design, it offers in addition to its inherent simplicity some other encouraging features. These include its apparent ability to capture energy beyond its nominal width, its potential for employing intelligent control, the fact that the seabed is used as a reference frame for power take-off and its apparent ability to self-limit its output and thus survive the worst wave conditions.

The cylinder is one of the devices in the UK which has recently received particular attention in order to bring its state of development up to that of the four earlier devices studied.

The Lancaster flexible bag. This device, invented by Professor French at Lancaster University has been developed recently in collaboration with Wave Power Ltd. It takes the form of a series of air bags mounted on a submerged hull [23], [24]. The device relies for its output on passing wave crests squeezing air out of the bags from low to high pressure ducts mounted on the hull. The flexible bag system was designed as an attenuator with the bows of the ship-like device facing into the incident waves, although recent versions assume an orientation between that of an attenuator and a terminator. One theoretical advantage of the system is that it is not a strongly resonant device; it can, in principle, give high efficiencies over a very wide bandwidth.

The 1979/80 version of the device is shown in figure 3.22. The two long bags are divided by flexible membranes (septa) and each compartment is successively compressed by a wave crest and deflated at a trough. This continues down the length of the device as each wave passes along it. The pulses of air fed into the high pressure air duct via non-return valves, are used to drive air turbines mounted in the centre of the device in a conning tower or at each end of the hull. The top flanges of an I-beam hull house the high and low pressure ducts possibly fabricated from glass-reinforced plastic (GRP), while the lower flanges contain water for ballast. In the most recent designs (1981), the I-beam has been replaced by a hull with a much broader beam.

At full scale, each unit would be made of mass produced prestressed concrete modules. Each bag, made of inch-thick fabric-reinforced rubber, would be divided into ten or twelve compartments which would be capable of independent operation if bags were punctured. The peak output of each device of this design has been estimated to be about 5 MW, although longer devices now being studied would give substantially higher outputs.

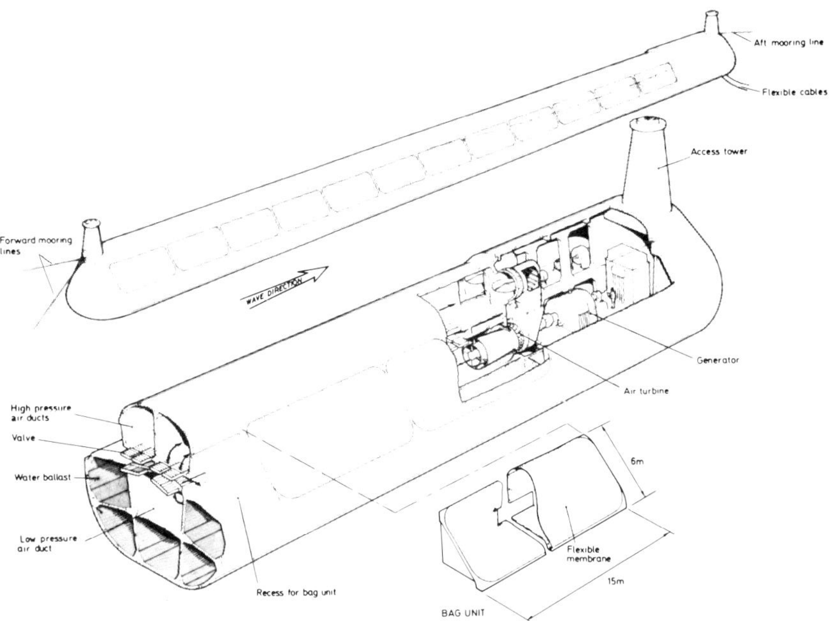

Figure 3.22 The Lancaster flexible bag wave energy converter. (Courtesy Mr J Platts, Wavepower Ltd and reproduced from reference [23] with permission of the BHRA.)

Initial tank tests on a small scale showed the feasibility of the device and more extensive testing is now being carried out. The device offers, in principle, the potential of cheap fabrication and has the advantage over some other floating devices in that its mooring forces and bending moments appear to be low. Work has proceeded on providing adequate pitch and roll stability and in investigating the best material to be used for the bags and septa. The precise shape of the bags under load is important. An S-shape [22] appears to be best (see inset in figure 3.22). The question of the integrity of the flexible materials remains one of the major unknowns in the design. For the projected design life of five years, the bags would encounter more than 20 million waves and would clearly require to resist damage from floating objects and general degradation including ultraviolet solar radiation. Such a performance is probably close to the limits of current technology but tests on quarter-scale bags are planned and this should yield much more information about the lifetimes that might be expected for the bags.

The Sea Energy Associates clam. The clam device, developed by Sea Energy Associates (SEA) (an industrially backed group centred on a team

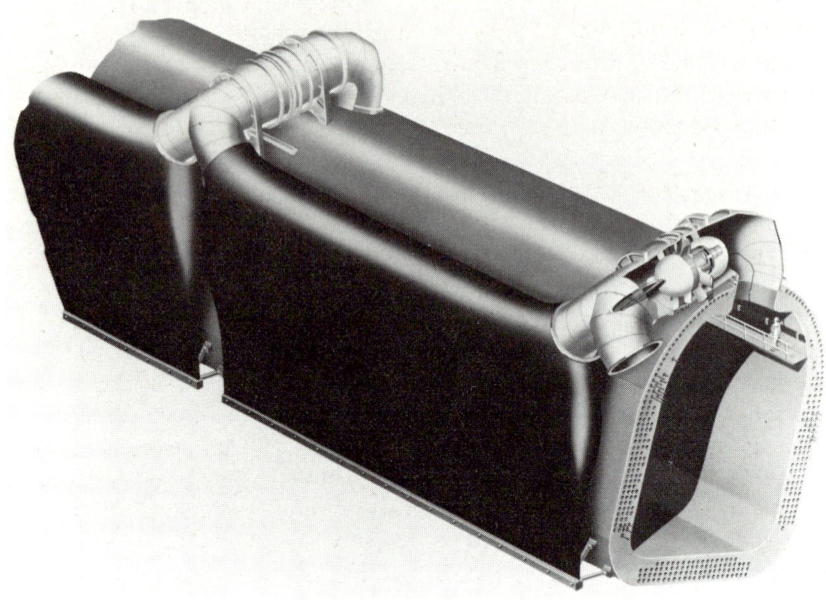

Figure 3.23 (a) Latest design of the SEA clam device and (b) impression of device in use. (Courtesy Dr N Bellamy, Coventry (Lanchester) Polytechnic.)

at Lanchester Polytechnic) since 1978, emerged as an offshoot of SEA work on the Salter duck [25]. In common with the duck, a floating spine is used as a frame of reference for the device. On the spine are mounted ten 30 m long, 15 m deep clam units which until recently were in the form of a steel flap connected to the spine at the bottom by a hinge (figure 3.23). As waves hit the front of the device, the opening and closing of the flap (by about 30°), pushes air through a series of large internal air bags. The air flow (in both directions) is used to drive a vertically mounted Wells turbine and is returned at low pressure back to the bags to close the circuit. All the clam units on a spine operate out of phase with each other, some opening and some closing at any time, and each clam has its own air turbine and alternator.

Considerable effort has gone into designing the rectangular section concrete spine. Long spines as envisaged by Salter need to be compliant in order to minimise bending moments. The SEA team have pursued a different philosophy, designing a short inflexible spine. Tests at 1/10 scale on Loch Ness and theoretical analyses are in good agreement in evaluating the bending moments which appear to be acceptable. There appeared at one time to be some danger that if the spine leaked it would waterlog and sink but this possible weakness has now been rectified.

The device, being essentially a non-resonant terminator, has to withstand considerable mooring forces but current designs envisage mooring at an angle of about 70° to the incident waves and allow the device to swing through 40° in either direction in order to maximise energy capture from waves approaching the device away from the prevailing direction.

The hinges, which until recently have been a major feature of the design, are standard on large earth-moving equipment and are used in very demanding environments. The fact that there were doubts about the ability of the hinge to survive in the conditions experienced at sea was the major reason for designing it out of the device. In the most recent designs the hinge has been replaced by a membrane which moves in and out in response to incident waves. As with the Lancaster system the long-term integrity of the bag system is yet to be proved, particularly in regard to abrasion, kinking and fatigue. Furthermore, the form of the bags will change with depth because of the variation of hydrostatic pressure and this may present problems of distortion. Because there are many independent units on a spine, loss of output from one clam would not drastically impair the performance of the system. Any unit could be independently switched off by venting its air system.

The clam and the Lancaster bag appear in many respects to be coming closer together in design.

Kaimei. Kaimei (Japanese for sea light) is a floating OWC attenuator device tested at large scale as part of the Japanese wave power programme [26],

[27]. The project has been supported by the IEA and experiments from the UK and USA have been carried out on the vessel. Figure 3.24 shows a photograph of the vessel moored in the Sea of Japan in 1978.

Figure 3.24 The Japanese Kaimei experimental ship.

The Kaimei vessel resembles a 2000 tonne ship moored head-on to the waves, but is in fact a buoy 80 m long and 12 m wide. The underside of the vessel is open and contains twenty-two chambers in which water oscillates and pumps air through to turbines mounted on the deck.

The first trials were carried out in the Sea of Japan in 1978–9 and despite anomalously low measured wave heights, useful operating information was obtained. The device was originally designed for a mean wave period of 6 s and the 80 m length was chosen for this reason. It became apparent that a more suitable choice of wave period would have been about 8 s with a consequent lengthening of the vessel. Some damage to valve flaps was incurred but generally the device survived well. Outputs from the generators mounted fore and aft gave about 60% greater output than those mounted at the centre. Mooring forces were found to be less than expected.

In August 1979, the second trials with the UK experiment (the US device was not available) commenced. A British-made air turbine appeared to operate well but overall the output from the buoy (some of which was taken to shore by submarine cable) was disappointingly low giving less than 1/50 of its expected average output. Many lessons were learnt including the need for improved positioning of the chamber openings and the need for a longer vessel. In some respects the unit was a success; two typhoons (one generating 10 m waves) were survived without major damage and repairs were carried out successfully. Results from the Canadian instrumentation on the vessel are still being analysed in detail.

Vickers oscillating water columns. The original Vickers device was in the form of a submerged duct which operated as a resonant point absorber [4]. To overcome the high structural costs often inherent in devices employing a large volume of air being compressed, the Vickers team have developed the original design into a family of devices [28]. These are still being developed.

One of the original Vickers duct designs and its operating cycle is shown in figure 3.25. It is a point absorber. A submerged tube is connected to a closed air reservoir at one end and the other end is left open to the sea. The length of the water column is chosen to be in resonance with the dominant wave frequency and the resonating column causes water to spill over into a reservoir. The air pressure in the reservoir determines the mean water column level, the damping applied to the water column oscillations, and hence the amount of water overspilling into the reservoir. The outflow of water from the reservoir, through a turbine, is controlled to maintain a high water level in the reservoir, minimising wasteful freefall into it. The difficulty, as mentioned above, is that both the air and water reservoirs need to be very large to achieve the right conditions, leading to a relatively large and expensive structure.

In the twin owc, the requirement for a large volume of air is overcome by linking the resonant air column to a second column adjacent to it. This is illustrated in figure 3.26. The concept can be either in the form of a bottom mounted terminator or an array of point absorbers. To ensure differential movement of the linked columns, the inlets are placed at different vertical heights—one primary inlet on the top of the device and the secondary inlet on the base. The primary inlet thus feels the full effect of passing waves whilst the secondary feels a much decayed pressure variation. The orientation of the inlets also produces a phase difference in the pressures acting on them due to surge wave motions. Power is extracted using a Wells turbine. A more efficient terminator using a reflector extending to mean surface level was designed in 1980. The most recent design of the twin column terminator (late 1981) does not use an auxiliary water column but relies instead on phase differences between cells. This device incorporates rectifying valves for each cell of the device which feed common high pressure and low pressure manifolds. Power is generated by axial-flow air turbines housed in special extensions of the plinth structure on which the device is mounted.

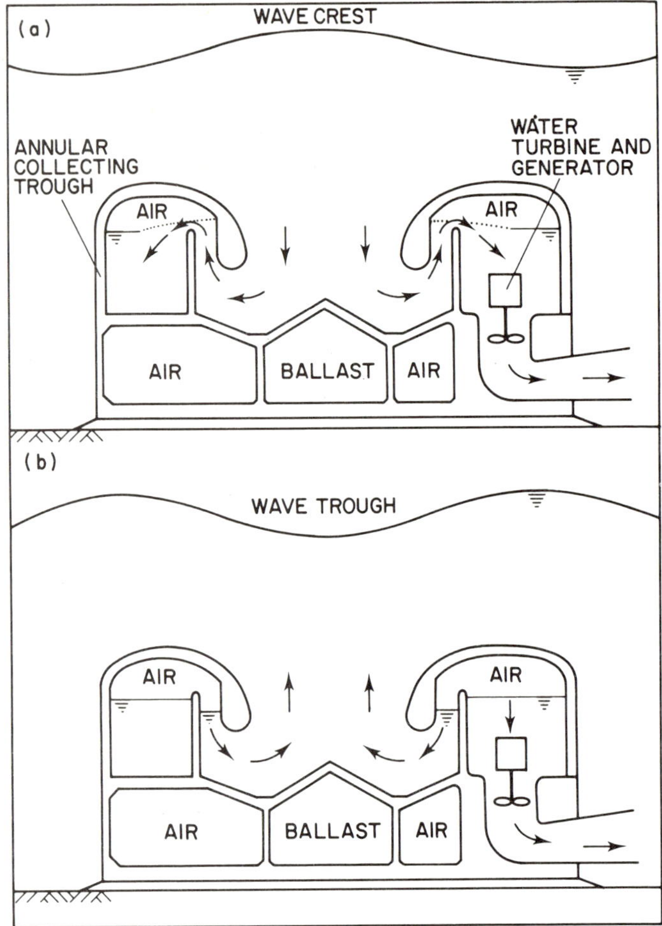

Figure 3.25 Design of original Vickers duct and illustration of its operating cycle. (a) Water flows into column, spills over the weir into the annular collecting trough and increases the air pressure. Water in the trough is forced by the excess air pressure through the water turbine. (b) Wave flows out of the column. The level in the trough falls below the top of the weir. Water continues to be forced through the turbine by excess air pressure. (Reproduced from *Energy Paper* 42, courtesy HMSO.)

The other main line of Vickers devices is submerged resonant attenuators, known as wave chambers. The most recent designs of this device comprise a long submerged chamber containing a free surface of

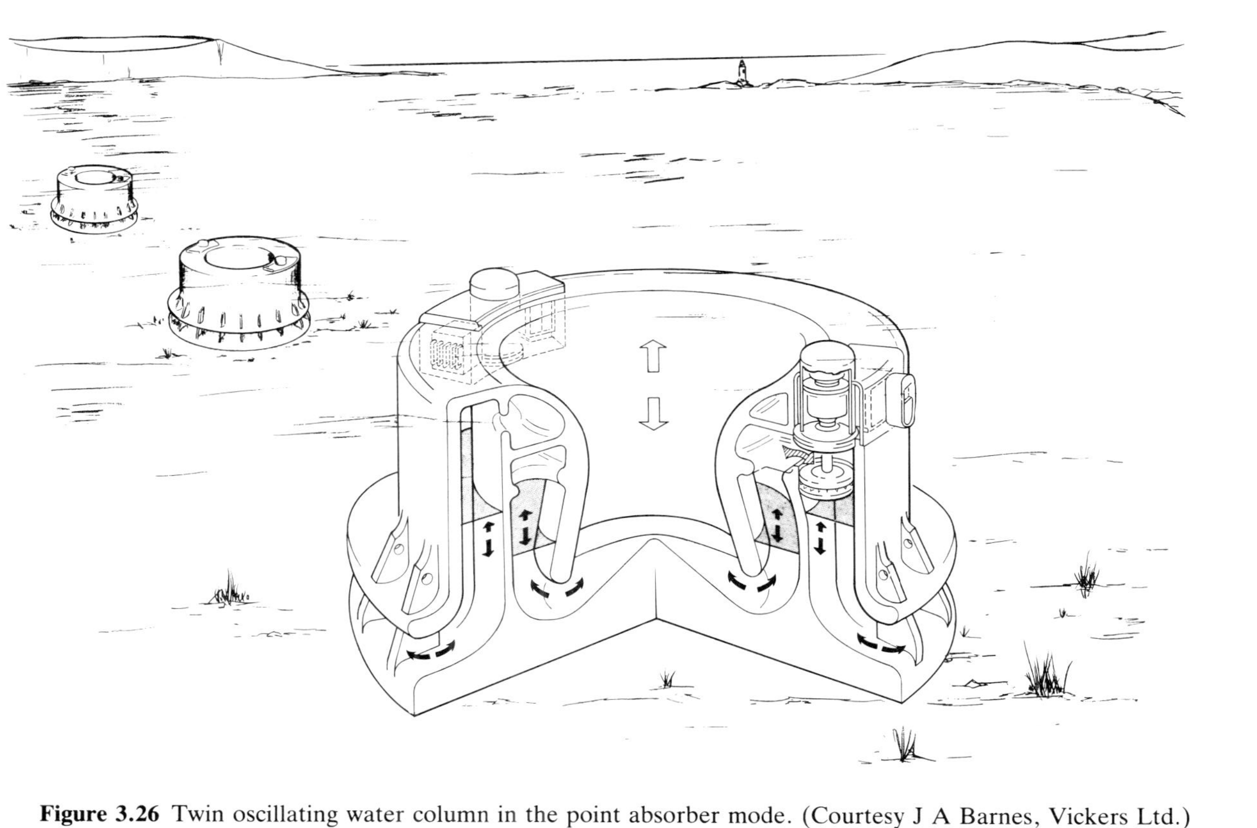

Figure 3.26 Twin oscillating water column in the point absorber mode. (Courtesy J A Barnes, Vickers Ltd.)

water, connected to open channels running along either side (figure 3.27). The chamber contains eight compartments, the extremities being attached to the seabed by precast concrete trestles. A central caisson piercing the surface at the centre of the device, like a submarine conning tower, houses a single axial-flow air turbine which is connected to common high pressure and low pressure mainfolds, providing a rectified air flow.

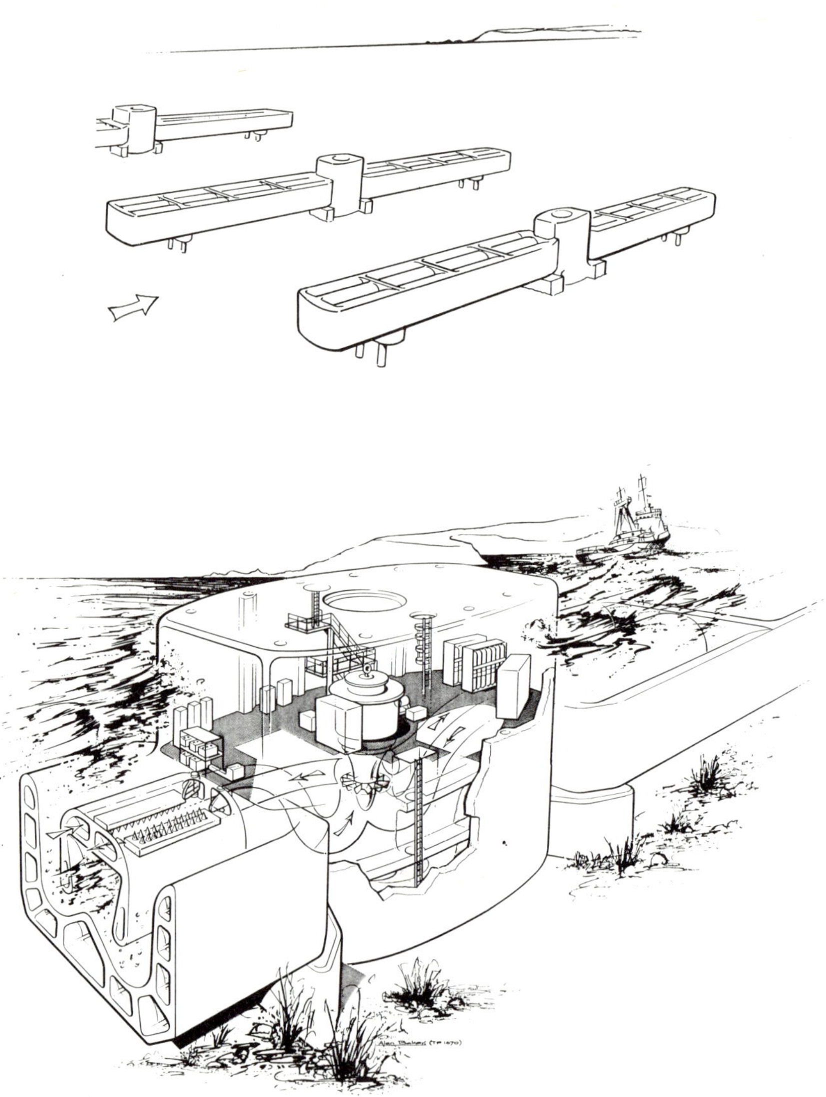

Figure 3.27 Vickers wave chamber (December 1981 design). (Courtesy J A Barnes, Vickers Ltd.)

Belfast buoy. This is an owc operating as a point absorber. The concept of the Wells turbine was originally developed for use with this device.

The original design consisted of two parts: an upper steel canopy supported on a steel buoyancy ring and a base caisson made of concrete and connected to the upper section by radial fins and a central steel shaft (figure 3.28). The device was ballasted so that it floated just below the water surface; the lower rim or buoyancy ring at such a depth that the action of the waves would drive water below the rim and in or out of the space below the canopy. This water motion would then force air through the vertically mounted turbine. In this floating design, it was thought possible that allowing the device to heave might augment the power output.

The concept has a number of attractions—mooring forces are relatively low and there are few moving parts. Simplicity has been a major aim of the development team.

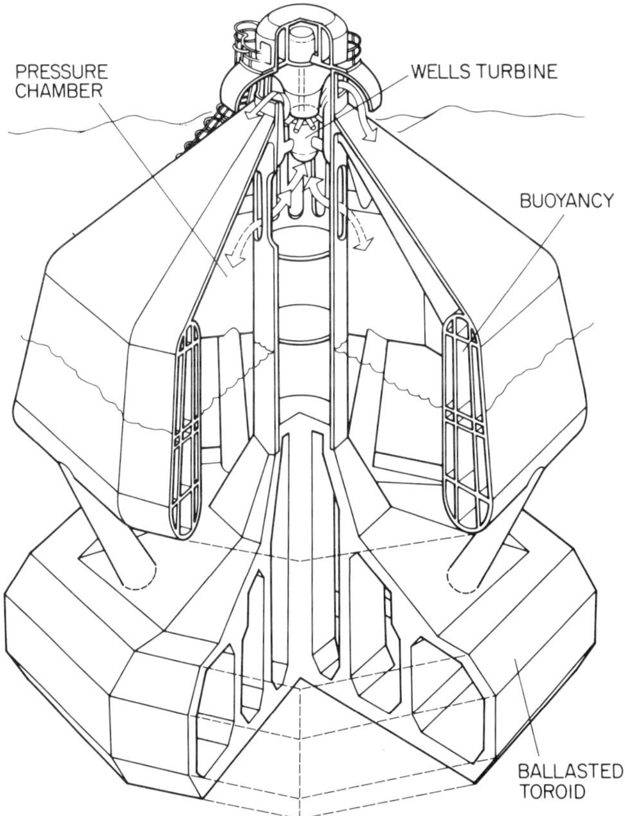

Figure 3.28 Original design for the Belfast buoy. (Courtesy The Queen's University of Belfast, Department of Civil Engineering.)

Recently, in order to reduce costs further, the buoy has been redesigned as a cylindrical concrete caisson attached to the seabed. Each unit, forming part of an array of point absorbers, would contain six water columns driving separate Wells turbines. These would be coupled to a common generator at the top of the device, above mean sea level.

The concept has been tested in wave tanks but a substantial model of the floating design employing a 4.5 m diameter fibreglass pressure chamber and generating 45 kV, was damaged in a storm on Strangford Lough in Northern Ireland in 1978. Unfortunately, this occurred before useful test results could be obtained.

The Royal Military College of Science triplate and the resonant raft (porpoise). The triplate, developed at laboratory scale at the Royal Military College of Science (RMCS) during 1978 [29], is designed to reduce cost by replacing structural material with a large volume of water. The principle is illustrated in figure 3.29. The plates B and C are fixed rigidly together and separated by half a wavelength, and this, in principle, ensures that they do not move horizontally. Plate A is a quarter wavelength in front of plate B (which reflects the waves) and thus this front plate moves horizontally against the frame of reference provided by B and C. In doing so, power can be extracted hydraulically.

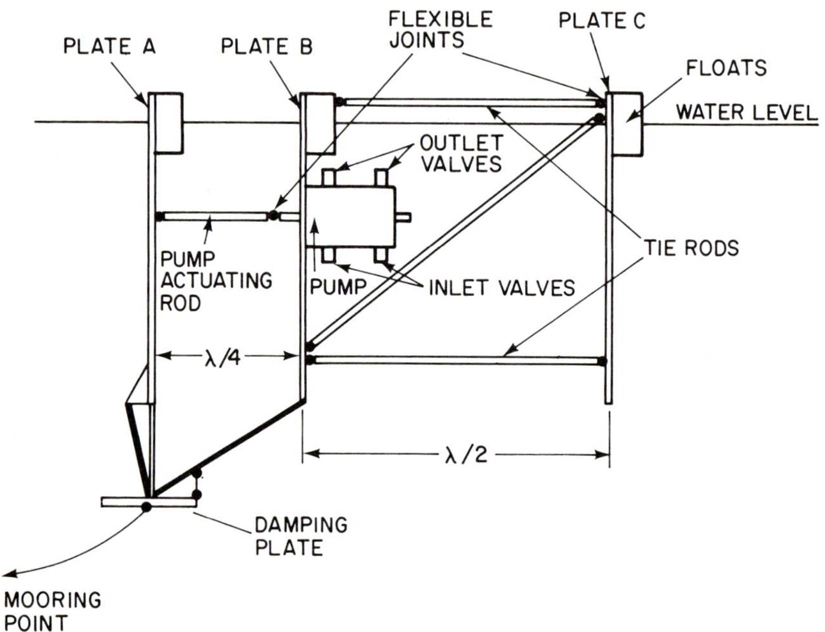

Figure 3.29 Triplate wave energy converter. (Courtesy of Professor Farley, RMCS Shrivenham and reproduced from reference [29].)

This elegant idea shares similar problems with many devices employing inertial frames of reference, particularly the large mooring forces. The tubular space frame is fairly complex and would experience a cyclic fatigue loading. Many hinges would be required in the structure and it is thus questionable as to whether the device could survive in the conditions to which it would be subjected.

The RMCS team have carried out tests on a small-scale version of another device called the porpoise [30]. This is in the form of a slender flexible beam or series of pontoons floating head-on to the waves. The structure is elastic and sufficiently thin to oscillate with the waves without damage. The raft is tensioned longitudinally by cables so that the structure buckles to enhance wave-induced vertical oscillations. These oscillations activate tube pumps incorporated in the structure which, it is suggested, can be intelligently controlled to ensure optimum operation.

Norwegian wave power devices. The work of Falnes and Budal on intelligent control of owc devices has already been discussed briefly. They first developed a heaving buoy in 1979 as part of the Norwegian wave energy programme which operated high pressure hydraulics. This was, however, abandoned because it was believed to be neither economically nor technically feasible.

In 1980, a new device was developed which is in the form of a conical buoy which can heave and drive an air turbine [31]. The buoy, being a point absorber, has a narrow bandwidth, and thus a computer fed with signals from a wave gauge is used to provide phase control. The system is being tested at 1/10 scale in a wave tank. A decision is likely to be made in late 1982 as to whether to build a large-scale version of this device for extensive sea trials.

Another Norwegian idea is to anchor platforms some distance from shore to act as a diffraction grating, funnelling water into a reservoir; the head of water built up in the reservoir in this way would drive a conventional hydroelectric station. It is not clear how seriously this rather ambitious scheme is being taken but there have been recent claims that it could be economic.

Lockheed dam-atoll. In 1979, the Lockheed Corporation in the USA announced a new device consisting of a dome-shaped structure, 75 m in diameter at full scale, floating just below the water surface [32]. The concept relies upon the fact that waves are focused by the shallow water around the dome and plunge into a 15 m diameter hole in the top of the dome. This water is guided by vanes into a swirling vortex in a central core where a turbine is housed. This is illustrated in figure 3.30.

It is not clear how this device would operate in real seas—particularly as the loss of energy on breaking at the dome and interference effects between waves may reduce power. A full engineering analysis is clearly required before this concept can be compared in cost and performance with the UK devices.

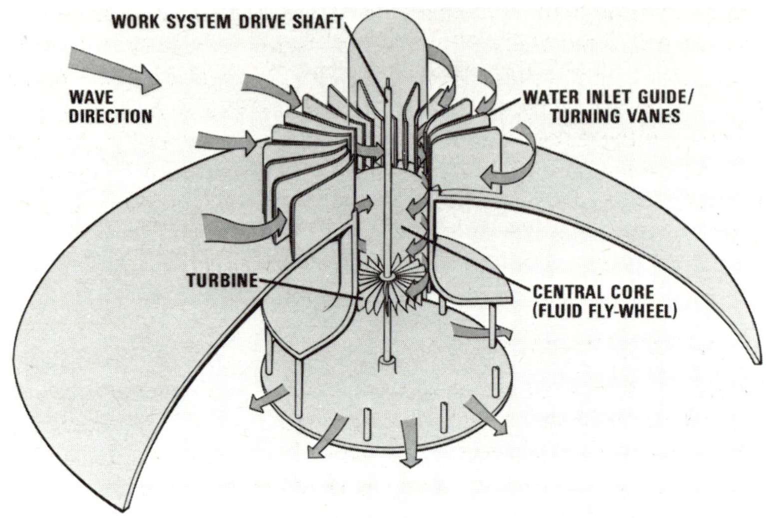

Figure 3.30 The Lockheed wave energy device. (Courtesy Lockheed Ocean Systems.)

The Mauritius ramp. A wave energy scheme for the island of Mauritius was studied in the early 1950s. This planned to use an existing coral reef as a foundation for a concrete ramp. Waves would run up the ramp and spill over into a reservoir formed between the reef and the shore. Originally it was planned to run a low head hydroelectric generator between the reservoir and the ocean. Because the storage in the reservoir would not be sufficient to maintain a continuous supply and because the head would be small (about 2 m) thus leading to low efficiency, Bott [33] suggested in 1975 that a second, higher reservoir should be constructed and water could be pumped to this by a hydraulic turbo-ram pump. The greater head of water (10–50 m) would allow conventional turbines to be used and peak loads to be met from the scheme. Economic evaluation of the project is complex because the scheme could be associated with commercial fish farming. The peak output would be about 5 MW.

In developing countries, where there is a suitable wave climate and particularly if a suitable shoreline topography exists, such schemes may prove a useful way of minimising oil consumption. More complex, high technology device concepts may demand technical resources which would not be available.

Other devices. Many other devices have been suggested. Work has been carried out at Sussex University on a solid cylinder moored beneath the surface and containing circular ducts through which water flows, while internal mechanical elements regulate the flow and absorb the wave energy. At Nottingham University a novel air buoy was designed at small scale involving a Savonius rotor. An alternative device described by the Lancaster team is the 'flounder'—a system of relatively small bodies of small positive buoyancy moving in surge and connected by tube pumps. Since none of these (or other devices omitted from this section) have been developed, nor look particularly promising, they will not be described in further detail in this section. Brief outlines of several of them are given in reference [34].

3.1.7 Structural response and materials

In designing wave energy conversion devices, as with ship design, it is essential to be able to predict the induced motions and loads on the structure, and various parts of it, for the full range of sea states. This then enables the structural response to be calculated and performance from the point of view of both fatigue and survival in extreme conditions to be evaluated. In practice this has been achieved in wave power research and development, both by theoretical and experimental means. The methods used have been extensively discussed in references [35] and [36].

Loads on the structure can be roughly divided into primary bending and twisting moments and direct and shear forces acting on major components of the structure. In addition, there are secondary loads such as hydrodynamic or inertial forces causing local deformations and stresses at hinges, mooring points etc. Linear response theory is used in studies aimed at solving the equations of motion which represent the mass, damping and stiffness from the structure, water, power take-off and mooring. Linear theory implies that the coefficients in the equation of motion are constant and the motions, loads, induced deformation and stresses are all proportional to wave height. This is generally considered to be an adequate approximation for assessing fatigue performance, but in analysing extrema, non-linear terms need to be taken into account and this is a more formidable problem.

In analysing structural response, i.e. deformations and stresses in a structure, fairly sophisticated finite element analysis computer programs are available. These would need to be used before a full-scale device could be built and operated. Already, using simpler analysis methods, it has been shown that the bending moments in conventional rigid spines as envisaged for the Salter duck, would be excessive and compliance needs to be built into the design.

Experimental work, either in a wave tank or in a scaled sea (e.g. Loch Ness or the Solent) has played a major role in allowing theory to be validated. Experiments also allow scale effects to be evaluated.

Study of structural design, both experimental and theoretical, also sets out to examine and design for the effects of breaking waves and impulsive slamming. In some locations, and in some parts of a device, forces arising from these motions can be of very considerable significance.

In order to achieve a long maintenance-free life, choice of materials for the main structure is clearly of great importance [4], [37]. Experience in the use of concrete at sea has been accelerated by the offshore oil programme. Concrete has advantages over steel with regard to corrosion resistance in the splash zone. However, unless series-produced using production line techniques, concrete would probably be too costly. If steel is used, corrosion resistant coatings and/or cathodic protection must be applied. In addition to marine corrosion, stray current corrosion and corrosion fatigue may also be important possible failure mechanisms and understanding of these phenomena is far from complete.

The extent of marine fouling (settlement and growth of marine plants and animals on the structure) will depend on location but could be important. It increases the weight, surface roughness, frequency of blockage of pipes, gates etc and can increase corrosion fatigue and brittle fracture. Removal can also damage protective layers, whilst non-removal makes routine inspection and maintenance more difficult.

As mentioned in §3.1.6 many devices have specific materials requirements. Use of rubber for bags in the Lancaster bag and clam are obvious examples. Use of glass-reinforced plastics, man-made fibres for mooring ropes and alloys for bearings and hinges need to be examined in the context of providing a long life in sea water.

3.1.8 Moorings

Unless seabed mounted, all devices need to be adequately moored and this presents formidable problems for large structures in some of the worst seas in the world [4], [38]. Thus much work requires to be carried out on providing safe moorings at an acceptable capital cost and with low maintenance costs.

Some compliance is necessary so that devices can reduce mooring forces by moving with the waves, but clearly compliance cannot be so great as to lead to collisions or crossed mooring lines. A corollary of this, is that compliance may decrease the available wave power resource by demanding large spacing between devices. Devices which operate beneath the surface, operate as attenuators or use long spines may present less of a problem from a mooring point of view.

The mooring line for a device could be a chain, steel-wire cable or man-made fibre rope. Chains tend to fail by abrasion at the joints, whilst steel cables need to be carefully protected against corrosion-fatigue. Little is known about the use of man-made fibre ropes for the type of duty demanded by wave power. Anchors would need to be larger than any currently in use; manufacture is not expected to be a particular problem, though testing and assessing their holding capability may be. Choice of anchor is clearly dependent upon the type of seabed to which the anchorage is made. In principle, the design of suitable shackles and swivels should not present major problems.

A considerable experimental and theoretical programme of research and development in this area is being pursued within the UK wave programme. For some devices, mooring may still present one of the greatest problems both in terms of producing designs at a reasonable capital cost and in minimising maintenance costs.

3.1.9 Reliability and maintenance—availability models

If wave power devices are to be sited many kilometres offshore in some of the roughest seas on earth, it will not be possible to man them or to carry out maintenance except during infrequent periods of calm. Thus reliability has to be high to minimise maintenance costs and maximise output.

Several studies of the reliability of various devices and transmission schemes have now been carried out as part of the UK programme and maintenance strategies have been formulated and costed. In some cases, it is estimated that they can constitute a very considerable part of costs when amortised over device life and can result in significant loss of output. Taylor [39] has recently estimated, for example, that the maintenance cost for a generalised wave device might be in the range 2–7 p/kWh. A device optimised for minimum maintenance offers the prospect of costs in the range 1–2 p/kWh.

The availability, A, of a component is a function of two variables, failure rate, f, and time to repair, t, and

$$A = (1 + ft)^{-1}.$$

The variable t, depends upon the actual time to repair, the delay due to inclement weather and any queueing for repair which is necessitated. Failure rate is also likely to be seasonal and to be correlated with inclement weather.

Models have been developed which seek to estimate the likely failure modes of devices, assume that these occur on a statistical basis and then cost lost output and the optimum number of repair teams, components held in stock and facilities such as support vessels, helicopters, workshops etc. The general conclusions of these models are that certain failure modes are likely to predominate—for example in control and cooling systems—and that these can be given particular attention in design, that 10% of the average annual output of systems might be lost during non-routine maintenance, and that, in general, reducing these losses by higher capital investment (e.g. redundancy in control systems) or higher operation and maintenance expenditure is likely to be the most advantageous strategy.

3.1.10 Output and transmission

Devices can produce output in different forms—electrical, chemical, hydraulic or as hot water—and each of these have been evaluated. The range of conceptual conversion and transmission possibilities is illustrated in schematic form in figure 3.31. The possibilities have been discussed in references [4] and [40]. It has even been suggested that wave power devices could be used to extract uranium from sea water.

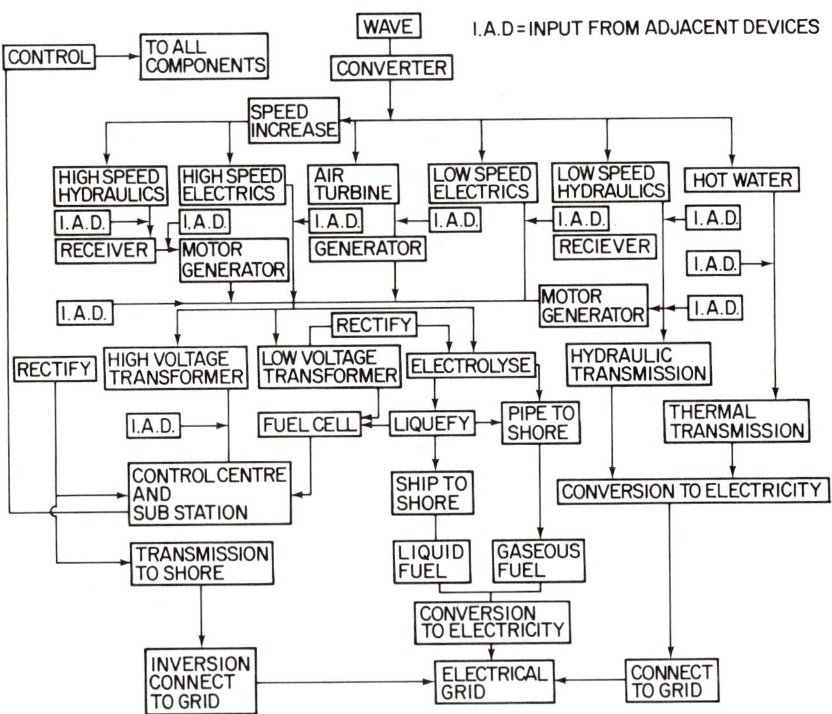

Figure 3.31 The range of conceptual conversion and transmission possibilities for wave energy systems.

The principle of thermal transmission is relatively simple, hot water being produced by a brake in the device, but is unlikely to find much application and has been effectively dismissed as a possible option. Hydraulic interconnection of devices may be very valuable in some device schemes, but hydraulic transmission to shore from ducks, rafts or cylinders would still need to be converted to electricity onshore and piping costs for the high pressures involved would be considerable. Production of hydrogen, ammonia or synthetic hydrocarbons on board the device looks costly. There are some problems in electricity generation, including, for floating systems, the design of flexible cables to allow for device motion, the mismatch between the availability of the wave power and requirements of the grid and the long distances from source to supply point (possibly involving environmental problems). Nonetheless, wave energy appears to be the best 'alternative' option in the UK.

If the electrical route is taken, the output of the device must be acceptable to the grid, i.e. reasonably smooth at a frequency of 50 Hz. However, typical output data from devices can be very spiky. Variations occur, for example, in the case of pneumatic devices twice during a wave, and wave heights vary such that peak power levels can be ten times the average level. This is illustrated for the Salter duck in figure 3.32. The output fed into the grid will depend upon many factors—the device itself, the number of essentially uncorrelated outputs being coupled together at the generator (diversity), any local storage and smoothing (such as the gyro on the duck or hydraulics) and the ratio of rated to mean power for the device. Furthermore, as with wind turbines, it would be uneconomic to rate generators and other equipment to convert peak outputs. A ratio of about three times the average power level is believed to be around the optimum for many wave power devices.

Preferred transmission schemes for electricity from wavepower have now been identified and costed [40], [41], [42]. It is still not certain whether AC or DC transmission is the best option but one of the presently favoured schemes assumes AC generation at an arbitrary frequency varying from machine to machine, each with locally controlled field forcing. Flexible cables then link the devices to the seabed on to a platform mounted transformer/rectifer station and these are series connected into a high voltage DC transmission to shore. An onshore inverter station then provides AC to the grid and achieves overall power control by current demand. Each inverter station takes an output of about 200 MW which it is assumed will be smoothed by the diversity of source and any local storage on the devices. A number of transmission schemes to supply this power to load centres in the UK have been evaluated and are shown for some possible geographical locations of wave energy converters in figure 3.33. The cost of transmission from the device to the supply point for a few gigawatts, has been estimated to be about 1–2 p/KWh, depending on

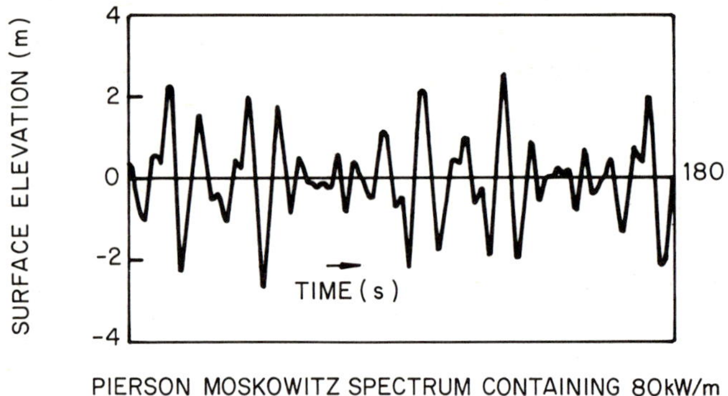

PIERSON MOSKOWITZ SPECTRUM CONTAINING 80kW/m

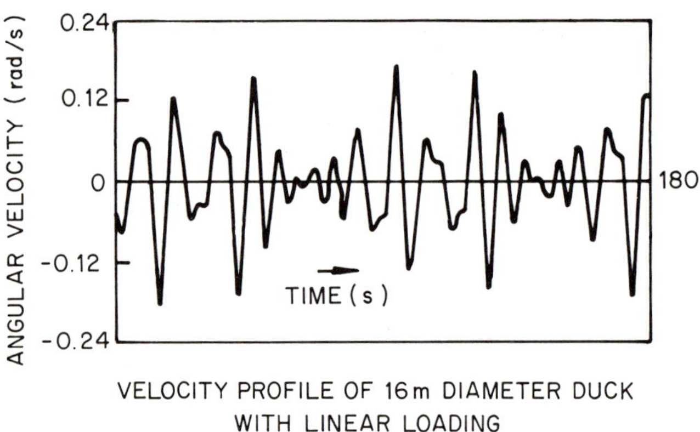

VELOCITY PROFILE OF 16m DIAMETER DUCK
WITH LINEAR LOADING

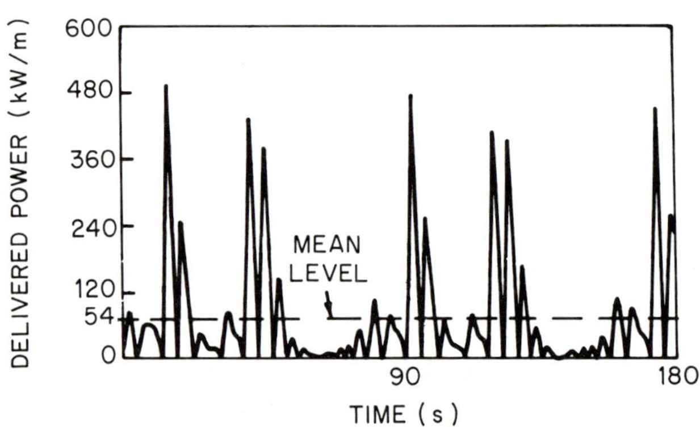

POWER DELIVERED BY DUCK WITH LINEAR LOADING

Figure 3.32 Variations in output from a Salter duck.

location. The costs clearly depend to a large degree on the amount of reinforcement required for the existing grid and this critically depends upon the wavepower capacity so these figures are very approximate.

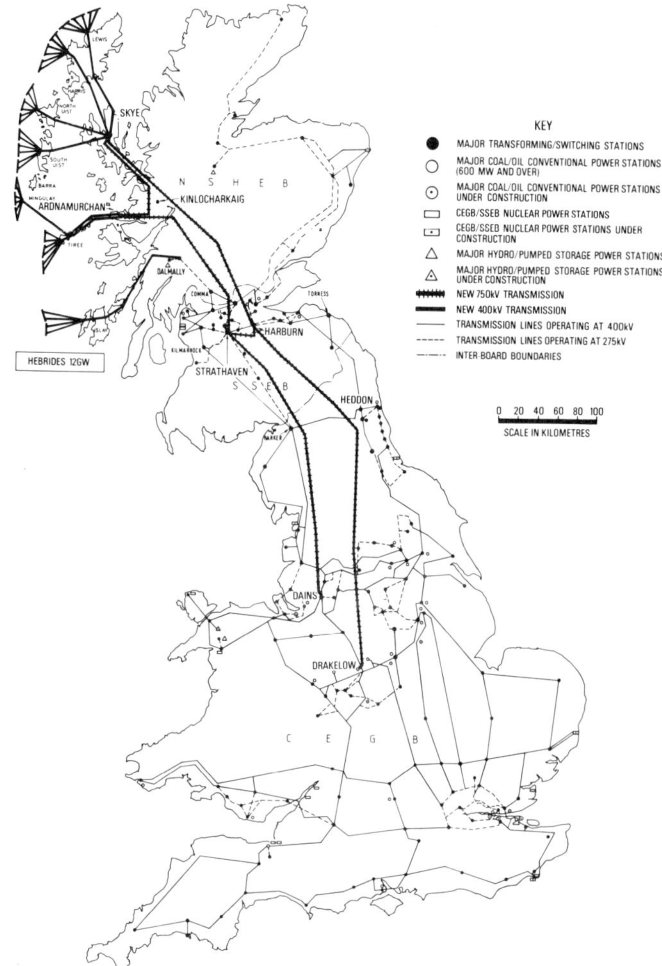

Figure 3.33 Transmission schemes for wave power in Great Britain.

System integration considerations, in terms of the fuel saving value and any capacity value for wind power have been reviewed in general terms in the last section of Chapter 2 and these should be generally applicable to wave power. Explicit calculations for wave power have not yet been carried out.

3.1.11 Environmental consequences

The environmental and social effects of wave power schemes are very complex [4] and will vary from location to location. The point should be made that, as with wind power, schemes would develop relatively slowly so catastrophic effects could be averted well in advance.

The consequences of developing wave power schemes can be characterised as effects on fisheries, navigation, the shoreline, population changes and the visual impact of electrical connections. The latter two are likely to give the greatest impact in areas such as northwest Scotland.

Studies have been carried out on the effects on certain types of fish and these effects seem to be quite small. The change in shoreline will depend upon the distance from shore of the converters, the bathymetry of the area and any diffraction effects arising from gaps between them. For northwest Scotland, at least, the effects are likely to be favourable. In poor weather, there will clearly be a danger of collision of large ships with the converters and they will need to be well marked. Reflected waves may lead to particularly hazardous waters for small boats close to the devices.

The main employment and social aspects would arise from construction and the provision of supply and maintenance bases. Construction may not be local but the bases almost certainly would be. These issues have not yet been studied in detail.

One of the chief objections and difficulties to wave power in northwest Scotland would undoubtedly be the environmental consequences of the electrical transmission lines spanning the highlands and islands.

3.1.12 Overall economics and resource

Various analyses of the costs of wave power have been carried out both by consultants to the DEn and by the various device teams [10], [43], [44]. Many of these are now out of date, however, and a further study of costs is currently proceeding at the time of writing (1982).

The available energy may be calculated from knowledge of the following factors:

(1) Power available to device. This will depend upon location and water depth. For the best UK location in northwest Scotland this may be about 50 kW/m as an annual average. This may be reduced by about 25% for some devices as a consequence of directionality.

(2) Mechanical efficiency. This varies from device to device, but is likely to be in the range 35–60%.

(3) Power conversion chain efficiency. Losses on converting the mechanical power to electrical power are likely to be in excess of 50%.

(4) Reliability. A reduction in the energy captured of between 10 and 30% has been assumed.

(5) Load factor. This again depends upon the device and its rating, but is unlikely to exceed 30%.

(6) Device spacing. Gaps for navigation will reduce the gross output from a line of devices by an estimated 20%.

These factors allow the total resource in the UK to be estimated and the total annual energy to be compared with the device capital costs, transmission costs and operation and maintenance costs in order to calculate the cost of energy in p/kWh for an assumed lifetime and discount rate (5% in the UK). Table 3.1, with data taken largely from [10], summarises the main features and most recently published costs of most of the devices discussed in this chapter, and it is clear that the costs of energy lie in the range 4–10 p/kWh in the view of the device teams, whilst independent studies suggest a range for the majority of devices from 5–15 p/kWh. When the first independent study of the various devices was carried out in 1978, costs were in the range 20–50 p/kWh [4] and the clear reduction in costs reflects the efforts put into design improvement.

As mentioned in the introduction, early estimates of the total resource were very optimistic. Using the figures in (1)–(6), Winter [2] has suggested that if wavepower devices were sited in the southwestern approaches to the UK, the northwest of Scotland and in the lower power density area off the northwest coast, a gross resource giving an average annual output of about 7 GW or an energy yield of about 60 TWh p.a. would be obtained. An independent CEGB analysis [3] has obtained a figure of 9.3 GW with a peak of 31 GW. There have been few attempts to analyse the world potential of wave energy but a rough indication of those parts of the world with wave energy potential is shown in figure 3.34.

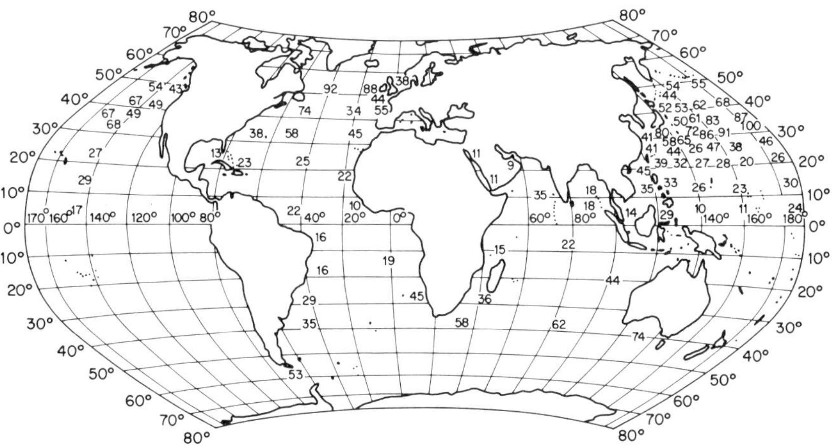

Figure 3.34 World-wide distribution of average wave power. Figures give estimates of power densities in W/m.

In conclusion, research in wave energy has made great strides in reducing costs and the best designs appear to be quite feasible in terms of the engineering required to build and operate them. Whether the early stage of development of wave power leads to the conclusion that there is potential for costs to be further reduced, or whether full-scale devices will again push up costs when the real engineering difficulties are tackled is clearly a matter for judgement. Costs for transmission, operation and maintenance and in some cases moorings, are more or less fixed and eat into the capital costs which would be allowed for the device itself if this is to be competitive, but it is to be stressed that much depends upon fossil fuel costs and the costs of competing technologies. Wave power remains an interesting insurance technology which in the view of the author needs to be further studied. In particular, new ideas should continue to be encouraged and experience gained of the problems which would require to be overcome in a wave power station.

In the UK, the direction of the wave power programme is likely to be decided in mid-1982, when a decision will be taken on whether to commit government funding to a prototype multi-megawatt demonstration device [45]. [In June 1982 the DEn concluded that a major sea trial should not be initiated at present. A limited programme of research to explore possible improvements in the technology is to be undertaken.]

3.2 Ocean thermal energy conversion (OTEC)

3.2.1 Introduction

One hundred years ago the French physicist D'Arsonval suggested that power could be extracted by exploiting the temperature difference between the warm surface layer and the cold deep water of tropical oceans (a principle illustrated in figure 3.35). Since the temperature difference will never be much more than 25 °C, efficiencies cannot be expected to exceed a few per cent. Nonetheless there is clearly a vast supply of solar heat energy available in this form which could, in principle, lead to a constant supply of electricity. The subject has recently been reviewed by Lavi [46].

In 1929 Claude, who had worked with D'Arsonval, built a prototype 22 kW OTEC system in a bay off the Cuban coast: Claude developed a simple cycle which used sea water admitted to a low pressure evaporator to provide steam to drive a turbine. The low pressure steam thus produced, was then condensed by contact with cold sea water in a spray condenser. Although the Claude cycle avoided the large heat exchangers used in the modern organic cycles to be described below, the cycle employed a large low efficiency turbine. Other problems with Claude's system were that a good vacuum was required involving good seals, that dissolved gases in the water required removal and that the electrical plant was located on land which entailed piping the cold water through tubes 2 km long. This led to significant warming of the water in transit. Notwithstanding the low efficiency and economic failure of the project, Claude's experiment remains a bold pioneer project which demonstrated that OTEC could be made to work. The system used by Claude is shown schematically in figure 3.36.

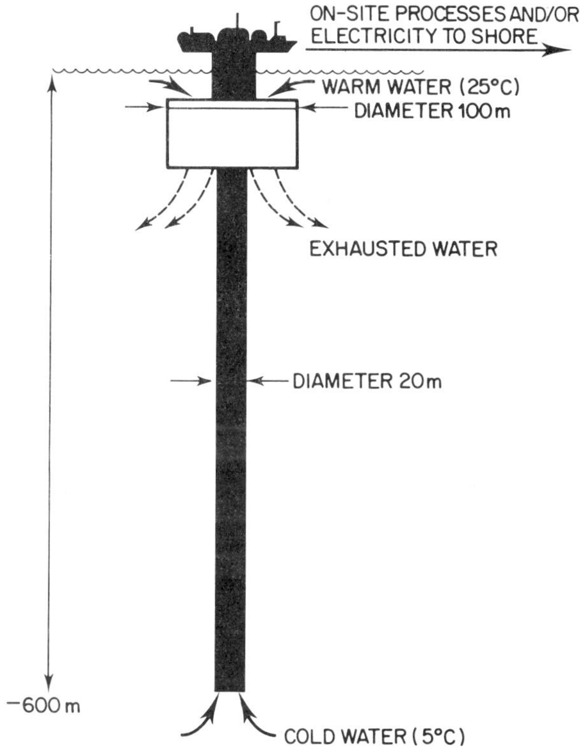

Figure 3.35 Diagram showing the principle of OTEC.

Two rather more ambitious projects using the same cycle were built by French engineers off the Ivory Coast in 1956. These were rated at 3.5 MW(e) each and exploited a temperature difference of about 20 °C between the surface and a depth of nearly 5 km. Problems of maintaining the pipeline and the large amounts of power used to operate the pumps and other ancillaries led to the abandonment of the project.

Recent attempts to exploit OTEC will be described in §3.2.7.

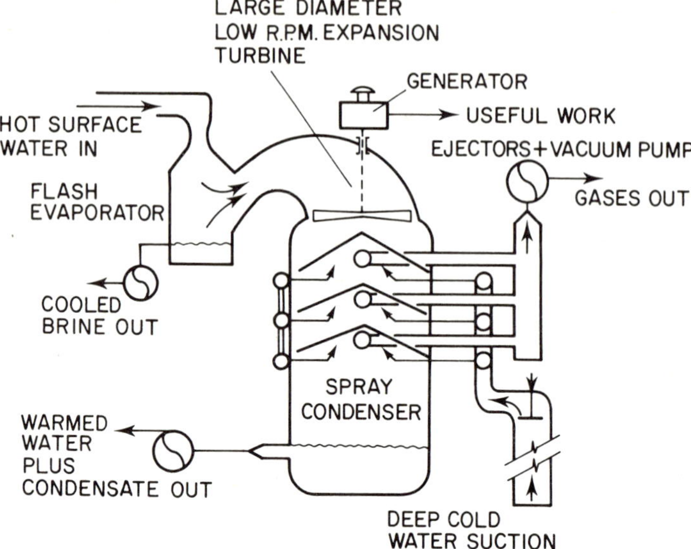

Figure 3.36 Claude's ocean thermal power plant. (Reproduced from *Solar Energy* by J R Williams (Ann Arbor Publications: Michigan) 1974.)

3.2.2 Resource and site requirements

The major factor in determining whether OTEC can be utilised as a source of power for a particular location is clearly that a sufficient thermal resource should exist. It is estimated that a temperature difference of at least 18 °C is required between surface water and water at a depth of 1000 m or more. A typical thermal gradient in the tropics is shown in figure 3.37. Prospective sites only exist in the equatorial belt from about 25 °S to 25 °N (see figure 3.38). Although ocean thermal energy is clearly not a potential resource for countries in northern Europe, it has, however, considerable potential for a number of developing countries with proximity to deep water. In some locations coastal contours have the effect of enhancing the thermal gradient to a very significant effect. Table 3.2 shows a list of some of the locations where a thermal gradient in excess of about 20 °C is thought to exist within 50 km of the coast. Estimates have put the practical global OTEC resource at over 20 GW; this is small compared with some of the other sources such as wind, geothermal and solar discussed in this book, but has a particular importance because of the high energy costs with which it must compete in many of the locations listed in the table. Some forms of OTEC system also have the capability of supplying fresh water where this is scarce.

Table 3.2 Some possible locations for OTEC with high gradients close to shore.

	Estimated temperature difference (0–1000 m)	Distance from shore to resource (km)
Indian and Pacific Oceans		
Comoros	20–25	1–10
Cook Islands	21–22	1–10
Fiji	22–23	1–10
Guam	24	1
Gilbert	23–24	1–10
Maldives	22	1–10
Mauritius	20–21	1–10
New Caledonia	20–21	1–10
Papua New Guinea	22–24	30
Philippines	22–24	1
Samoa	22–23	1–10
Seychelles	21–22	1
Solomon Islands	23–24	1–10
Sri Lanka	20–21	30
Vanuatu	22–23	1–10
Latin America and Caribbean		
Bahamas	20–22	15
Barbados	22	1–10
Cuba	22–24	1
Dominica	22	1–10
Dominican Republic	21–24	1
Grenada	27	1–10
Haiti	21–24	1–10
Mexico	20–22	32
Saint Lucia	22	1–10
Saint Vincent	22	1–10
Trinidad	22–24	10
Virgin Islands	21–24	1
Africa		
Ivory Coast	22–24	30
Kenya	20–21	25
Mozambique	18–21	25
Nigeria	22–24	30
Tanzania	20–22	25

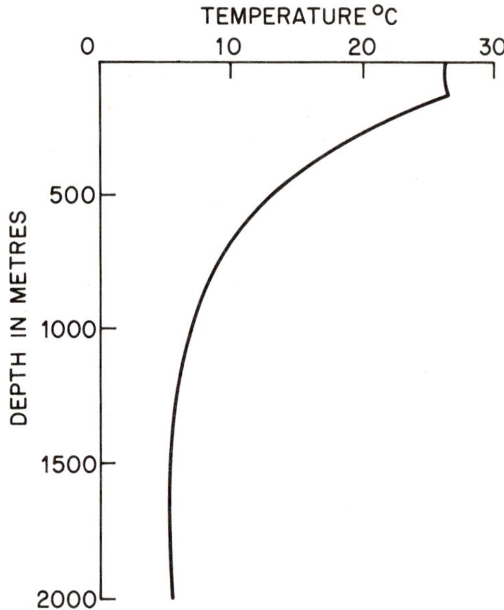

Figure 3.37 A typical thermal gradient in the tropics.

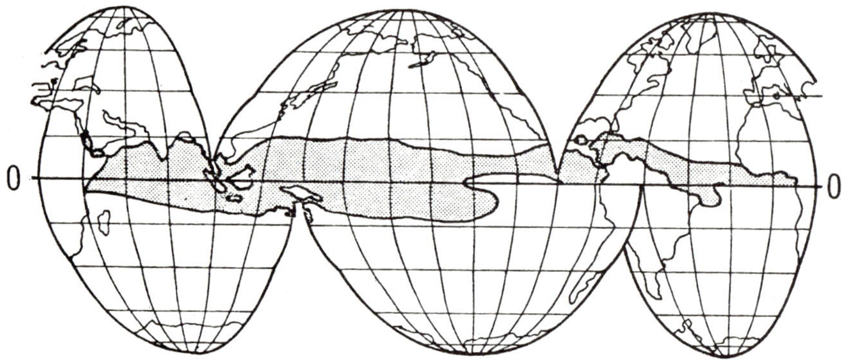

Figure 3.38 Possible regions of utilisation (shaded) for OTEC. (Courtesy Lockheed Missiles and Space Inc.)

3.2.3 *Modern OTEC plant designs*

Power plant. In the early 1970s two approaches were suggested to improve the relatively simple cycle employed by Claude. These were the controlled flash evaporation system and the indirect vapour cycle system [47], [48].

The first of these, illustrated in figure 3.39 avoids the problems of sea water corrosion and trapped air and has the additional advantage of being able to produce fresh water. Warm sea water is allowed to flow down a series of parallel widening tubes and as the pressure drops with depth, the water evaporates. The low pressure steam flows through a large low pressure turbine and into a condenser where it is cooled and condensed by the deep water. Although overcoming some of the disadvantages of the Claude cycle, the use of the large and inefficient low pressure steam turbine is still a major drawback, as also is the requirement for a vacuum and the need to continuously draw off non-condensable gases. This type of cycle has been studied in some detail by Westinghouse [49] and by SERI in the USA. It has been argued that the open cycle system may eventually become cost effective because the structural shell may act both as pressure vessel and structure, supporting and connecting all cycle components and providing flow passages for liquids and vapours. Furthermore, such a system may employ a single, large, low-speed turbine rather than a large number of small turbines.

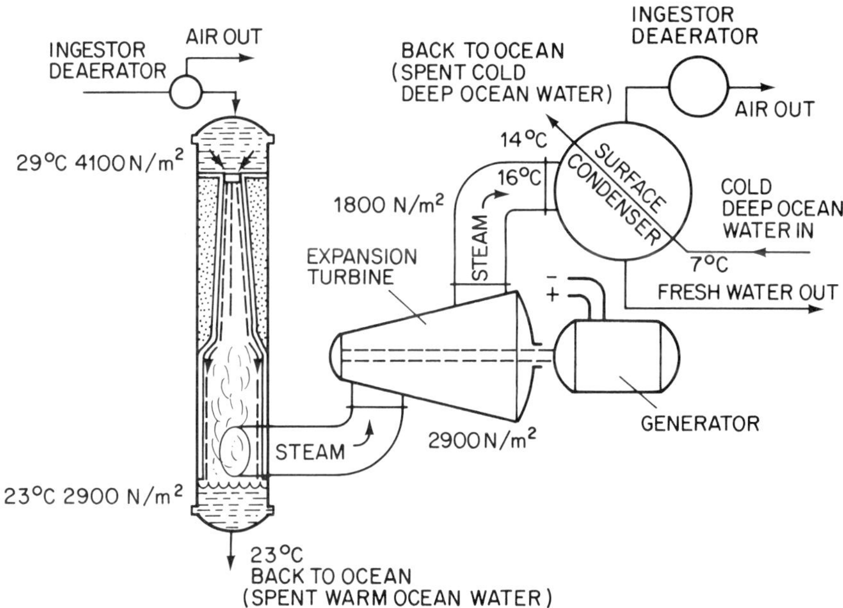

Figure 3.39 Controlled flash evaporation ocean thermal power plant. Water temperatures and pressures in °C and N/m² respectively. (Reproduced from *Solar Energy* by J R Williams (Ann Arbor Publications: Michigan) 1974.)

The indirect vapour cycle, illustrated in figure 3.40 employs a boiler but uses higher pressure working fluids with a consequent reduction in turbine size and an increased efficiency. It is argued that in OTEC devices, because of the lower temperatures and pressures used than in conventional steam plants, tube walls can be made of thinner and cheaper materials, thus reducing plant costs to some degree. This design of OTEC plant appears, at present, to be the favoured system and has been studied in some detail by teams at Massachusetts and Carnegie–Mellon Universities and subsequently by consortia led by Lockheed and TRW Systems Group. It operates as a simple Rankine cycle, but because of the low temperature difference between source and sink, superheaters, reheaters and feed heaters are not used. Various working fluids have been considered including ammonia, isobutane, propane, refrigerant 11 and various mixtures of organic chemicals. In determining which of these is the best, cost, environmental acceptability in the event of leakage or fire and compatability with structural materials have to be considered. Cost is largely determined by the heat transfer coefficients for condensation and evaporation and by the vapour density. Because of its relatively high heat transfer coefficient, ammonia allows smaller evaporator areas than the other working fluids considered and also appears to allow the use of a smaller condenser. Against this must be set the toxicity and flammability of ammonia and the fact that it cannot be used with conventional copper–nickel condenser tubes thus requiring the use of titanium or possibly aluminium. Nonetheless, ammonia appears to be the favoured fluid. The Lockheed OTEC design which uses ammonia is shown in figures 3.41 and 3.42.

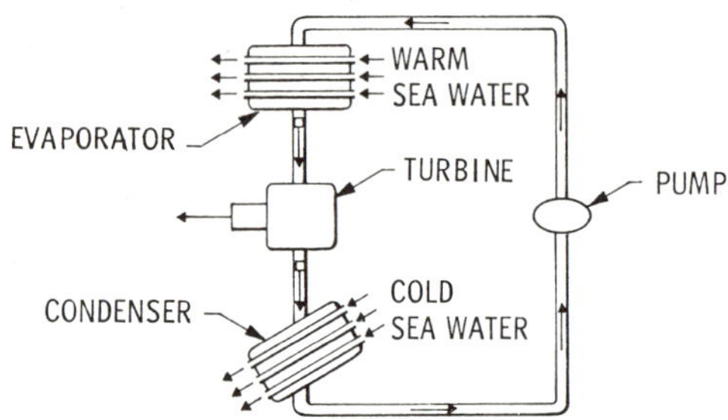

Figure 3.40 Illustration of the closed OTEC cycle. (Courtesy Lockheed Missiles and Space Inc.)

The Lockheed system is rated at 265 MW and the TRW system at 100 MW (net). There is no consensus as to the optimum size of unit. A study by the Electrical Power Research Institute (EPRI) [50] suggested a unit size of 50 MW with a theoretical OTEC power station comprising eight such units, giving a maximum power output of 400 MW. The same study showed that for a temperature difference of 22 °C, an efficiency of just over 2% would be achieved compared with a Carnot efficiency of nearly 7.5%. The losses are mainly due to the need for finite temperature differences across the evaporator and condenser and the utilisation of about 25% of the generated power to operate auxiliaries such as water pumps. Although it has been pointed out that OTEC can, in principle, generate continuous power output without regard to the vagaries of the weather, there is, nonetheless, a seasonal variation in the marine temperature gradient at most locations. A typical example of this is illustrated in figure 3.43. Roughly speaking, power output varies as the square of the temperature difference.

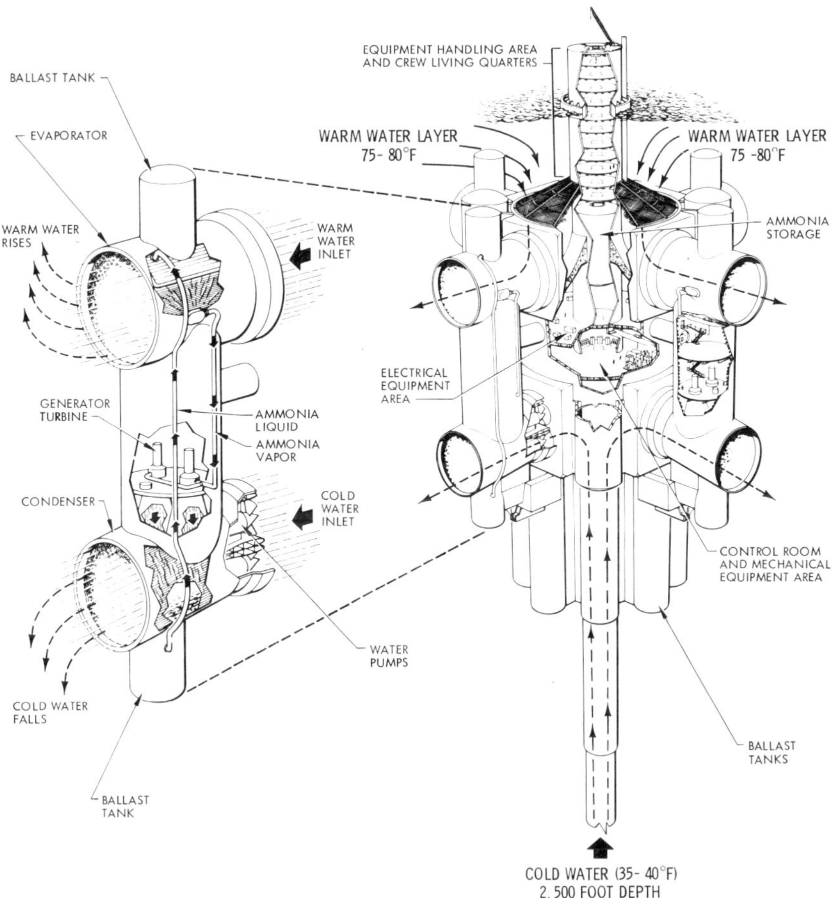

Figure 3.41 The Lockheed OTEC design (see also figure 3.42). (Courtesy Lockheed Missiles and Space Inc.)

Most of the closed cycle plant is fairly conventional. Heat exchangers are the most expensive items of plant and most designs are of the shell and tube type with sea water on the inside and ammonia outside. Attempts have been made to enhance the heat transfer in these components by using fluted surfaces and specially roughened surfaces. Titanium tubes 100 μm

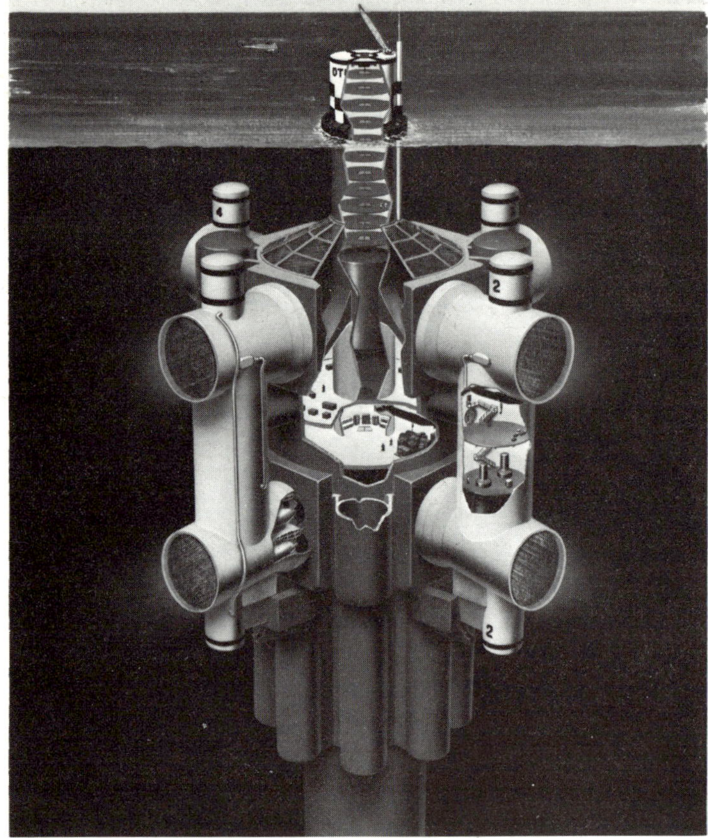

Figure 3.42 The Lockheed OTEC system (see also figure 3.41). (Courtesy Lockheed Missiles and Space Inc.)

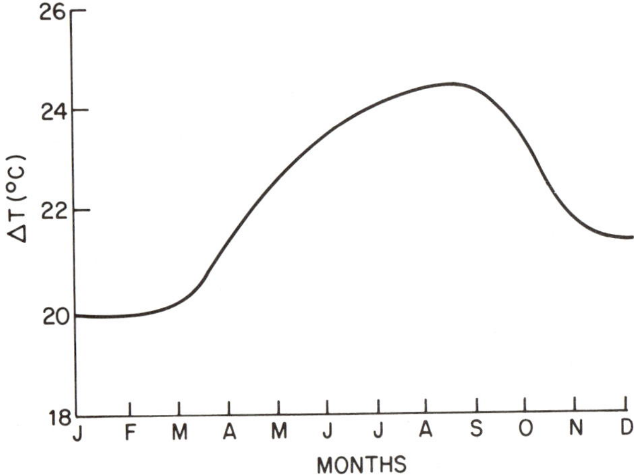

Figure 3.43 Typical seasonal temperature variation in the tropics.

thick are envisaged since titanium offers good corrosion and erosion performance despite its high cost.

A number of evaporator designs have been considered (figure 3.44); pool boilers, thin film evaporators and spray evaporators have been particularly favoured. In pool boilers, the evaporator tubes are surrounded by a large pool of ammonia, evaporation occurs at the tube surfaces and the vapour is collected at the top. Efficiency tends to be limited by the hydrostatic pressure of the ammonia. In the other two systems, the ammonia forms a film or series of drips on the tubes; high efficiency is only achieved when the surface of the tubes is mainly covered with evaporating ammonia. Thin film evaporators appear to offer the greatest potential for OTEC systems [50].

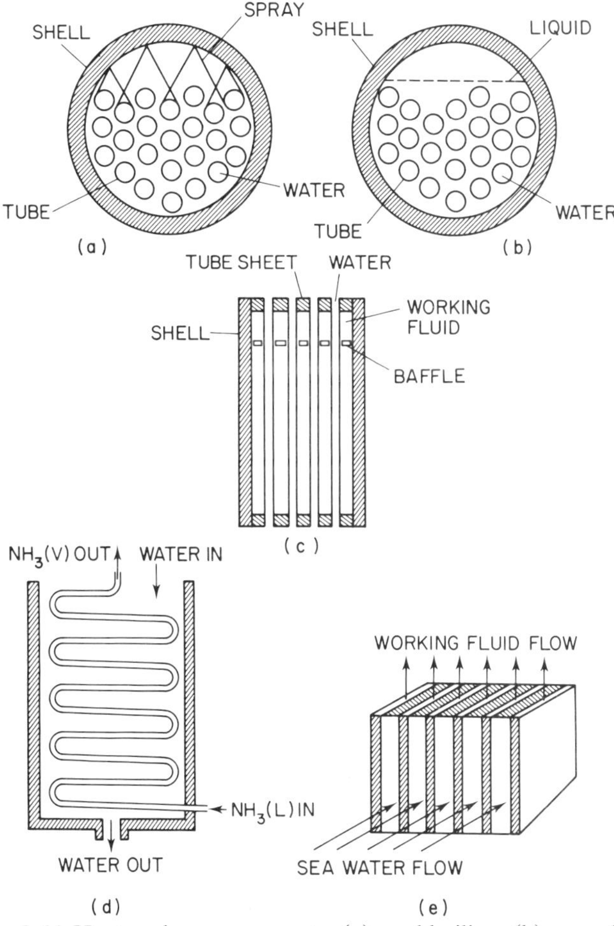

Figure 3.44 Heat exchanger concepts: (a) pool boiling; (b) spray film; (c) vertical falling film; (d) folded pipe 'trombone'; (e) plate 'compact'. (Courtesy Professor A Lavi, reproduced with the permission of *Energy* from the June 1980 issue.)

Ammonia turbines are of largely conventional design, usually employing only a single stage. Speed control is achieved in single stage devices by a simple by-pass system to the condenser. Pumps to deliver the large volumes of sea water to the heat exchangers would present difficulties, since no suitable pumps of more than one half the required size for a 50 MW unit have yet been manufactured. Other auxiliaries include systems to achieve adequate protection against biofouling. These include chlorination, ozonisation, foam cleaning and mechanical cleaning. Protection against biofouling is essential in order to achieve efficient heat transfer and chlorination appears to be the favoured method.

One of the main problems with OTEC systems (as with wave power) may be in achieving acceptable reliability. Although sited in calmer water, the huge number of tubes in the heat exchangers (up to 400 000 in a 50 MW unit) clearly presents a formidable reliability problem. Automatic methods of sealing ammonia leaks have been proposed.

In summary, from the plant point of view, closed cycle OTEC looks feasible in principle and employs largely state-of-the-art technology, although the large heat exchangers and pumps may be difficult to fabricate. Corrosion and biofouling appear to be the main operational problems.

Advanced concepts. Two radically different methods of utilising the thermal gradient have been suggested and these will briefly be considered here. They are the foam OTEC system [51] and the mist lift cycle [52].

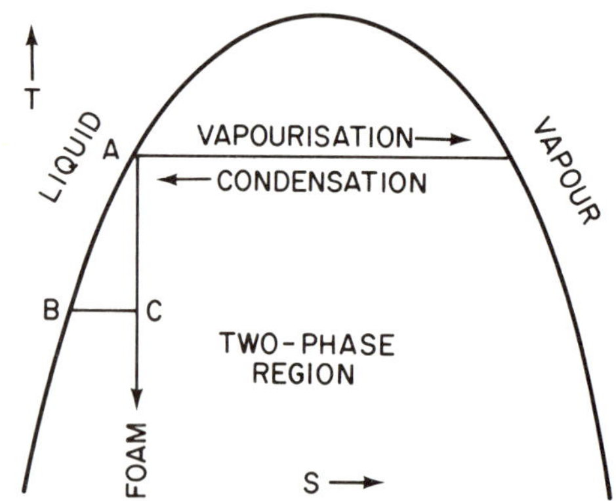

Figure 3.45 Temperature–state diagram for a foam OTEC system. (Courtesy Professor C Zener and reproduced with the permission of *Energy* from the June 1980 issue.)

Figure 3.45 shows the thermodynamics of the foam concept. If an element of water has its pressure isentropically lowered by dP, from P_A to

P_C, the enthalpy released is given by the expression

$$\int_{P_C}^{P_A} \frac{\mathrm{d}P}{\varrho}\,.$$

This is approximately given by the area ACB in the figure and this itself approximates to the expression

$$\tfrac{1}{2}(C_p/T)\,\Delta T^2$$

where C_p is the specific heat of water at constant pressure.

In the foam OTEC system as envisaged by Zener and Greenstein [51], an evacuated dome is built on a ship and warm surface water is sucked into the dome. A generator forms foam which rises a hundred metres or more and is then broken into liquid and vapour. The head of water generated is then used to drive a water turbine. The vapour is condensed at the surface by cold water from depth. According to the proponents of the scheme, 1 GW could be generated from a 400 m diameter system operating with a head of about 200 m. The principle is illustrated in figure 3.46.

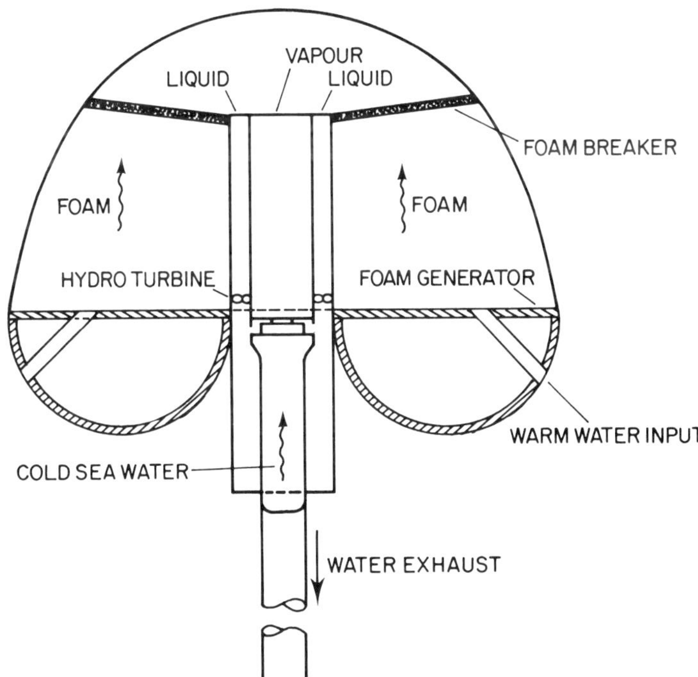

Figure 3.46 Conceptual diagram of a foam OTEC plant. (Courtesy Professor C Zener and reproduced with the permission of *Energy* from the June 1980 issue.)

The reason for employing foam is that evaporation takes place from the bubble surfaces. Negligible loss of enthalpy occurs in providing the heat to cause this evaporation because the heat need only travel short distances. Since the vapour and liquid travel physically together, the two phases are tightly coupled both thermally and mechanically. Zener and Greenstein, from early experimental work, believe that such a system is feasible on a commercial scale. Like other open cycle systems non-condensable gases would require to be removed from the sea water.

In the mist lift cycle [52], the principle is similar (figure 3.47) but instead of foam, a very fine mist of droplets suspended in their vapour is used.

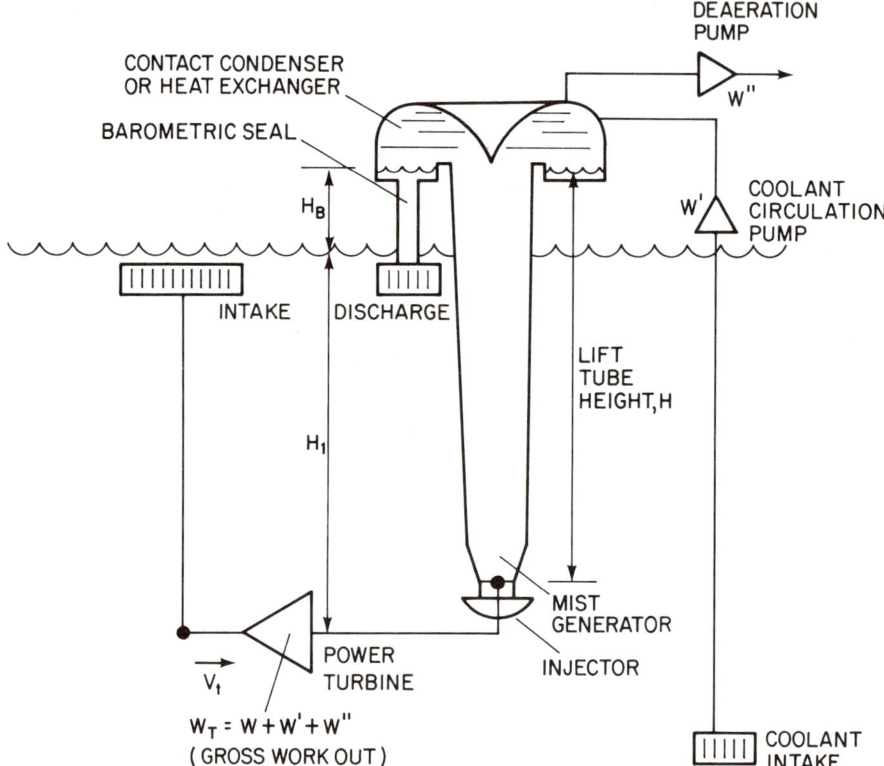

Figure 3.47 Schematic diagram of a mist lift OTEC plant. (Courtesy Professor F Charwat and reproduced with the permission of *Energy* from the June 1980 issue.)

These two advanced systems are currently being explored but are unproven and have not yet been costed in any detail.

Platforms. The technology required to support the power plant, provision of the deep cold water pipe and construction and mooring are all somewhat similar to those examined in the wave energy section of this chapter. Although the average conditions experienced by an OTEC plant will clearly be much less demanding than those on wave power plant, extreme conditions may not be that dissimilar. Many of the locations where OTEC is feasible are within the hurricane belt and survival against such conditions would need to be ensured.

Several types of hull have been proposed for OTEC systems including barge or ship, semi-submersible and submersible systems. The last two of these hull types have been used for oil exploration but present problems of access and provision of space. Furthermore, the size of a large OTEC system would be greater than state-of-the-art technology and there would be problems in providing an adequate connection to the cold water pipe. On the other hand submersible and semi-submersible hulls should provide good resistance to extreme conditions. Various elaborations on these designs have been proposed including spar, sphere, disc and tubular truss constructions. The spar is in the form of a vertical cylinder piercing the surface, the sphere and disc both semi-submersed. Each of these are designed to minimise motion and thus reduce mooring problems, but at the large scale envisaged, fabrication would involve new technology and present formidable problems.

The most satisfactory hull type appears to be that of the barge or ship constructed from steel or concrete. This type of structure has been used successfully in oil exploration and the technology is largely available. Because it is a surface system, large mooring forces might be expected in some conditions, but unlike wave power the OTEC ship would probably be manned and could employ supervised dynamic stabilising systems. This would involve a loss of power output and it is probable that a combination of mooring and dynamic positioning would be used. Moorings are in some respects simpler, but in other respects more demanding, than for wave power devices. The need to minimise forces on the cold water pipe presents an important limitation on the freedom of movement of the structure. The depth of water in which the OTEC system must operate means that anchoring would probably need to be achieved using gravity rather than drag anchors. Whatever method is employed, the experience with wave energy shows that mooring such massive structures raises formidable problems. In principle, it has been suggested that an OTEC ship or barge might 'graze' for the best thermal gradient.

Cold water pipes probably represent the greatest difficulty in OTEC technology. In the Lockheed design, the reinforced concrete pipe starts at the bottom of the submarine platform 200 m below the surface and telescopes down a further depth of 600 m or so. The outside diameter varies from about 40 m at the top to 30 m at the bottom. In the TRW plant, the fibre-reinforced plastic cold pipe is 15m in diameter and extends to a depth of nearly 1300 m; the EPRI baseline unit has a 1000 m long inlet.

These structures must withstand various loads including drag due to currents, differential wave forces, platform motion, gravity and internal flow. Although theoretical modelling of these loads has been carried out, understanding of the response to the full range of possible operating conditions must await results from prototype installations. Connection to the hull will clearly also be a major problem and although various universal and ball joint systems have been proposed, they are novel at the sizes required for large OTEC systems.

3.2.4 Utilisation of OTEC

Getting power ashore from OTEC systems is similar in many respects to transmitting power ashore from wave energy schemes. Floating OTEC systems, like floating wave power devices, will require the use of flexible cables. In a few locations where sufficiently deep water of high thermal gradient can be found close to shore, or a suitable reef, it is possible that the electricity producing plant could be sited onshore and thus problems of electrical transmission reduced or eliminated.

Various studies [48] have been carried out into uses other than electricity generation for OTEC energy or into combined uses. These include using the cold nutrient sea water to grow biomass or to feed aquaculture, provide fresh water, produce high value chemical feedstocks or to process raw materials. An advanced concept study of a design for 'a tropical grazing OTEC plant-ship' to produce liquid ammonia at sea has been carried out by a consortium involving Johns Hopkins University and several US companies. European studies [53] have also selected several uses (in addition to electricity production) for further study and these are listed below.

(1) De-salination of warm water leaving the electricity producing plant.

(2) Electrolysis of water to produce gaseous hydrogen followed possibly by liquefaction to facilitate transport.

(3) Synthesis of ammonia or methanol combining gaseous hydrogen with nitrogen or carbon dioxide.

(4) Production of aluminium.

(5) Aquaculture and biomass production.

The choice of end use clearly depends upon the location of the plant, the distance from electricity loads and the demand for the various raw materials. This is similar to the analyses carried out for wave energy, which in a UK context favoured the production of electricity.

Most of the processes being studied are within existing technology. Production of methanol would rely upon finding an economic source of carbon and the chemical process advocated is non-standard. Biomass production would introduce problems arising from separation of the cold water from the OTEC hot water intake.

3.2.5 Environmental factors

OTEC would be unlikely to lead to severe environmental effects. In many respects it will have a similar impact to wave energy but without effects upon the shoreline on the lee of devices. Clearly, impacts would be strongly site-specific but might include the following.

(1) Damage to marine organisms from entrainment in the device.

(2) Effects due to ocean water mixing. This would depend upon the size and number of devices in a particular area. It could, in principle, lead to changes in local climate as well as affecting marine species.

(3) Chemical pollution due to biocide discharge, working fluid leaks or other chemical release. This would clearly be a greater environmental threat if industrial production were associated with the plant.

(4) In open cycle systems, the release of non-condensable gases such as CO_2 may produce detrimental effects if OTEC were ever to be introduced on a very large scale.

(5) Effects on marine life on reefs or in open water. Spawning and migrational impacts may be particularly evident.

(6) Impacts arising from construction, maintenance and shore installations.

(7) Hazard to shipping.

3.2.6 Economics

Like geothermal plant, OTEC could theoretically supply electricity continuously and would not depend upon the vagaries of the weather for its output. In terms of electricity production it is clearly comparable with other high initial capital cost, low fuel cost plant (such as nuclear or hydro). Many of the tropical locations where OTEC might be feasible rely heavily upon imported oil and it may, if technically proven, find early application in some of these areas. One difficulty may be in finding the necessary local expertise to support this advanced technology.

Analysis of the economics of OTEC has been carried out in Europe and the USA. A consortium of European companies (EUROCEAN) have estimated that for a first 100 MW(e) plant operating for 30 years at a rather optimistic availability factor of 90%, electricity could be produced from a 20 °C temperature gradient for 5.4 c/kWh (at 1977 prices). The fixed charge rate, which includes operation and maintenance costs was taken to be 16%. The total capital cost of just over \$2600/kW was distributed between the various components as illustrated in table 3.3 which has been adapted from tables given in references [45] and [53]. Figures are also given for the Lockheed and TRW designs which claim slightly lower electricity costs. A sensitivity analysis of the effect of thermal gradient on costs suggested that a decrease in temperature difference to 16 °C would increase the cost of electricity to about 13 c/kWh, whilst increasing the difference to 22 °C would lower the cost to 4 c/kWh. It should be noted that the estimates do not include transmission to shore and onwards. Production of ammonia was also believed to be economic in the European study.

Table 3.3 Comparison of OTEC cost estimates (adapted from reference [52]).

	TRW	Lockheed	Japan	EUROCEAN
Net production (MW(e))	100	160	74	100
Design ΔT (°C)	20	18.5	21	20
Working fluid	Ammonia	Ammonia	Ammonia	Ammonia
Cost year (US$)	74/75	74/75	76	76/77
Cost of heat exchangers ($/kW)	798 (44%)	1479 (58%)	—	700 (40%)
Structure	490 (27%)	468 (18%)	—	480 (27%)
Cold water pipe	198 (11%)		—	48 (3%)
Mooring	15 (1%) (dynamic)	194 (8%)	—	50 (3%)
Pumps	66 (4%)	163 (6%)	—	120 (7%)
Turbines	37 (2%)	78 (3%)	—	130 (7%)
Other direct costs	208 (11%)	182 (7%)	—	230 (13%)
Total direct costs	1812	2564	—	1757
Indirect costs	290	63	—	879
Total capital costs ($/kW)	2102	2627	2900	2636
Projected costs for mature design (from [45]) (1980$, $\Delta T = 22.2$, 100 MW plant)	844	1015	—	—

The EPRI study [50] based on the 400 MW prototype with a power plant life of 15 years and a range of availabilities from 50–90%, estimated that, at the highest availability, electricity could be produced from a 22 °C temperature difference for 6.4 c/kWh (at 1978 prices). Total capital costs of $3000/kW were split evenly between power plant and platform. Annual operation and maintenance costs of the manned vessel (assumed to be one of several sharing some facilities) were estimated to be about $50/kW p.a.

Again, the costs of bringing power ashore were not included. A fixed charge rate (which excluded operation and maintenance costs in this case) was taken as 15% per year. The authors emphasised that the main unknowns in estimating costs were the heat exchangers, platform, mooring, deployment of the plant, assumed lifetime and operation and maintenance. Much lower costs (less than 2 c/kWh at 1975 prices) have been estimated for large complexes (3 GW(e) by Curto [54]).

Economic analysis using a UK test discount rate of 5% but assuming lower availability figures would suggest that costs for OTEC electricity could be less than half those estimated for the best UK wave power devices. It is difficult, however, to determine whether costings and energy production estimates have been evaluated with equal rigour for both technologies. It is probable that OTEC is closer to commercial realisation than wave power but formidable problems have yet to be solved. In any event, it is unlikely that OTEC and wave systems would compete in many areas of the world because the climatic conditions driving each technology are clearly widely different and, as mentioned previously, OTEC holds no prospects for supplying electricity to the UK.

3.2.7 Current OTEC projects

The most intensive experiments on OTEC are being carried out in the USA, where $30M per year was being spent at the end of the 1970s. Development work has been proceeding on heat exchangers, ammonia turbine design, materials evaluation and design of the cold water pipe and its associated technology. Work on submarine cables for high voltage DC transmission has also been underway since 1977.

A recent practical demonstration of OTEC was achieved off the coast of Hawaii in the summer of 1979 using a 50 kW plant known as mini-OTEC (see figure 3.48). The project was funded by the state of Hawaii and a group of companies including Lockheed, Dillingham Corporation of Honolulu and Alfa-Laval of Sweden. Lockheed, who previously received funding from the National Science Foundation, engineered the power and ocean systems while Alfa-Laval supplied the titanium plate-type heat exchangers used in the experiment. The system was built from standard components on board a barge on loan from the US Navy. Cold water was obtained from a 600 m pipe which also formed part of the mooring.

Mini-OTEC operated for about 1000 hours in its first trials, and produced 15 kW of usable power. The other 35 kW was used to power the associated plant. The major role of mini-OTEC was to confirm the engineering calculations but other significant information was also obtained. For the site off Hawaii, for example, it was found that biofouling, entrainment of debris and trapped gas in the cold water were much less of a problem than had been anticipated. Further tests of the system are planned.

Figure 3.48 The mini-OTEC power plant at its operational site off the Kona coast of the Island of Hawaii. Mini-OTEC was funded by the state of Hawaii, Lockheed Missiles and Space Company, Sunnyvale, California and Dillingham Corporation, Honolulu, with Alfa-Laval of Sweden contributing the two titanium heat exchangers.

The US national programme funded by the Department of Energy has until recently provided for research using a further engineering test facility, OTEC-1. A view of this floating OTEC facility is shown in figure 3.49. OTEC-1 was constructed by TRW Inc. and Global Marine Developments and the facility was built by converting a 26 000 ton tanker. It was first used for OTEC studies off Hawaii in 1980. The vessel was planned for use as a research facility which would not generate electricity but could be used to study corrosion, biofouling and heat exchanger performance. The system was closed cycle incorporating a 1 MW titanium tube-and-shell heat exchanger which could simulate 10 MW of capacity. Three cold water pipes each 850 m in length and 1.2 m in diameter were built into the hull. Originally, it was planned to test five heat exchanger configurations (two shell and tube and three flat-plate) but before extensive testing could be carried out cut-backs in the US programme led to its early retirement.

The results of OTEC-1 were to be used to plan larger units which it was intended to build in the 1980s, including a 40 MW unit and eventually a 400 MW system for operation in 1990. Apart from Hawaii and Puerto

Rico, OTEC on a large scale could find application in many areas off the South Atlantic coast of the USA, around Florida for example.

During 1980, public laws were enacted similar to those discussed for wind power which it was hoped would hasten the commercialisation of OTEC. One such law established a national goal of 10 000 MW of OTEC capacity by the year 2000 and directed that two or more pilot plants be constructed to produce 100 MW by 1986. A further law declared OTEC plants seaward of the high water mark to be subject to maritime law and established a $2 billion fund to provide ship mortgages for OTEC demonstration plants. The tax credits and legal obligation on utilities to purchase electricity from renewable sources mentioned in Chapter 2, refer equally to OTEC.

Figure 3.49 The OTEC-1 ship. (Courtesy US Department of Energy.)

Unfortunately, as part of the budget cuts introduced by the Reagan administration it seems likely at the time of writing (mid-1982) that federal funding for OTEC in the USA will be very severely cut. It thus seems possible that large demonstration projects will either be curtailed or dropped altogether from the US solar programme. As with the similar legislation concerning wind energy, it seems unlikely that federal funding for commercialisation will proceed, and further development in the USA now seems to be firmly in the lap of individual states such as Hawaii, utilities and other industrial companies. There are encouraging signs that companies such as General Electric and Basic Resources Corporation may proceed with 40 MW pilot plants in collaboration with the state of Hawaii.

The major European programme on OTEC is being carried out by a consortium of European companies called EUROCEAN comprising mainly Swedish, Dutch and Italian firms. They see a potential export market for the technology. France has an independent OTEC programme. EUROCEAN are carrying out development work aiming at a pilot plant of 10 MW, with full-scale components so that no further scaling-up problems will be encountered on moving to full scale. It is anticipated that at the end of the two year planning stage the plant will take three years to build. This will allow completion in about 1983. The estimated cost is $50M (1978 costs). The UK does not have an OTEC programme.

The French programme involves a three-phase, eight year plan to get a working OTEC system by the mid-1980s. Component testing costing $5–6M will take three years and construction of the 1–10 MW plant costing $20–30M will take a further three years. The project is two-thirds funded by the French government and one-third by French industry.

As part of the Japanese 'sunshine project' a vigorous OTEC programme is being pursued [55]. A 1 MW system to test components is planned for 1985 and will cost an estimated $6M. A 10–25 MW engineering test plant is planned for 1990, followed by a 100 MW full-size station towards the end of the century. In the meantime 700 W and 50 kW trial systems have been designed and, in the case of the smaller system, tested. The 50 kW plant is expected to be in operation in 1982 and this will utilise power station cooling water to maintain a 30 °C temperature gradient between the top and bottom of the 500 m cold water pipe.

Because of the geographical location of the OTEC resource, as discussed in §3.2.2, it has application in many developing countries [56]. Reports and feasibility studies have been carried out for Curaçao in the Caribbean, Tahiti, US Virgin Islands, Puerto Rico and Nauru in the South Pacific. Consideration is being given to schemes in the Ivory Coast, India and Jamaica. Details of some of these are given in table 3.4.

3.3 Other concepts for obtaining energy from the ocean

There are two further methods of extracting power from the sea which have been examined for their feasibility. These are the extraction of power from salinity gradients and from ocean and river currents.

3.3.1 Salinity gradients

In order to obtain energy from salt water, two solutions of different concentration must be available. Such a salinity pair might be formed from sea water, saline lakes (e.g. the Dead Sea, Great Salt Lakes) or brines left over from salt manufacture, coupled with a very low concentration source such as river water. The energy which can, in principle, be extracted from the two solutions is directly proportional to absolute temperature and the logarithm of the ratio of their concentration (activity ratios). The idea of extracting energy from salinity gradients was revived in the USA and Israel in the mid-1970s [57] and has been extensively reviewed by Schlechter [58].

The methods suggested for extracting the energy are the reverse of those used for desalination. These include using the osmotic pressure difference between sea water and fresh water (22 atmospheres), the difference in electrochemical energy which mainfests itself across an ion exchange membrane (reverse electrodialysis) and the difference in vapour pressure between salt and fresh water.

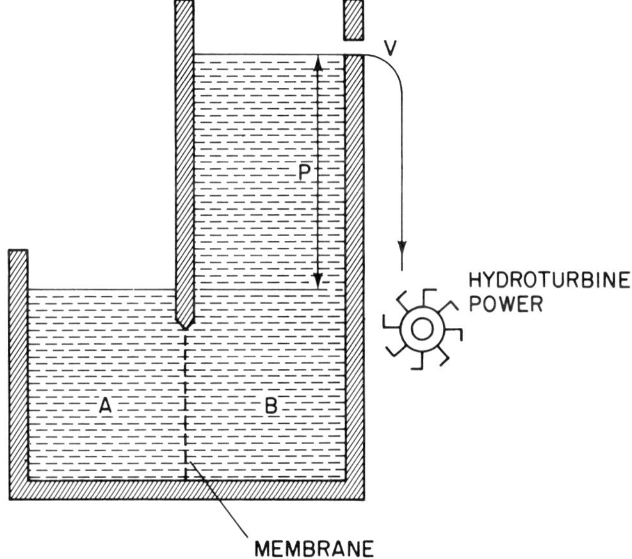

Figure 3.50 An osmotic cell. Water from the dilute solution A will enter the concentrate compartment B and raise its level. This difference in height can be used to operate a hydroturbine. (Courtesy Dr A Schlechter and reproduced from reference [58] with the permission of UNITAR and Pitman Publishing Inc.)

The free energy lost when a volume, V, of pure water is mixed with a larger volume is PV, where P is the osmotic pressure of the solution. Thus

Table 3.4 OTEC project studies (from reference [56]).

Project characterisation	Caribbean island		Caribbean island		Pacific island		
	United States Virgin Islands	Floating	Land based	Floating	Land based		Floating
Study contractor	Westinghouse Electric Corporation (United States)	Johns Hopkins APL (United States)	EUROCEAN	Netherlands Hollandische Beton Group	CNEXO†	Todden Sekkei Co., Tokyo Electric Power Service Co.	Sunshine Project (Japan)
Plant type	Hybrid closed	Closed cycle	Closed cycle	Closed cycle	Closed and open	Closed	Closed
Working fluid	Ammonia	Ammonia	Ammonia	Ammonia	Ammonia Sea water	R–22	R–22 or ammonia
Proposed location	United States Virgin Islands	Puerto Rico	Curaçao	Curaçao	Tahiti Nauru	Nauru	Okinawa
Net plant power capacity (MW(e))	2–5	33.8 net onshore	3	10	3–15	0.1 (100 kW) Total demonstration	1
Capital investment ($M 1980)	50–70	206 (including 10 per cent profit, 10 per cent contingency)	9.1 for power alone, 17.7 for combined ODA plant	73.0	60 for 3 MW, 150 for 15 MW	—	50, including research

Table 3.4 Continued.

Project characterisation	Caribbean island		Caribbean island		Pacific island		
	United States Virgin Islands	Floating	Land based	Floating	Land based		Floating
Unit cost ($/kW)	—	6110	9100	7300	20 000 for 3 MW, 10 000 for 15 MW	—	
Electrical cost ($M/kWh)	—	66–83	70 internal calculation for plant	61	175 for 3 MW, 112 for 15 MW	—	
Production cost at present ($M/kWh)	170	80–100	75	75	175		
Output	1 Fresh water 2 Electricity	1 Electricity	1 Electricity 2 Fresh water 3 Aquaculture	1 Electricity	1 Electricity 2 Fresh water	1 Electricity	1 Electricity
ΔT (mean) (°C)	21	22.4	20	20	23		

† Le Centre National pour l'exploitation des oceans (France).

river water flowing into the ocean at a rate of $1\,m^3/s$ could theoretically generate 2.2 MW of power. A device which might, in principle, extract some of this energy is shown schematically in figure 3.50. Water from the dilute solution in compartment A is separated from the concentrated solution in compartment B by a semi-permeable membrane. Water from the dilute solution will enter the concentrate compartment and raise its level and the difference in height which is achieved can' then be used to drive a water turbine. It is claimed that practical devices to harness this energy could use existing membranes and for a river water/sea water couple could generate electricity for 20 c/kWh [59] but this would appear to be very speculative. According to Loeb *et al* [60] costs would be lower for more concentrated sources such as the Dead Sea. The major technical problem would appear to be to achieve high enough fluxes across the membrane.

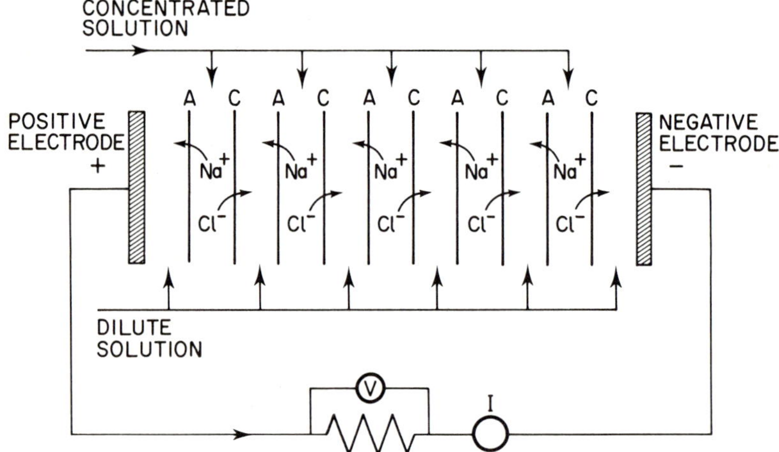

Figure 3.51 Reverse electrodialysis. The membranes marked A will allow only positively charged ions to pass and those marked C will allow only negatively charged ions through. Electric power can be obtained at the electrodes. (Courtesy Dr A Schlechter and reproduced from reference [58] with the permission of UNITAR and Pitman Publishing Inc.)

The electrodialysis technique would use specially treated membranes which allow only ions of one polarity to cross, leading to the establishment of a voltage across the membrane [61]. An illustration of the type of device which might be used is shown in figure 3.51. If concentrated brine is allowed to flow between pairs of plates in the cell and fresh water allowed to flow between the adjacent pairs, positively charged Na^+ ions will drift through the selective membranes marked A while the chloride ions, Cl^-, will drift through the selective membranes marked C. The voltage obtained is dependent upon the number of membranes in the device (being typically 0.2 V/membrane), the absolute temperature and the ratio of the concentrations of the solutions. To maximise the current, the internal resistance of the device must be reduced to the lowest possible value and thus the channels and membranes must be made as thin as possible. If membranes as thin as 0.005 cm and channels as thin as 0.01 cm could be developed, it is estimated that electricity could be generated from a sea water/fresh water couple for about 100 c/kWh. Mass production of membranes would, it is claimed, be expected to bring these costs down to 10–20 c/kWh [58]. Very large areas of membrane would be required, necessitating the use of a large amount of pumping power and this may call into question the practicability of such devices.

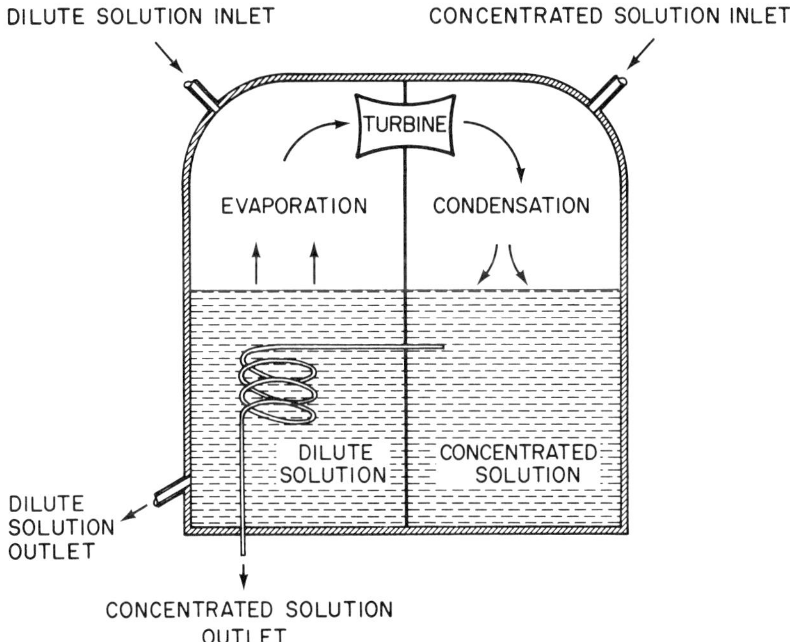

Figure 3.52 Vapour pressure differential turbine. The heat of condensation is returned to the dilute solution by means of the heating coil. Both solutions are continuously charged. (Courtesy Dr A Schlechter and reproduced from reference [58] with the permission of UNITAR and Pitman Publishing Inc.)

If a dilute and concentrated solution of brine are connected by a vacuum, the dilute solution will evaporate and condense into the concentrated solution. The transport of vapour in this way can be used to do work. However, the process would rapidly lead to cooling of the evaporating solution, thus lowering its vapour pressure, and the evaporation would anyway tend to equalise the concentrations and bring the process to a halt. A possible device which might overcome these problems is shown in figure 3.52. In this device [58], the heat generated by the condensing vapour is returned to the evaporating dilute liquid via a thin heat-conducting wall. This maintains the two solutions at almost equal temperature and if the liquid in each compartment is continuously changed, the concentration difference can be kept nearly constant. Because the fluid leaving each compartment will be of a different concentration to that entering, it has been suggested that a multi-stage device could be developed. Electricity is generated by a turbine between the two compartments. This type of device is even less developed than the reverse electrodialysis and osmotic techniques.

It is difficult to estimate the size of the resource which might be available from salinity gradients. The greatest potential clearly exists where salt is manufactured, where large salt deposits (above or below ground) can be found in the vicinity of a large river or coast or where there exist special geographical features such as salt lakes or depressions (e.g. the Qattara depression in Egypt) which could be used as giant evaporating pans. The global resource from rivers emptying into the sea has been estimated to be about 2.5×10^9 kW but how much of this is recoverable is clearly very questionable.

No commercial devices yet exist to exploit this possible form of energy and most studies are still at the feasibility stage. Research work is proceeding in the USA, Japan and Israel. The latter country is particularly interested because of the large potential resource offered by the Dead Sea. Salinity gradients could, in principle, be combined with a hydro scheme if the Mediterranean were to be connected to the Dead Sea by a tunnel.

3.3.2 Ocean and river currents

A 2 m/s current gives a power density of over 4 kW/m^2 and thus tidal streams and river currents, if they could be exploited, might offer a sizeable energy resource. Fraenkel and Musgrove [62] have, for example, shown that five tidal flows in the UK (North Channel between Ireland and Scotland, the Pentland Firth, English Channel, Orkney–Shetland Channel and Alderney Race) taken together yield an average power of over 16 GW. The authors make the point that if 10% of this could be exploited it would make a significant contribution (about 5%) to UK electricity needs. It could prove to be valuable for island communities. Wyman and Peachey [63] concluded that a total power of 10–15 GW might be captured in the UK. Elsewhere in the world such ocean currents could also provide substantial amounts of energy. It has been estimated [64], for example, that the Florida current in the USA constitutes a gross resource of 25 GW. However deep water resources such as this may be technically even more difficult to exploit.

The underwater equivalent of a windmill has been suggested as the best method of exploiting tidal streams. Vertical axis machines have, in general, advantages over other alternatives in that they have a high efficiency compared with drag dependent devices, sweep a rectangular cross section which allows shallows to be better exploited and have a low solidity thus reducing costs. On the other hand, where the flow is largely bi-directional, a horizontal axis screw may be preferable [63]. Since the dominant stresses are hydrodynamic (rather than centrifugal) for a given design and water velocity, stresses are independent of size. Thus machines of 100 m diameter or greater could be contemplated. Musgrove and Fraenkel have suggested that the rotors would be suspended mid-way between the surface and seabed so as to reduce wave induced loadings and the loss of output and shear effects resulting from the reduced flow near the seabed. A 100 m diameter rotor giving about 4 MW in a 2 m/s stream would rotate at about 7 rpm, so it is suggested that in common with some wave power systems, hydraulic pumps might be used and the compressed seawater converted centrally to electricity. The output from tidal streams offers the advantage that it is predictable.

This concept clearly has many of the difficulties which are being tackled in wave power studies, including materials and corrosion problems, moorings, fouling and power take-off. Its particular weaknesses would appear to be the hazard to shipping if such submerged devices were used on a large scale and the difficulties of inspection and maintenance [63]. The possible use of ocean currents is being studied in the USA and Japan.

A more practicable idea may be to use vertical axis windmill-type structures for power generation or irrigation in underdeveloped regions. It may be possible to exploit the concept in such rivers as the Nile, Niger, Euphrates and Indus which cross arid regions where energy sources are scarce and costly. Problems of fouling, cavitation of blades, gradients due to shallows and river banks and the disturbance of flow patterns and hazard to boats may still militate against such ideas, but simple experimental studies [62] have confirmed that river currents might be exploited with some prospect of being economic in favourable locations.

References

There are very few good general references on wave power which are up to date. Among the best are the article by Glendenning (1980 *Proc. IEE* **127** 301–7) and references [4] and [20] below.

1   Denton J D, Glanville R, Gliddon B J, Harrison P L, Hotchkiss R C, Hughes E M, Swift-Hook D T and Wright J K 1975 *CEGB Res.* **2** 28–40

2   Winter A J B 1980 *Nature* **287** 826–9

3   Rockingham A P, Taylor R H and Walker J F 1981 Offshore wind and wave power—A preliminary estimate of the resource *Proc. 3rd BWEA Workshop, Cranfield* (Bedford: BHRA) pp63–70

4   Wave energy 1979 *Energy Paper* 42 (London: HMSO)

5   Count B M (ed) 1980 *Power from Sea Waves* (London: Academic) pt II

6   Coulson C A 1955 *Waves* (Edinburgh and London: Oliver and Boyd)

7   Count B M and Robinson A C 1976 *CEGB Report* R/M/N895

8   Glendenning I 1977 *Appl. Energy* **3** 197–222

9   Glendenning I and Count B M 1976 *Proc. RSA Symp. on Renewable Sources of Energy* (London: Royal Society of Arts) pp50–81

10 Cottrill A 1981 *Offshore Eng.* January issue 25–36

11 Salter S H 1980 *Proc. IEE* **127** 308–19

12 Bellamy N W 1979 Wave power experiments at Loch Ness *Proc. IEE Future Energy Concepts Conf.* (London: IEE) pp167–9

13 Cockerell C, Platts M J and Comyns Carr R 1978 The development of the wave contouring raft *Proc. Heathrow Conf. on Wave Energy, Heathrow Hotel* ed P Quarrell, Energy Technology Support Unit (available from HMSO) pp7–16

14 Platts M J 1979 The development of the wave contouring raft *Proc. IEE Future Energy Concepts Conf.* (London: IEE) pp160–3

15 Rance P J 1978 The development of the HRS rectifier *Proc. Heathrow Conf. on Wave Energy, Heathrow Hotel* ed P Quarrell, Energy Technology Support Unit (available from HMSO) pp49–53

16 Elliot G and Roxburgh G 1981 Wave energy studies at the UK National Engineering Laboratory *Proc. BHRA Int. Symp. on Wave and Tidal Energy, St John's College Cambridge* paper J3 (Bedford: BHRA) pp269–81

17 Long A E and Whittaker T J T 1978 The Belfast device *Proc. Heathrow Conf. on Wave Energy, Heathrow Hotel* ed P Quarrell, Energy Technology Support Unit (available from HMSO) pp61–5

18 Grant R J, Johnson C G and Sturge D P 1981 Performance of a Wells turbine for use in a wave energy system *Proc. IEE Future Energy Concepts Conf.* (London: IEE) pp117–23

19 Falnes J and Budal K 1978 *Norw. Marit. Res.* **6** 2–11

20 Count B M 1979 *IEEE Spectrum* September issue 42–9

21 Evans D V 1976 *J. Fluid Mech.* **77** 1–25

22 Shaw T L, Watson P J, Sullivan J A, Cure J R and Armitage R N 1981 Options for power transmission to shore from the Bristol

cylinder *Proc. BHRA Int. Symp. on Wave and Tidal Energy, St John's College Cambridge* (Bedford: BHRA) paper J2 pp261–8

23 Platts M J 1981 Engineering development of the Lancaster flexible bag *Proc. BHRA Int. Symp. on Wave and Tidal Energy, St John's College Cambridge* (Bedford: BHRA) paper J4 pp282–97

24 Chaplin R V and French M J 1980 *Power from Sea Waves* ed B M Count (London: Academic) pp381–98

25 Bellamy N W 1981 A second generation wave energy device—the clam concept *Proc. IEE Future Energy Concepts Conf.* (London: IEE) pp127–33

26 Butterworth J 1978 The Kaimei project *Proc. Heathrow Conf. on Wave Energy, Heathrow Hotel* ed P Quarrell, Energy Technology Support Unit (available from HMSO) pp45–8

27 Ohno K, Masuda Y, Kai G, Miyazaki J and Ohbayashi S to be published *Proc. Int. Conf. on the Management of Oceanic Resources, 1981* (to be published by the Engineering Committee on Offshore Resources, Institution of Civil Engineers)

28 Drew S D 1981 Progress towards a submerged oscillating water column device *Proc. BHRA Int. Symp. on Wave and Tidal Energy, St John's College Cambridge* paper X5 (Bedford: BHRA) pp385–96

29 Farley F J M, Parks P C and Altmann T H 1979 A wave power machine using three vertical floating plates *Proc. IEE Future Energy Concepts Conf.* (London: IEE) pp118–22

30 Farley F J M, Parks P C and Grimshaw J F 1981 Porpoise, the buckling resonant raft *Proc. BHRA Int. Symp. on Wave and Tidal Energy, St John's College Cambridge* paper X4 (Bedford: BHRA) pp371–83

31 Budal K, Falnes J, Iversen L C, Hals J and Onshus T 1981 Model experiment with a phase-controlled point absorber *Proc. BHRA Int. Symp. on Wave and Tidal Energy, St John's College Cambridge* paper G1 (Bedford: BHRA) pp191–206

32 Higgins T P 1979 *Marine Tech. Soc. Int. Conf. on Ocean Energy, New Orleans* pp86–90

33 Bott A N W 1979 Electro/mechanical aspects of the Mauritius 'passive' type wave energy project *Proc. IEE Future Energy Concepts Conf.* (London: IEE) pp81–5

34 Quarrell P (ed) 1978 *Proc. Heathrow Conf. on Wave Energy, Heathrow Hotel* (Energy Technology Support Unit, available from HMSO)

35 Count B M (ed) 1980 *Power from Sea Waves* (London: Academic) pt III

36 Count B M 1979 The absorption of energy from ocean waves *Proc. IEE Future Energy Concepts Conf.* (London: IEE) pp96–9

37 Hudson J A 1978 Materials aspects of wave energy converters *Proc.*

Heathrow Conf. on Wave Energy, Heathrow Hotel ed P Quarrell, Energy Technology Support Unit (available from HMSO) pp99–104

38   Hancock R 1978 Mooring and anchoring of wavepower devices *Proc. Heathrow Conf. on Wave Energy, Heathrow Hotel* ed P Quarrell, Energy Technology Support Unit (available from HMSO) pp95–8

39   Taylor R J 1981 Wave energy: the influence of maintainance/repair requirements *Proc. BHRA Int. Symp. on Wave and Tidal Energy, St John's College Cambridge* paper C2 (Bedford: BHRA) pp99–104

40   Glendenning I 1978 Generation and transmission *Proc. Heathrow Conf. on Wave Energy, Heathrow Hotel* ed P Quarrell, Energy Technology Support Unit (available from HMSO) pp79–90

41   Bishop H W 1980 *Electr. Rev.* **207** 29–31

42   Mayes R P and Eunson E 1979 Initial thoughts on the transmission implications of large wave power complexes *Proc. IEE Future Energy Concepts Conf.* (London: IEE) pp406–13

43   Clark F J P 1979 *Raporteurs Report, IEE Future Energy Concepts Conf.* (London: IEE)

44   Grove-Palmer C O J 1981 *Power from Sea Waves* ed B M Count (London: Academic) pp441–9

45   Grove-Palmer C O J to be published *Proc. Int. Conf. on Management of Oceanic Resources, 1981* (to be published by the Engineering Committee on Offshore Resources, Institution of Civil Engineers)

A more recent account of several of the devices is given in *Proc. of a Workshop on Wave Energy, Maidenhead 16–18 December 1979* published by Energy Technology Support Unit, AERE, Harwell, for the Department of Energy.

For the reader requiring further introductory reviews on OTEC, reference [46] provides an excellent overview. The paper by Schlecter [58] likewise deals comprehensively with the subject of salinity gradients.

46   Lavi A 1980 *Energy* **5** 469–80

47   Anderson J H and Anderson J H Jr 1966 *Mech. Eng.* **88** 41–6

48   Chereminisoff P N and Regino T C 1978 *Principles and Applications of Solar Energy* (Michigan: Ann Arbor Science Publishers)

49   Coleman W H 1980 *Energy* **5** 493–502

50   EPRI 1979 *EPRI Report* ER-1113-SR

51   Zener C and Greenstein M 1980 *Energy* **5** 503–10

52   Charwat A F and Ridgway S L 1980 *Energy* **5** 511–24

53   Griekspoor W and Van Der Pot B J G 1978 OTEC in Europe *Proc. Int. Symp. on Wave and Tidal Energy, BHRA, Canterbury* paper D2 (Bedford: BHRA) pp11–22

54 Curto P A 1980 *Energy* **5** 529–38
55 Kamogawa H 1980 *Energy* **5** 481–92
56 UN General Assembly 1981 *Report of the Technical Panel on Ocean Energy (Preparatory Committee for the UN Conference on New and Renewable Sources of Energy)* (New York: UN)
57 Loeb S 1975 *Science* **189** 654–5
58 Schlecter J 1981 *Proc. Unitar. Conf. on Energy Resources* (London: Pitman)
59 Norman R S 1974 *Science* **186** 350–2
60 Loeb S 1975 *Science* **189** 654–5
61 Pattle R E 1954 *Nature* **174** 660
62 Fraenkel P L and Musgrove P M 1979 Tidal and river current energy systems *Proc. IEE Future Energy Concepts Conf.* (London: IEE) pp114–17
63 Wyman P R and Peachy C J 1979 Tidal current energy conversion *Proc. IEE Future Energy Concepts Conf.* (London: IEE) pp164–6
64 *Proc. MacArthur Workshop on the Feasibility of Extracting Useable Energy from the Florida Current* 1974

4

Tidal energy

4.1 Introduction

The harnessing of tides to provide energy has a very long history. Tidal mills were used in the Middle Ages, particularly in Britain and France. The first serious publications on the matter date from the early eighteenth century. Schemes to build enclosures, and designs of specialist turbines were examined in France, Germany and Britain at the end of the nineteenth and beginning of the twentieth century. A brief history of tidal schemes has been given by Charlier [1].

Tidal energy schemes are sometimes confused with river hydro schemes. They have many features in common but there is, however, a fundamental difference between them. Hydro schemes utilise the flow of water arising from variations in topography and the consequent downward acting terrestrial gravitational forces. Tidal energy arises from the upward acting lunar gravitational force which results in a cyclical variation in the potential energy of water at any point on the earth's surface. Frequently, these variations are strongly amplified by topographical features such as the size and shape of estuaries.

In this chapter, the characteristics of the tidal variation will be discussed, some of the considerable number of possible ways of utilising tidal energy will be presented and the technology which is required will be briefly reviewed. Large tidal schemes have important environmental effects and these will be mentioned in general terms as will system integration and costs. The major scheme which has so far been built (La Rance in Brittany, France) and some of those which have been the subject of detailed evaluation (Severn Estuary in the UK, Bay of Fundy in North America) will be examined. Finally, the further potential for tidal energy in the UK and elsewhere in the world will be briefly assessed.

The recent assessment of the prospects of tidal power from the Severn Estuary [2] provides a comprehensive discussion of many of the factors affecting tidal schemes and the interested reader is referred to both volumes of this study for a deeper analysis of the matters discussed in this chapter. Many of the references given in this chapter are accounts of work carried out as part of this study.

4.2 Characteristics of the tides

Sea level in most places varies approximately sinusoidally on a 12.4 hour cycle and the peak-to-trough amplitude is known as the tidal range. Further variations are also imposed on this semi-diurnal pattern as a result of the complex gravitational field arising from the relative positions of the earth, sun and moon. The most important of these is the spring–neap cycle which varies over a period of about fourteen days. The ratio between the maximum spring tide and minimum tidal range at neap, can be as great as three to one. This pattern is further modified by the fact that the moon's orbit is elliptical. Longer term variations are less significant, but because the power which can be extracted varies roughly as the square of the tidal range they may not be negligible. The semi-annual cycle caused by the inclination of the moon's orbit to that of the earth may cause a variation in tidal range of about 10%. Even longer cycles of up to 1600 years occur, the most important of these has a period of about nineteen years and leads to a variation in tidal range of about 4%.

In mid-ocean the tidal range is typically about one metre, but frequently this is strongly modified by coastal contours. In estuaries, where the tidal range may be as great as 10–15 m at the most favourable sites, the amplification arises from proximity to resonance. In a simple model of a channel of uniform cross section, resonance occurs when the length of the channel is close to one quarter of the length of tidal movement. In real terms, this simple approximation is complicated by variations in width and depth. In particular, frictional losses resulting from water moving over the seabed and banks introduce damping. It is important that a full under-standing of the characteristics of particular prospective sites should be obtained because the construction of a dam across an estuary may itself move the system nearer to or further away from resonance [2] and as discussed further, below, there is some prospect that measures can be taken to approach resonance more nearly in some estuaries.

4.3 Basic modes of operation of a tidal scheme

There are two principle ways in which power can be generated from a tidal estuary. The first of these is to use a single basin which involves building a barrage at a suitable point along the estuary, installing turbines and generating electricity when the tide is ebbing or flooding or in both directions. The second approach is to build a two-basin scheme in which the generation can be retimed to take place always during the hours of peak demand by using the basins alternately. In both types of scheme, but particularly the second, pumping can be arranged to maintain output during the lower tidal ranges by further raising or lowering the basin levels.

4.3.1 Single-basin schemes

These may operate in three basic modes.

Ebb generation. The incoming water is allowed to flow through sluices and the turbine passageways. At high tide, these are then closed and when the water on the seaward side of the barrage has ebbed sufficiently for a large enough head to have been established, the entrained water is allowed to drive the turbines. Flow stops when the water in the basin behind the dam reaches such a level that the head difference becomes less than the minimum required to operate the turbines. When the sea level becomes equal to the level in the basin, the sluice gates are opened and the cycle is repeated. The relationship between water levels in the estuary and the basin and the timing of the power produced is illustrated for such a scheme in figure 4.1.

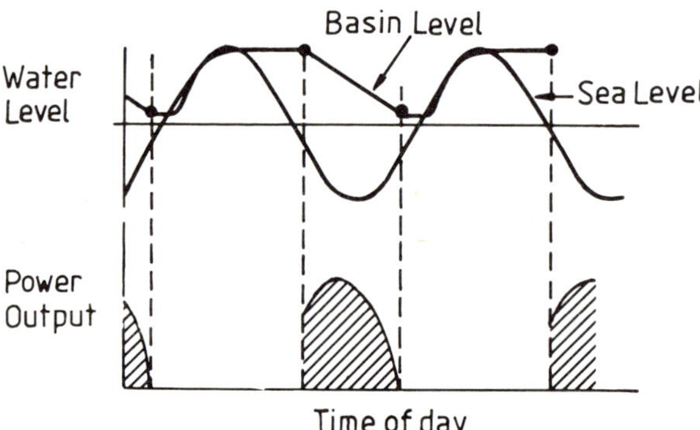

Figure 4.1 Schematic diagram of water levels and power outputs for an ebb generation scheme.

In practice, the operation of this relatively simple scheme is quite difficult to optimise. For example, it would be possible to operate an ebb generation scheme in such a way that a very large number of turbines were generating when the head difference between the sea and the basin was at a maximum, giving a short burst of power. Alternatively, a small number of turbines might be operated for the longest possible time for which there was a sufficient head difference. These extreme options are illustrated in figure 4.2. In practice, the timing of the output and its relationship to the demand and the differences in capital cost between various schemes largely determine the optimum strategy. This will be discussed further in §4.5.

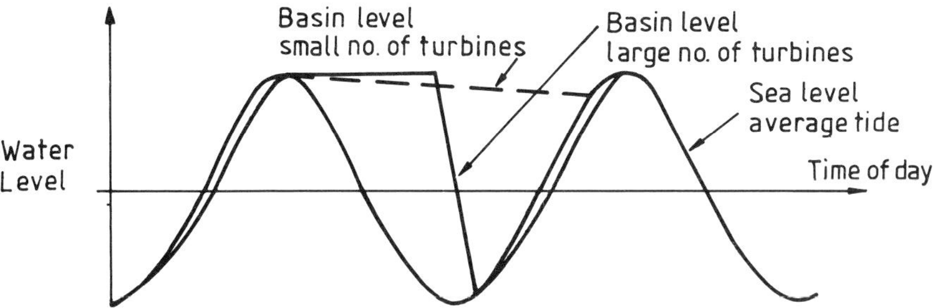

Figure 4.2 Hypothetical operating cycles.

Ebb generation schemes should usually start operating when the difference in level between the basin and sea is about half the tidal range. At first, the turbines are operated at maximum efficiency in order to conserve water and keep a high head. As the tide recedes faster and the need to conserve head becomes less important, the turbines are usually controlled to operate at maximum power (rather than energy). The flow through the turbines can be regulated in a number of ways and these will be discussed in §4.6.

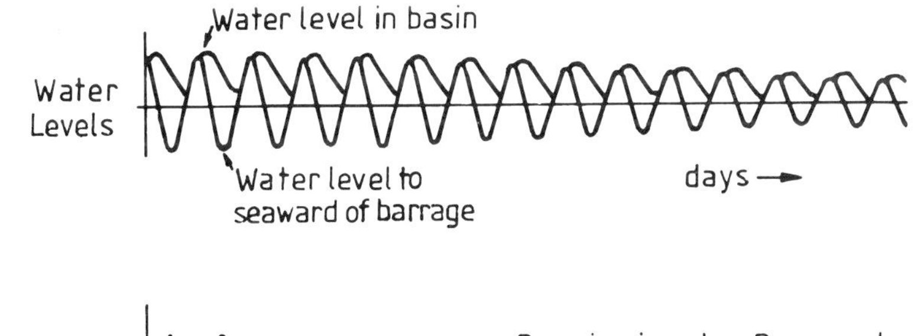

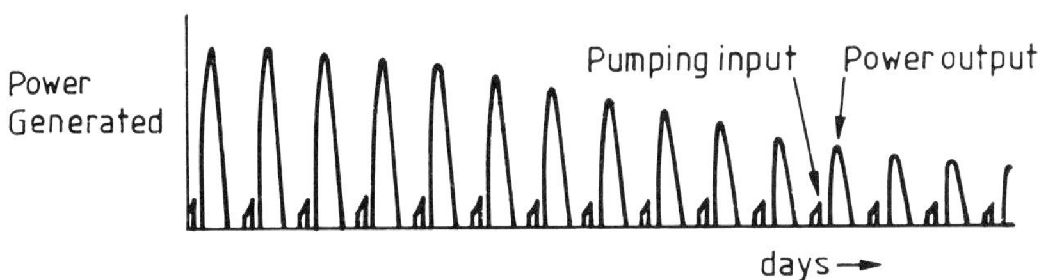

Figure 4.3 Ebb generation with pumping at high tide.

In principle, more energy can be produced by an ebb generation scheme if the turbine generators are designed to motor so that water can be pumped into the basin at high tide (see figure 4.3). The energy gain arises because this pumping would take place at low head and the extra water would be used to generate electricity when the head is greater. In practice, such an apparently attractive scheme does not always provide a net benefit. Firstly, pump turbines are less efficient and more costly and secondly, the energy for pumping must frequently be drawn from the electricity supply system at times when demand is high and costly low-merit steam plant is being used.

Flood generation. In this configuration, operation is the reverse of ebb generation. The sluices and turbine passageways are closed against the incoming tide so that the water level rises on the seaward side of the barrage. When a sufficient head has been built up, the turbines are brought into operation and continue until the water in the basin has reached about mid-tide. The relationship between water levels on each side of the barrage and the timing of the power produced is shown in figure 4.4. Flood generation schemes are generally less favoured than ebb generation for two reasons. Firstly, the estuary above the barrage is kept at low-tide for prolonged periods with possible adverse ecological effects and with re-duced access for shipping. Secondly, the energy produced is less than in an ebb generation scheme. This latter consequence arises because the surface area of the estuary decreases with depth, leading to a more rapid reduction in the head of water when generation commences and thus the volume of entrained water which can be used to drive the turbines is reduced.

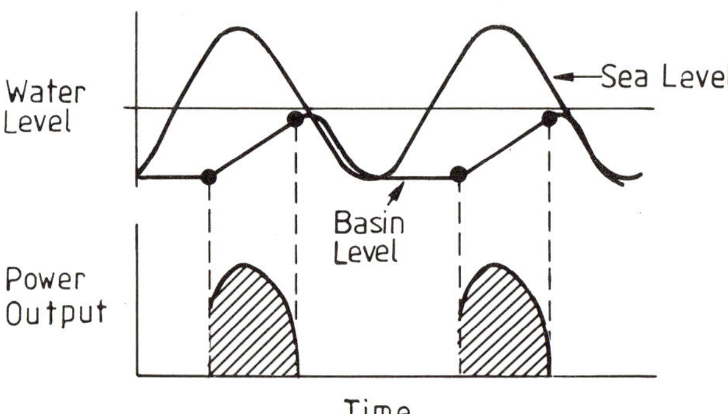

Figure 4.4 Schematic diagram of water levels and power outputs for a flood generation scheme.

The possibility of combining a flood generation scheme with pumping, is generally unattractive for the same reasons which were discussed in the section on ebb generation.

Two-way generation. The water levels and power output for two-way generation are shown in figure 4.5. Towards the end of flood generation, the sluices are opened to entrain a head of water behind the dam and are then closed. When the level of water on the seaward side of the barrage has dropped sufficiently, water is released through the turbines in the ebb generation mode.

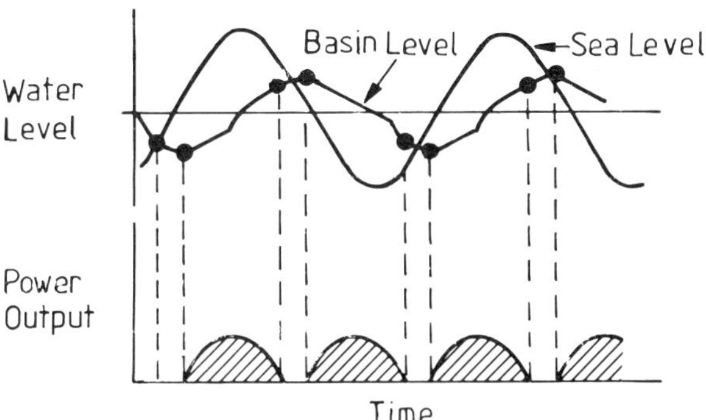

Figure 4.5 Schematic diagram of water levels and power outputs for a two-way generation scheme.

Generating electricity on both flood and ebb tide does not lead to a major increase in energy yield over a simple ebb generation scheme for a number of reasons. Because of the need to achieve a useful head during the previous flood generation phase there is a sizeable reduction in the initial water level for the ebb generation part of the cycle. Furthermore, the phase cannot be taken to completion because it is necessary to open the sluices and reduce the level of water in the basin in preparation for flood generation. The same applies in reverse during the flood generation phase. There are also other disadvantages. A larger number of turbines are required to achieve the same power output and in some schemes, this extra number may be difficult to accommodate. The turbines are less efficient and more complex if they are required to operate in two directions and the turbine water passage has to be longer. This also leads to a more expensive design and larger caissons are required to house the turbines. As in the case of flood generation, the reduction in the maximum water levels in the basin means that navigation for large vessels is adversely affected.

There are, however, some merits in two-way generation. Because water levels in the basin remain closer to mean sea level, some environmental problems present in other schemes are minimised. Power is produced in four blocks rather than two, thus increasing the firm power contribution from the tidal scheme (although possibly increasing the amount of cycling on other plant in the system). For the same energy output the peak power generated is less than for the other modes of operation so that transmission costs may be reduced by achieving a higher utilisation of lower capacity lines.

As with ebb and flood generation on their own, the association of pumping with a two-way scheme does not appear to increase its attractiveness.

4.3.2 Double-basin schemes

In the simplest form of double-basin scheme the two basins operate as independent ebb-and-flood-generation schemes with output characteristics similar to those of two-way generation. Like single-basin schemes the power produced varies according to moon time and this frequently does not match the demand for electricity. Double-basin schemes, which usually include provision for pumped storage, enable the periods at which electricity is generated to be retimed. There are many possible methods of operating such schemes; two of these will now be described.

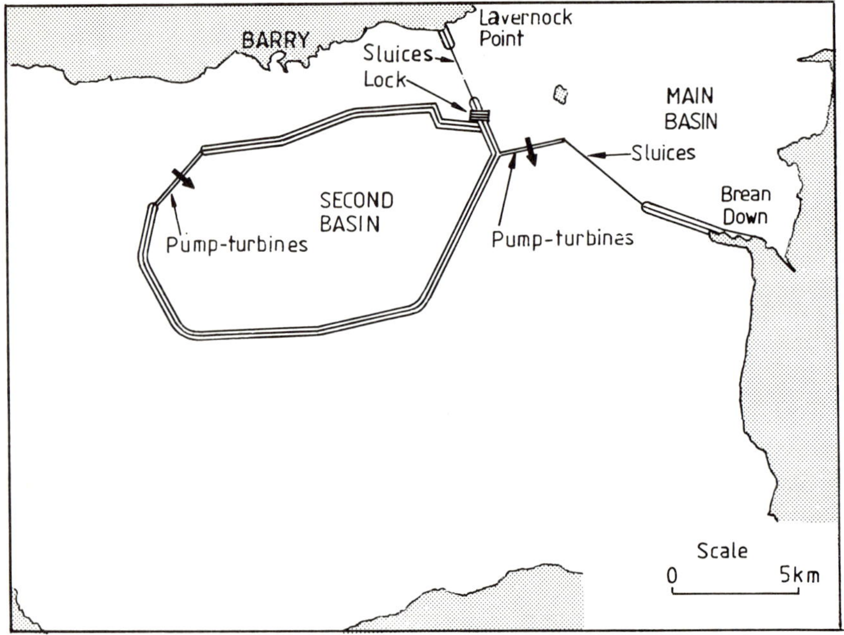

Figure 4.6 The two-basin pumped storage scheme.

Figure 4.6 shows a form of two-basin scheme proposed by Shaw for the Severn Estuary [3]. The main barrage is similar to a single-basin design except that it has fewer sluices. The principle of operation is that turbines in the main enclosure pump from the sea into the basin so that the water level is high prior to a period of high electricity demand. The second basin, built in deep water, utilises pump turbines with about the same electrical capacity as those in the main basin, but operating at higher heads. The level of water in this basin is always kept below the level of the tide and thus energy can be generated at any time during periods of demand so as to make good any deficit from the main enclosure turbines. Because the output from both basins can be arranged to be zero during the night when demand is low (and indeed energy would be imported from the grid in order to pump), the scheme is frequently referred to as 'constrained'. The variations in water level and output for such a scheme are shown in figure 4.7.

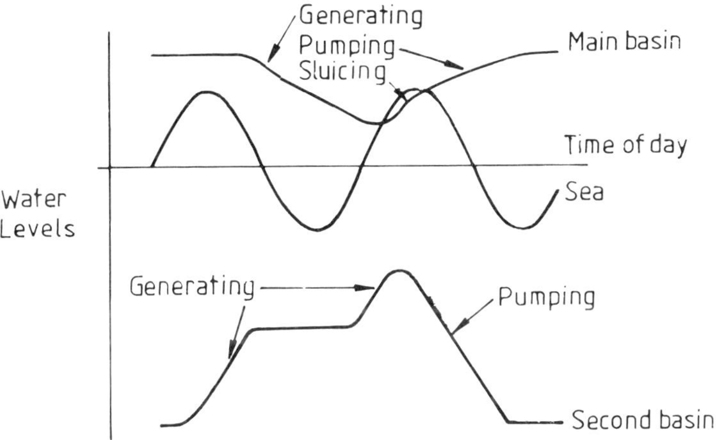

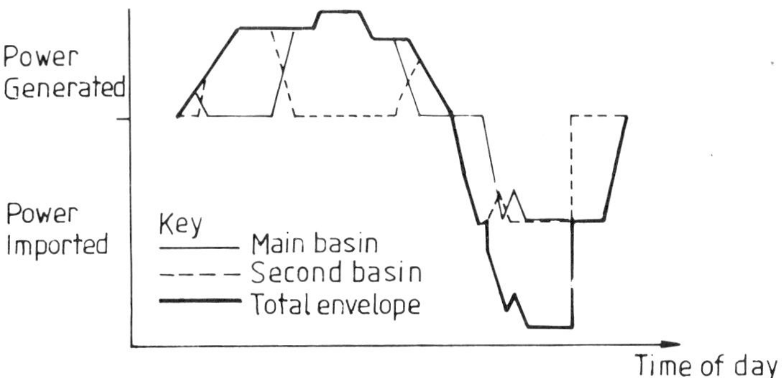

Figure 4.7 The constrained two-basin scheme. Schematic variation of water and power levels.

Another method of operating a double-basin scheme [2] is referred to as 'unconstrained' operation. In this case, the main barrage is operated in a normal single-basin ebb generation mode. When this basin is unable to supply power during the day, output from the second basin is used to make good any deficits and to ensure that a constant output is generated all day. This output level is sometimes lower than the maximum that can be achieved by the main basin alone and any spare energy generated is used to pump out the second basin. Energy generated at night by the main enclosure turbines is also used in this way. At spring tides extra energy can be produced from the second basin, whilst at neap tides energy has to be imported from the grid in order to pump this basin to its lowest level. This is illustrated in figure 4.8. It should be noted that, in principle, the same barrage could operate in whichever of these modes is most advantageous at a given time. For part of a week for example, operation could be in the unconstrained mode and for the rest of that week could be constrained in output.

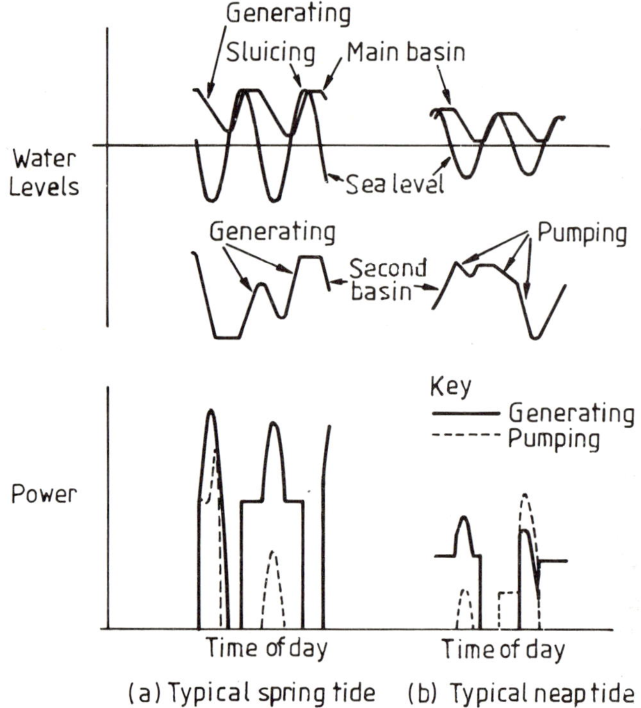

Figure 4.8 The unconstrained two-basin scheme. Schematic variation of water and power levels.

Optimisation of any double-basin scheme is much more complicated than for single basins. Additional factors which have to be considered include the relationship between output and the utility demand curve, the trade-off between turbine efficiency and pump efficiency, the number and size of turbine pumps in each basin, the area of the secondary basin and the optimum water levels.

The major additional cost is clearly in providing the civil works and turbines for the second basin. The pump turbines in the second basin would also be more expensive because of the high pumping head and the need for double regulation to maintain efficiency over the wide ranges of head. Against these extra costs must be set the value to the electricity supply system of providing storage and retiming the output. The value of such storage depends critically on the mix of plant in the system. For example, if sufficient base load nuclear plant was available to meet demand at night, the value of tidal electricity generated at such times would be greatly reduced and storage might be attractive. However, this storage could be obtained just as effectively from inland schemes (such as pumped hydro) which are designed to be used by the system as a whole and which may be cheaper.

The main environmental consequence of two-basin schemes operated in the constrained or unconstrained mode, would be that at periods of low demand (for example at night), water levels in the main basin would remain at or above high tide levels. This could be advantageous for shipping but in some cases may lead to problems of drainage and may adversely effect the ecosystem of an estuary.

It is difficult to generalise about the relative advantages of different barrage schemes because their value depends upon the details of the power system into which they are connected, environmental considerations and the costs of construction. These all vary from site to site. The recent study of the feasibility of a Severn Barrage concluded that ebb generation would have the highest benefit to cost ratio and double-basin schemes the lowest.

4.4 Enhancing output by modifying estuarial resonance

It has been suggested, for example by Count [4], that power at neap tides might be increased by as much as 60% in tidal schemes if flow through the barrage could be controlled in order to move the estuary closer to resonance. This can, in principle, be achieved by maintaining a phase difference between the barrage head and the flow. This would seem to imply that pumping is required, but it has been suggested by Robinson [5] that closing the sluices over part of a cycle while a head builds up and then releasing it over a controlled time may be able to achieve the same effects.

The theory of this process has recently been examined in some detail [5], [6] for model rectangular estuaries where there is no friction. These idealised analyses suggest that power could be enhanced if the sluices are opened to produce a tidal surge twice during each tidal cycle for a time t, when $t = 2l\,(gh)^{-\frac{1}{2}}$, where l is the length of the basin, and h the depth below mean water level. For flood generation this would need to take place close to high tide outside the barrier and for ebb generation close to low tide. The principle, showing the effect of reflection in establishing the optimum condition, is illustrated in figure 4.9.

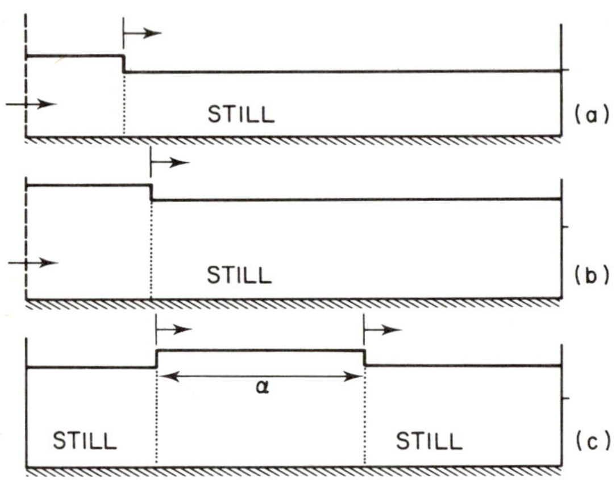

Figure 4.9 Surge propagation in an enclosed basin. (a) Barrier open, positive surge propagating away from the barrier. (b) Barrier still open, surge has reflected once at the basin end and once at the barrier. (c) Barrier closed, negative surge now follows the positive surge, with water flow in region α which propagates up and down the basin. (Courtesy Dr I S Robinson, University of Southampton and reproduced from reference [5] with permission of the BHRA.)

These simple models effectively treat the dynamics of the tides, the barrier flow characteristics and the operating cycle as a single system. They are, at present, very simple and assume that the instantaneously produced surge height is very small compared to h, that the sea level is not altered by the surge and that no negative surge is reflected seaward from the barrier. These assumptions are in addition to the simplifications about shape and friction. Furthermore, environmental effects may preclude such measures being employed in real estuaries.

Nonetheless, the possible gains are considerable and it will be important to test these ideas in accurate models of real estuaries and to attempt to develop realistic operating strategies which might increase the power produced. The ideas will probably not be applicable in shallow estuaries where surges will be dissipated rapidly but may be important in deep and short fjord-like estuaries.

4.5 Optimisation of single-basin schemes

Mention was made in §4.3 of some of the additional problems which must be overcome in order to optimise the design of a two-basin scheme. These extra degrees of freedom make optimisation very difficult and will not be discussed further here, but an outline will be given of the method which is used in order to arrive at an optimum configuration in the simpler case of a single-basin tidal scheme. The questions have been discussed recently with special reference to the Severn Estuary by Wheeler [7]. Environmental and social factors are clearly of great importance but these are normally considered separately and will be discussed in §4.10.

Choosing the most cost effective scheme is an iterative process and may be carried out to various levels of sophistication. The methodology described by Wheeler does not, for example, involve running a model of the electricity supply system into which the barrage output is to be integrated, although this would certainly need to be done for a detailed evaluation. It is only applicable to deciding between alternatives which are broadly similar in design and output. The three main variables which enable the unit cost of energy to be determined are the annual energy output, the cost of the scheme and the timescales for construction.

The energy which is likely to be produced is normally calculated using computer models [8]. The simplest of these, known as flat-surface models, assume that the water on the seaward side of the barrage is that given by tide tables, whilst upstream it is assumed that the water surface is flat with flow through the barrage immediately changing the level of water in the basin. The models do not predict the changes in tidal range seaward of the barrage due to a particular scheme, because dynamic effects are not included. Models which take such effects into account are much more expensive to run, so what is frequently done is to use a dynamic model to calculate the reduction in tidal range and energy output for one scheme on a given barrage line and then to explore the effect of changing such variables as the number of turbines and sluices by using a flat-surface model.

Where a number of alternative barrage lines are possible the first objective is to attempt to reduce the choice to a workable number. This may be achieved by using a dynamic model to explore the tidal range reduction for a number of possible lines and then using interpolation to estimate the reduction at the others. The number of turbines and sluices for each line can then be studied using a flat-surface model and a consistent estimate based on experience of turbine diameter and capacity.

Having chosen perhaps two lines for more detailed analysis, the next phase of the optimisation is to make a provisional choice of turbine diameter and submergence. This is performed by choosing the maximum size of proven turbine which can be transported to site and fixing the submergence to a value which optimises energy output against the civil engineering costs. A conservative value of submergence is normally chosen in order to reduce possible cavitation in the turbines.

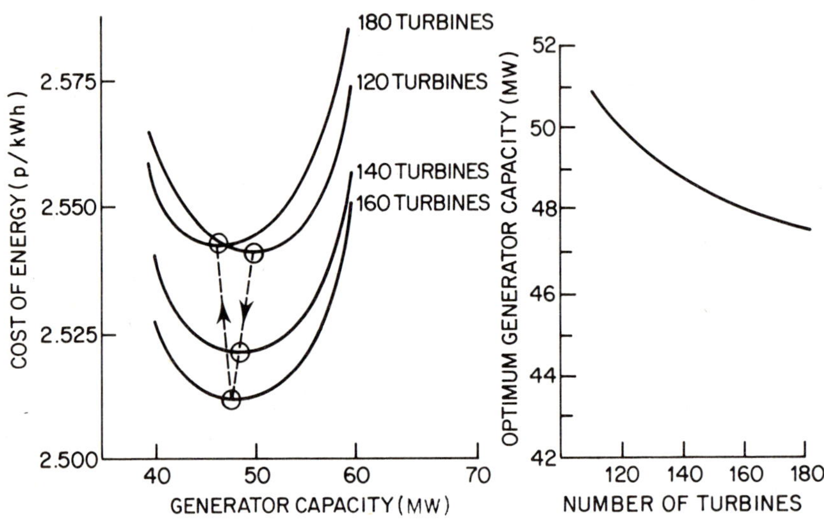

Figure 4.10 Effect of turbine numbers on optimum generator capacity. (Courtesy Mr S J Wheeler and reproduced from reference [7] with permission of the BHRA.)

Generator capacities are then fixed since the optimum for a given barrage line appears to be relatively insensitive to the number of the turbines (figure 4.10). The variation of annual energy output can be calculated from a flat-surface model. In optimising capacity, transmission costs as well as generator costs for various sizes need to be included in the analysis. To optimise the number of turbines and sluices, the unit cost of energy is calculated for a range of possibilities. It is important to ensure that the numbers can be accommodated in water which is sufficiently deep. In practice, it is usual to choose a reasonable number of turbines and then to calculate the cost of energy as a function of output for varying numbers of sluices. If this is repeated for different turbine numbers, a family of curves is obtained (figure 4.11), the envelope of which allows the minimum unit cost of energy to be determined. What is really required to optimise the numbers is a curve of marginal unit cost against annual energy output. In practice, the marginal cost is found to rise sharply as the numbers increase and the value of energy falls, so the optimum is usually close to the number of turbines and sluices which give minimum cost per unit.

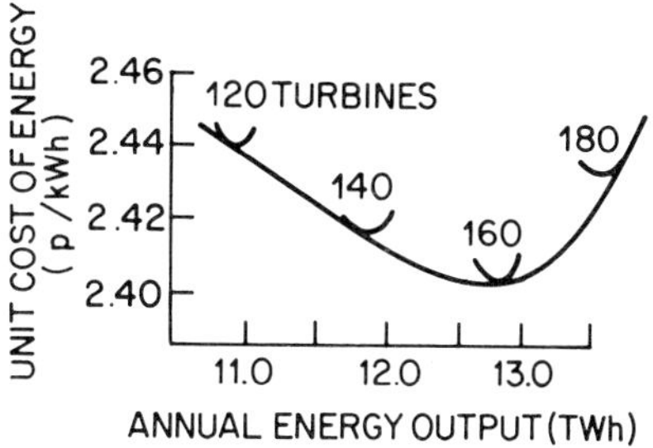

Figure 4.11 Unit cost of energy for a typical barrage line. Each 'sub-curve' shows sluice numbers varying between 120 and 165. (Courtesy Mr S J Wheeler and reproduced from reference [7] with permission of the BHRA.)

The most suitable barrage line should be apparent when the unit cost of energy is plotted against the annual energy output. A typical plot taken from recent studies of a Severn Barrage (see §4.11.2) is shown in figure 4.12. In order to be absolutely certain that a minimum in the cost to benefit ratio has been obtained more detailed studies including system modelling need to be carried out and further iterations of the optimisation process may be required. The procedure discussed here and summarised in table 4.1 adapted from a similar table in [7] provides, however, a useful first approximation in determining whether a given line in an estuary is likely to be attractive.

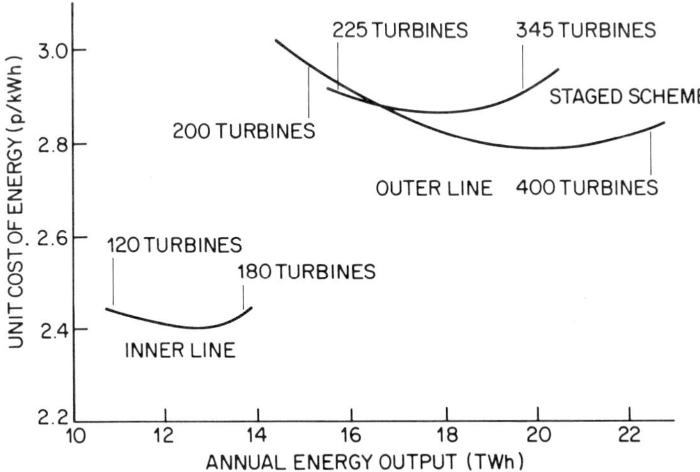

Figure 4.12 Energy costs for lines shortlisted in the Severn tidal power study. The curves shown take no account of the value of energy from the different schemes. (Courtesy Mr S J Wheeler and reproduced from reference [7] with permission of the BHRA.)

Table 4.1 Optimisation procedure (from reference [7]).

Operation	Recommended performance criterion	Recommended method for calculating energy output	Comments
Selection of short-list of barrage lines	Unit cost of energy	'Flat surface' model	A dynamic model may be used to investigate the range reduction at a few barrage lines
Selection of turbine diameter and submergence	Unit cost of energy	'Flat surface' model	Because of uncertainties in the turbine cost and cavitation criterion a degree of subjective judgement required
Optimisation of generator capacity	Unit cost of energy	Dynamic model in conjunction with 'flat surface' model	Use dynamic model to calculate the output from one scheme on each barrage line. Use 'flat surface' model to show the effect of changing generator capacity
Optimisation of turbine and sluice numbers	Unit cost of energy	Dynamic model in conjunction with 'flat surface' model	Dynamic and 'flat surface' models used in same fashion as in previous operation

4.6 Turbines

There are now three types of hydroelectric plant which are generally considered suitable for a tidal scheme. These are the bulb-type, rim turbine and the tubular turbine. These horizontal axis types have now virtually replaced the vertical axis Kaplan turbine which was costly and required deep water. These turbines and their methods of regulation will now be discussed. References [9] and [10] provide more detailed reviews of the subject.

4.6.1 Bulb turbines

This device was patented by Escher-Wyss in 1933 and is now well developed and widely used. It is of the type used, for example, at La Rance and chosen as the lead machine in the study of the Severn Barrage.

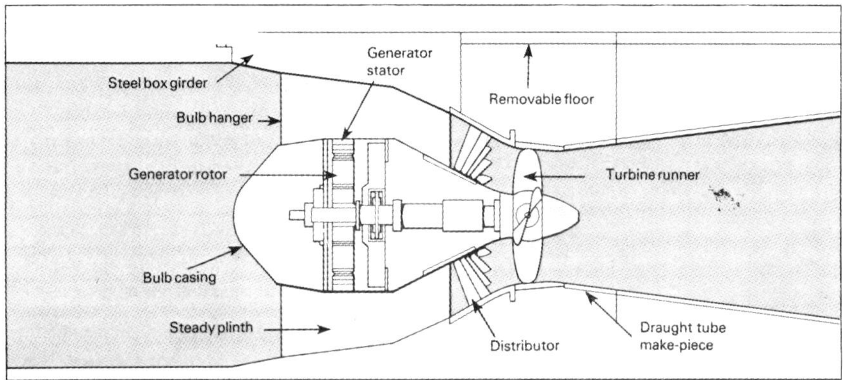

Figure 4.13 Bulb turbine. (Reproduced from *Energy Paper* 46, courtesy HMSO.)

Figure 4.13 shows the layout. The generator is enclosed in a steel bulb within the water passage. Normally this bulb is upstream of the turbine runner which is directly coupled to the generator. The bulb turbine is compact and self-contained but has a number of disadvantages including poor access to the generator, difficulty in providing generator cooling and low inertia, thus limiting its use in weak grids where electrical instability may result. Low head turbines operate at low rotational speeds and unless the generator speed is increased by gearing, the bulb diameters are relatively large. Because this determines the length of the convergent section of the water passage, a longer device also results from this limitation.

4.6.2 Rim generator turbines

This turbine was invented in 1919 and in the 1940s Escher-Wyss built a number of such turbines of small diameter for use in German river hydro schemes. In the last decade Escher-Wyss have developed larger diameter rim machines under the name of 'Straflo'. As the name implies, the rim generator rotor is mounted on a rim fixed peripherally to the runner blades as shown in figure 4.14. Because the generator is no longer situated in the waterway, the water flow is straightened and losses are minimised. Other advantages include the absence of a drive shaft, easy access to the generator, the ability to use a larger diameter generator than in the bulb type (which leads to simple generator cooling and greater inertia). Set against these are the difficulties in sealing the generator from the water passage and in providing a reliable method of supporting the generator.

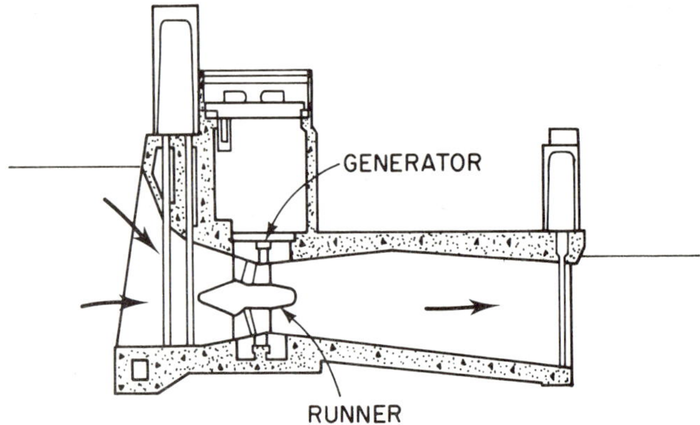

Figure 4.14 Rim generator turbine. (Reproduced from reference [9] courtesy Mr J L Carson and *Power* magazine.)

The application of Straflo turbines to large tidal schemes will depend on the successful development of these machines in the very large sizes required. A 7.6 m diameter, 20 MW machine is shortly to be tested at Annapolis Royal in the Bay of Fundy. This experiment and details of the design of this large turbine have recently been discussed by Focas [11] and Miller *et al* [12].

4.6.3 Tubular turbine

This type of turbine is shown in figure 4.15. The generator is again located outside the waterway. To avoid turning the water through a large angle, the drive shaft is usually set at an angle to the horizontal. The machine offers the advantages of easy access to the generator and room to add a gearbox to increase the generator speed and thereby reduce its size and cost. The tubular turbine also offers high rotational inertia.

The disadvantages are that alignment problems can result from the long drive shaft, the power house tends to be fairly long, and thus costly, and the drive shaft has to operate in cross flows which can result in complex bending stresses and vibration.

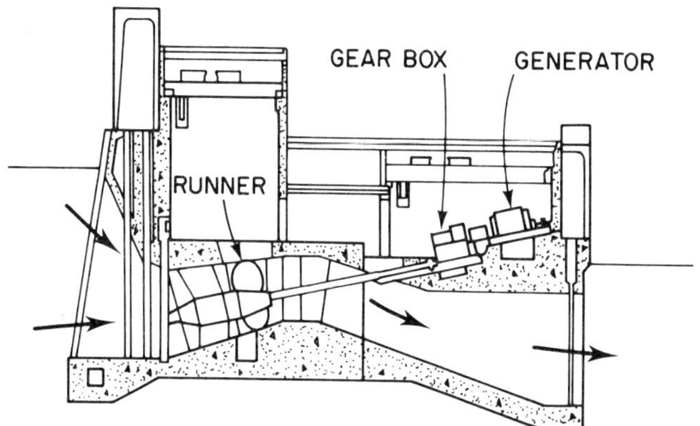

Figure 4.15 Tubular turbine. (Reproduced from reference [9] courtesy Mr J L Carson and *Power* magazine.)

4.6.4 Regulation of the turbine

Whichever of these turbines is chosen for a tidal scheme the output generally has to be regulated. The only situation in which regulation is not required is when the head remains nearly constant during operation as it does on some run-of-river hydro schemes. Regulation to achieve maximum efficiency over a range of operating conditions can be achieved in three basic ways [13].

(1) Varying the opening of the distributor guide vanes.

(2) Varying the angle of the runner blades.

(3) Varying the speed and torque by controlling the generator excitation and connecting to the system by a DC link.

To achieve the highest efficiency with constant speed operation, both the guide vanes and runner blades must be variable. This is known as double regulation and entails extra capital costs.

Figure 4.16 shows a typical set of curves of energy output as a function of rotational speed for various methods of regulation. Although less efficient, variable runner blades are frequently used because this method of control is mechanically simpler, minimising the number of parts which move relative to each other. On the other hand, a downstream gate is normally required to initiate or stop the flow and to avoid runaway in the event of sudden loss of load. For the Severn Barrage study, adjustable guide vanes were chosen as the 'lead' option.

In tidal schemes, variable speed operation increases the energy produced because the speed is then well-matched to the head at all times. Also the use of a DC link avoids the need for synchronising and problems of electrical stability. These benefits have to be offset against the cost of rectifying the output from the generators to DC and inverting back to AC at grid frequency. Furthermore some losses (about 1%) are incurred in the conversion process.

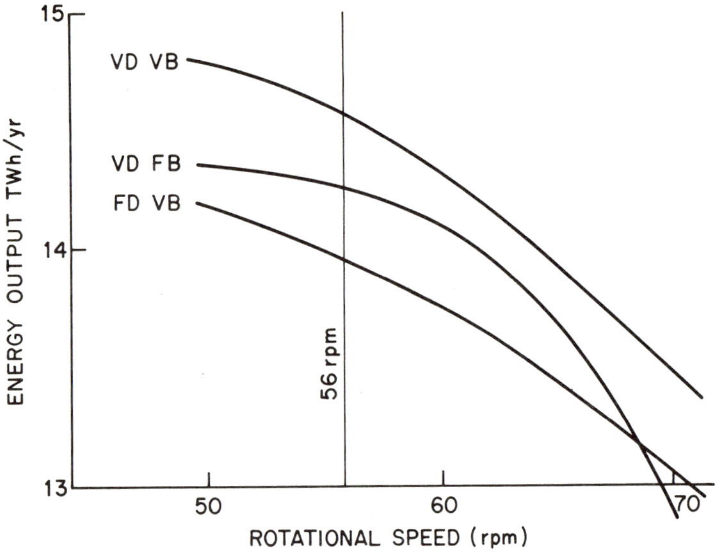

Figure 4.16 Energy output for fixed speed generation for inner barrage. Key: V, variable; F, fixed; D, distributor; B, blades. 60 MW synchronous generator capacity limit. (Reproduced from *Energy Paper* 46, courtesy HMSO.)

The consensus appears to be that, provided sufficient water depth is available, the largest size turbine possible should be used for a given barrage scheme. In the case of the Severn Estuary study, 9 m diameter bulb turbines have been provisionally chosen and these would drive synchronous generators rotating at about 56 rpm and rated at 45 MW. Generator design has recently been discussed by Gardner and Jervis [14]. Transmission is not discussed here, because it is largely scheme-dependent. Arnold [15] has discussed the interconnection to the grid of the Severn Barrage.

4.7 Civil engineering

For most tidal schemes the total cost is dominated by the civil engineering costs of constructing the barrage. In the case of the Severn Barrage, a breakdown of costs is given in table 4.2 (see §4.9). In this section, a brief description will be given of modern construction techniques employing caissons, the technology used to place these in an estuary and possible methods of construction for embankments.

The Rance tidal barrage was built by constructing a coffer-dam [16]. This technique is costly (30% of the Rance costs) and time consuming and strong currents can be generated in the gaps in the coffer-dam before closure is completed. In more recent schemes which have been the subject of detailed study, prefabricated concrete structures known as caissons would be used. These were first utilised for constructing the wartime Mullberry Harbour.

The typical building sequence for a barrage in a large estuary is first to place in position any ship locks, followed by turbine caissons, sluice caissons and embankments.

4.7.1 Turbine caissons

Figure 4.17 shows one design of turbine caisson proposed for the Severn Estuary. It would weigh about 100 000 tonnes and would be built in a special construction yard and floated to the barrage site. The structure is double-skinned to afford protection from internal flooding in the event of damage, and each bay is watertight below the high water mark to ensure that flooding would be contained within a small part of the structure. It is designed to withstand the highest likely unbroken or broken wave forces, the impact of small vessels and the large forces exerted during the operation of towing the structure into position.

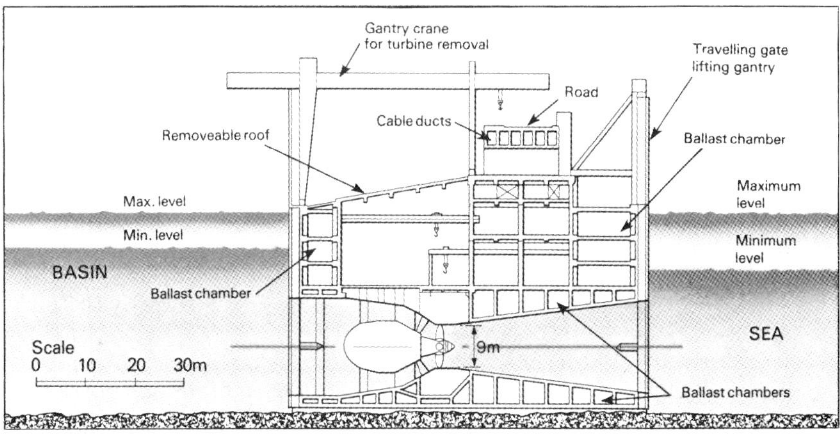

Figure 4.17 Bulb turbine caisson. (Reproduced from *Energy Paper* 46, courtesy HMSO.)

There are several possible designs. The one shown in figure 4.17 is formed from box sections thus utilising materials and space efficiently. Other designs take a spherical or cylindrical form in order to utilise fully the inherent strength of the concrete in compression. In all designs, stability is enhanced by mounting the turbines at a low level with lighter equipment in the upper bays. To allow relatively easy removal the Severn Barrage designs envisage hanging bulb turbines from a steel box girder structure forming part of the caisson and providing torsional restraint from below by a concrete plinth. In most designs of caisson, cellular voids at low levels are used for ballasting during float-out and for fine ballasting during the difficult task of positioning the structure. The upper ballast chambers allow adjustments to trim to be made during float-out when the caisson must also be dynamically stable against wave action.

Design of turbine caissons has recently been discussed in detail [17], [18].

4.7.2 *Sluice caissons*

These are much smaller and less complicated structures than turbine caissons. They hold the gates required to fill and empty the basin during operation and provide channels for water flow during closure of the estuary.

There are three basic types of gate [2] and these are illustrated in figure 4.18. Among the features which are desirable in the design of a sluice gate for a tidal scheme are the ability to operate under a differential head, provision of good access for maintenance and ease of replacement, low operating costs, the capability of supporting variable water levels when the turbines are inoperative, ability to withstand wave attack and ease of release if the gates are jammed closed.

The vertical lift gate has the advantage of allowing a large discharge rate because of the tapered shape of the flow passage and it satisfies most of the above criteria except that de-watering of the complete sluice passage is required if the gate is jammed closed. It is a proven design and is probably the most favoured device for deep water. Radial gates provide easier passage for migrating fish and in some designs most of the moving parts are kept above water level. As with the vertical lift gate, de-watering of the sluice passage is required if the gate is jammed closed. It is probably the more favoured design in shallow water. The flap gate has a number of disadvantages. These include lack of control over the sluicing operation, a tendency to produce seabed scour by deflecting the water flow downwards, inability to support variable water levels with the power plant inoperative, generally poor access for maintenance, a considerable vulnerability to wave damage and a lack of provenness in large-scale operation. However, it has very low operating costs and jammed gates can be removed fairly easily.

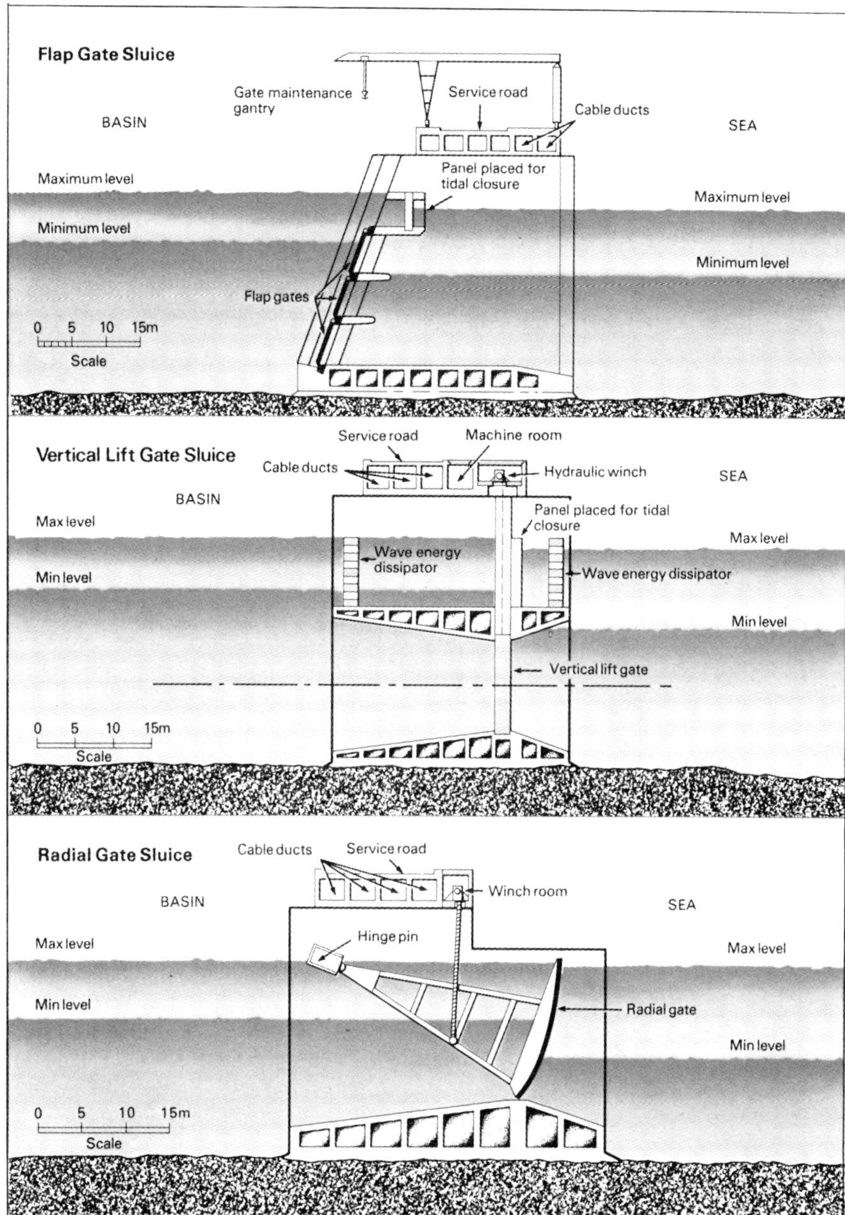

Figure 4.18 Types of sluice caisson. (Reproduced from *Energy Paper* 46, courtesy HMSO.)

4.7.3 Towing and positioning of caissons

The technology for towing and positioning caissons on to their foundations has been largely developed for North Sea platforms. These are larger than turbine caissons but tidal forces are clearly less in the North Sea so the difficulties involved in positioning may be comparable.

Positioning would be achieved during the slack water window at neap tides and the length of time available (typically about an hour) would depend upon the forces which can be resisted. Clare and Oakley [19] have recently given a critical review of three methods which might be employed in the Severn Estuary. In the static method, the caisson is held by anchors and winched into position and then flooded. For large caissons in a tidal estuary, the slack water window is unlikely to be long enough for this technique to be employed. The lift barge method employs a U-shaped barge which holds the caisson between its hulls and the whole assembly is towed into position by a fleet of tugs. When roughly positioned, legs are deployed which can be used to lift the barge clear of the water and the caisson can then be moved accurately into its final position. The technique is probably best used for small caissons, because for large structures the loads on the barge crane beam become very high, stresses on the caissons at their lifting points may be unacceptable and the forces on the barge legs may limit the size of caisson which can be manoeuvred. The dynamic method is probably most favoured. In this method, the caisson is ballasted until it achieves the lowest possible bottom clearance during towing and is then pushed or drawn, by four or five tugs, against its neighbour at the start of the slack water window. It is then lowered on to temporary foundations by flooding. The method gives coarser alignment than some of the other methods but should be acceptable. In the Severn Barrage study [2], it was suggested that hinged links would be used between the tugs and the caisson in an attempt to decouple some of the motion of the tugs. The preparation of foundations for caissons in the Severn Estuary has been reviewed by Campbell [20].

4.7.4 Embankments

These can represent more than a tenth of the cost of a barrage. They can be constructed using well-tried, conventional technology. The main techniques available are the dumping of rock-fill using bottom dump or side dump barges, end tipping using bulldozers, dumping from large cranes or overhead cableways, or embedding caissons and building the embankment around them. Some more novel techniques are now being suggested which might find application in building tidal barrage embankments. These include building a steel framework across an estuary from which sand may be dumped between perforated steel sheets (the Eiderdamm method) or slip-forming the embankment. This has recently been suggested by R C H Russell of the Hydraulic Research Station in the UK as a possible method for building the embankments of a Severn Barrage [2].

The factors which need to be considered in choosing methods of construction apart from cost, provenness and the timescales involved, are the dependence on tide and weather, effects on navigation and the materials requirements (including transport and environmental effects at the source of supply). Barge-dumped rock-fill and end-tipped rock-fill appear in most cases to be the most favoured currently available methods.

Figure 4.19 shows a section of the design of embankment proposed for a Severn Barrage.

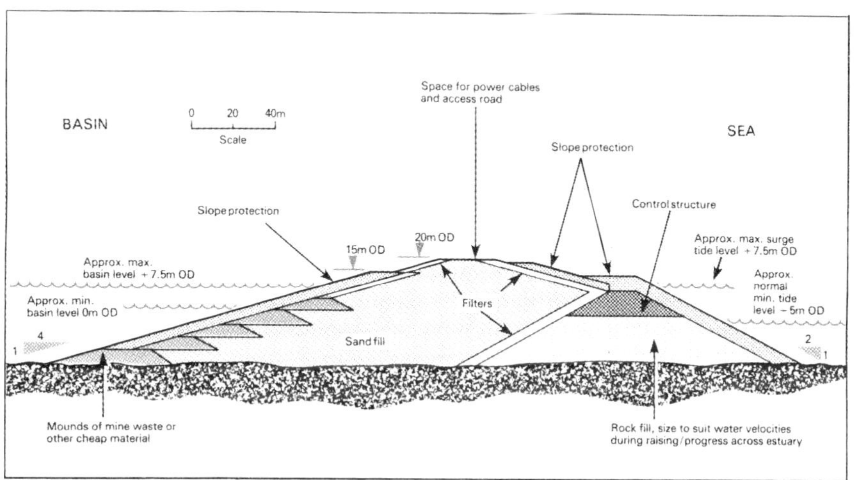

Figure 4.19 Cross section of the embankment design proposed for the Severn Barrage. (Reproduced from *Energy Paper* 46, courtesy HMSO.)

4.8 Assessing the value of tidal schemes

The general methodology used in assessing the value of renewable energy sources to a utility was discussed in Chapter 2. The points made there apply equally to tidal energy. There are, however, two additional points to note.

(1) Some potential schemes are very large and it is therefore of great importance to make a thorough assessment of their value as fuel savers or as providers of firm power for various future plant mixes and levels of demand. Whereas wind, wave and solar power can be built up slowly and any adverse system effects observed, this is not true for large tidal schemes.

(2) The output from a tidal barrage can be predicted. This does not add to the firm power which a scheme can offer (the Severn Barrage stage one

scheme (7.2 GW) has been estimated to offer about 1 GW of firm power) but it should simplify its accommodation into the system. Unlike the other sources which rely for their output on the vagaries of the weather, large unexpected changes of output from a barrage would only result from a major electrical fault. For very large schemes, however, the generation of 'blocks' of power means that large amounts of conventional plant would require to be two-shifted† in order to cope with the changes in tidal output. The capability of a system to respond to these changes depends upon its size and plant mix. The systems analysis of the Severn Estuary has recently been discussed by Brown *et al* [2].

4.9 Cost of tidal energy

The cost of a particular tidal scheme depends critically on local conditions. As a guide to the breakdown of costs between various components of a barrage, table 4.2 shows the percentages of total expenditure ascribed to ship locks, embankments, caissons, turbines and transmission for the Severn Barrage 'lead' single-basin scheme and a much larger outer barrage scheme.

Table 4.2 Percentage contribution to costs of Severn Barrage schemes.

	Inner (7.2 GW)	Outer (12 GW)
Locks	8%	4%
Embankments	13%	13%
Caissons	41%	39%
Turbines	25%	29%
Transmission	13%	15%

In both of these cases the cost of turbines and caissons contributes over two-thirds of the total cost and this is probably fairly typical.

The cost of the electricity generated is again strongly site dependent. In studies of many major schemes, however, the cost of electricity appears to compare favourably with the projected cost of electricity from fossil fuel plant, but is higher than the projected cost of nuclear power.

4.10 Environmental effects

The effect of a tidal power system on the environment clearly depends upon the location of the scheme. These matters will be discussed more fully

† Two-shifting is the scheduling of a power station such that it supplies power twice a day, e.g. at the morning and evening periods of peak demand.

later in §4.11 with reference to several actual or potential schemes. It is possible, however, to list some of the effects which need to be examined closely before tidal schemes are undertaken. Environmental effects may result from changes in water levels, changes in water flows and velocities, changes in sedimentation, the physical presence of the barrage and the effects of construction activities both close to the site and as a result of supplying the necessary materials. The impacts can include [2] those on: ports and navigation; facilities for recreation; water quality; sea defences; agriculture and land drainage; visual amenity; industry and employment; opportunity of providing an estuary crossing; bird life; migratory fish; the ecosystem in the estuary and its margins.

4.11 Studies of the operation and potential of major tidal schemes

It has already been pointed out that the feasibility, economics, mode of operation and environmental impact of tidal schemes depend strongly on location. The rest of this chapter will therefore be taken up with a discussion of the largest operational tidal scheme in the world at La Rance in Brittany and two other projected schemes—the Severn Estuary in the southwest of England and the Bay of Fundy in North America. The last two have been the subjects of detailed appraisals. Finally, in the last section, a number of other possible locations in the UK and around the world which have been considered for tidal schemes will be mentioned.

4.11.1 La Rance tidal scheme

The Rance Barrage near St Malo in Brittany was constructed between 1961 and 1967, the first output being achieved in 1966. It is the only operational tidal power scheme in the world with the exception of a small (400 kW) scheme in the USSR and a 500 kW scheme in China. The position of the Rance Estuary is shown in figure 4.20, the large tidal range (up to 12 m at spring tides) is thought to result from the coastal topography of the Cotentin Peninsula.

The tidal scheme at Rance was built as a 240 MW prototype to gain experience for the possible construction and operation of a much larger complex of such schemes on the Brittany coast. The massive Iles de Chaussey scheme is illustrated in figure 4.21. This could, in principle, generate up to 15 GW. Although built as a single-basin scheme, La Rance was designed to provide the maximum flexibility of operation, with double-action pump turbines able to augment the filling or emptying of the basin and the ability to generate and pump on the ebb and flood tides.

 Tidal energy

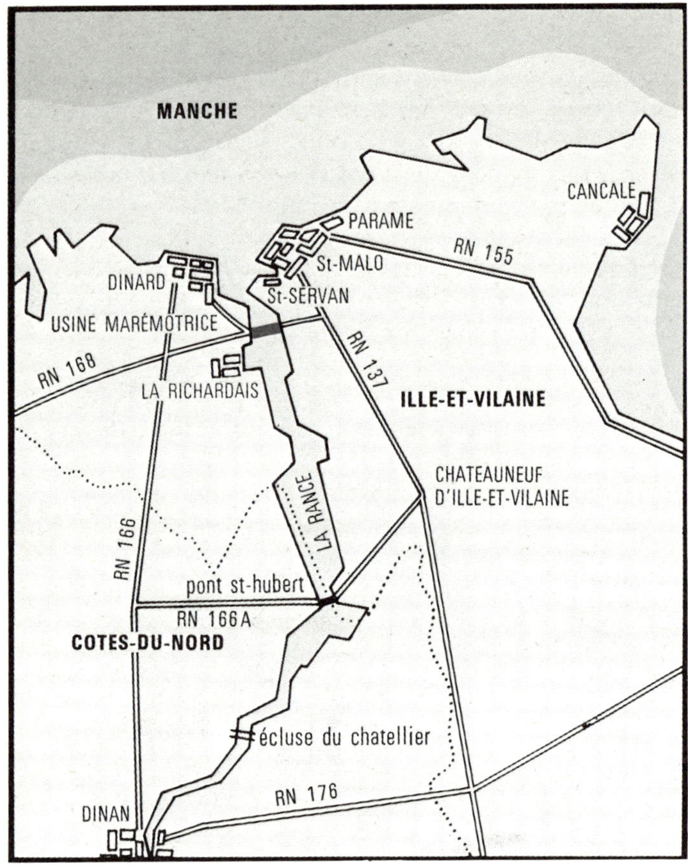

Figure 4.20 Position of the Rance barrage. (Courtesy Electricité de France.)

Reviews of the project have been given by Cotillon [16] and more recently by Banal and Bichon [22] and some details of the performance of the barrage are summarised in table 4.3. Figure 4.22 shows the layout of the barrage and figure 4.23 shows an aerial view. On one bank of the estuary there is a 65 m long, 13 m wide ship lock. Adjacent to this is the power house which is 390 m long and 13 m in width. This houses twenty four 10 MW bulb turbines each with a runner diameter of 5.35 m. These are reversible, horizontal shaft machines with adjustable blades and movable guide vanes. Each turbine is coupled to an air-cooled generator with an air–water heat exchanger and the whole assembly is supported in a 53 m long conduit by stay vanes and prestressed tie-rods. Each unit can turbine or pump in either direction. The power house is connected to a

163 m long embankment which was constructed from rock-fill and which joins on to a natural feature called the Chalibert rock. From the rock to the shoreline at Pointe de la Briantais, there is a 115 m long sluice structure which houses six gates each 15 m wide and 10 m high and capable of passing nearly 10 000 cubic metres of water per second at mid-tide with a head difference of about 5 m. The construction of the barrage was carried out from inside a specially constructed coffer-dam.

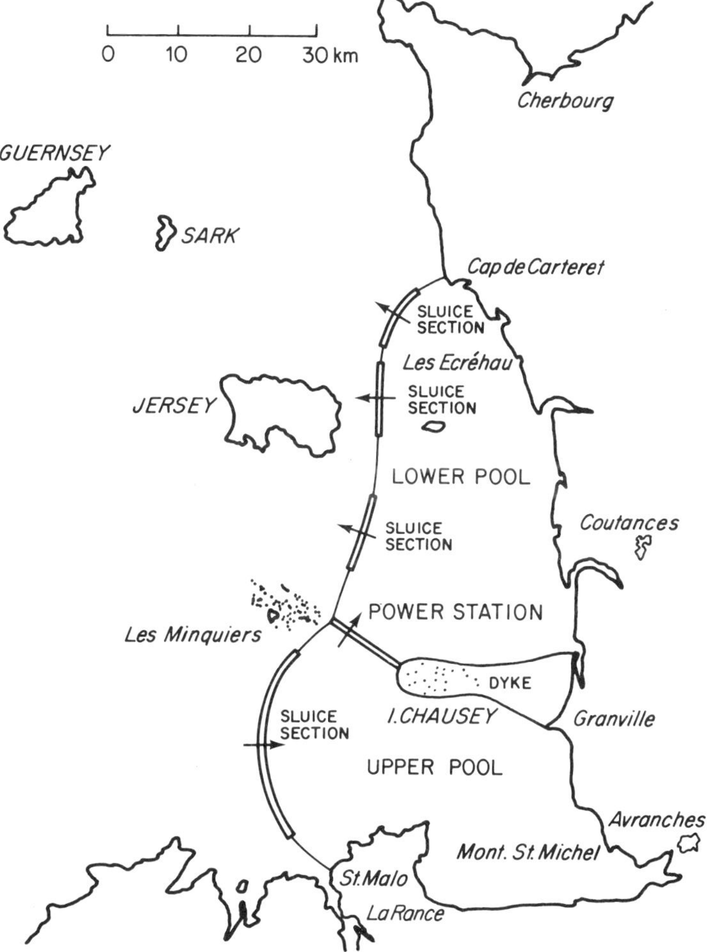

Figure 4.21 The Iles de Chaussey tidal power project.

The turbines have now been immersed for more than 100 000 hours and have operated for about 85% of this time. A major engineering problem was encountered in 1975, when lugs holding the generator stators in position in the bulb housings were found to have been weakened by severe stresses during start-up for pumping. These were removed for strengthening over a period of more than four years. With the exclusion of this type of fault, the availability of the plant has at times been more than 90% (see table 4.3) and by timing engineering work to periods of low output at neap tides, the energy availability has been in excess of 95%.

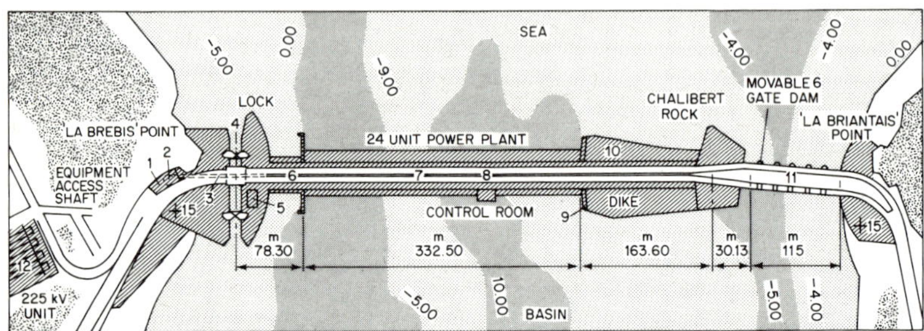

Figure 4.22 Layout of the Rance barrage. 1, access building to large rooms at +16.65 m level (marine charts). 2, shaft for descent to rooms at level −7.00 m, diameter 12 m. 3, access gallery at −7.00 m, passing under the lock, about 80 m long. 4, navigation lock, lock chamber: 65 × 13 m, invert at +2.00 m. 5, administrative building and main access to the plant. 6, Equipment disassembly bays and maintenance shops. 7, 24 bulb-unit bays, distant 13.30 m from each other. 8, control bay. 9, wall at plant end, constituting the retaining wall of the rock-fill dike. 10, rock-fill dike. 11, six sluices equipped with 15 × 10 m gates. 12, line departure unit, three 225 000 V lines. (Courtesy Electricité de France.)

The position of the barrage is such that wave loadings are negligible but much experience has been gained in corrosion protection. A successful cathodic protection system using about 20 kW of power has been used on the runner blades and other submerged metal parts and it is thought that in future schemes this could reduce the need for special steels.

The choice of 10 MW capacity units was made at La Rance in the 1960s, partly because of the constraints of using the available technology which meant choosing turbines of 5 m diameter and partly as a result of a deliberate policy of ensuring the greatest flexibility in operation. A head of 5.65 m was chosen with a design flow of 6600 m³/s because these figures correspond to a tide halfway in range between an ordinary neap tide and a mean tide. The choice of this rated head (the rated head is the minimum head required to achieve maximum output) and the proportionally lower area of sluices has the consequence that the optimum mode of operation of the Rance scheme is different from that suggested below for the Severn

Table 4.3 Generation figures for the La Rance scheme (from reference [16]).

	1968	1971	1972	1973	1974	1975 Actual	1975 Supposed	1976	1977
Gross output (GWh)	426	509.5	560	567.5	606	524		444.5	462
Input (GWh)	24.5	42	59.5	62	91	35.5	62	—	—
Plant consumption (GWh)	7.5	8.5	8.5	8.5	8	7		6.5	6.5
Net output (GWh)	394	459	492	497	507	481.5	571	438	455.5
Availability factor (%)	77	94	93.9	95.2	95.5	85.75		72.2	75
Assumed net output with a 95% availability factor (GWh)				497	507	510	509	480	480
Pumping gains (GWh)	44	74	108	112	120	47	112	—	—
Net output without pumping gains (GWh)				385	387		497	480	480

Barrage. In essence, the French scheme is believed to operate at its most economic in the following modes:

(1) For spring tides, two-way generation is favoured (primarily because of the choice of rating for the machines).

(2) For intermediate tides (7–11 m range) either flood tide generation or direct pumping from sea to basin is favoured.

(3) For neap tides direct pumping from sea to basin is sometimes an attractive option to supplement generation.

Figure 4.23 Aerial view of La Rance tidal scheme. (Courtesy Electricité de France.)

The most economical mode of operation is clearly not always synonymous with achieving maximum output. To achieve this, direct turbining with appropriate pumping is advocated for a range of less than 11 m, whilst for higher ranges two-way operation without pumping achieves the maximum generation of energy.

Environmental and amenity considerations for this relatively small scheme have been evaluated and taken seriously. Changes in sedimentation and effects on the ecosystem in the estuary have not been of major significance in this particular case. The provision of a ship lock was originally included because certain naval vessels were believed to require access to the lower reaches of the estuary. In the event, this was not the case, but the number of pleasure craft has increased and, in retrospect, the size of the lock should have been greater. The provision of a road across the barrage has shortened a major route and has been popular. Bringing the turbines up to speed rapidly (they can all be brought on line in parallel in less than two minutes) has occasioned some difficulties with the generation of waves along the shore thus causing potential hazards. This is therefore carefully controlled.

The decision to build La Rance was made because of uncertainties about the future of nuclear power and to help the development of a less prosperous region of France. Notwithstanding the rapid growth in nuclear power in France, attention is now being paid once more to the feasibility of building further tidal schemes in Brittany. It has been estimated [2] that if the Rance scheme were built today, it would cost about £160M and the cost of electricity would be just over 3 p/kwh. This is cheaper than the expected cost of oil-fired generation, similar to that of coal although higher than the cost of nuclear. Although La Rance has had technical problems, its overall performance has been very encouraging. Much has been learnt about the real problems of operation and about large bulb turbines. It has recently been reported [2] that with the benefit of hindsight the design would not be changed much were it to be repeated, except that the 160 m length of embankment might be replaced by more turbines.

4.11.2 The Severn Barrage

The earliest known proposal for a barrage across the River Severn was made in 1849 and has been attributed to Thomas Telford. This, however, was to improve navigation to inland canals. A reproduction of a drawing of this barrage has been given in [2]. The first major proposal suggesting that a barrage might be constructed in the Severn Estuary to generate electricity was made in 1918. In 1933, the Brabazon Committee reported on the proposal—known as the English Stones Scheme—which envisaged a barrage from Sudbrook to Redwick (see figure 4.25). This scheme favoured a barrage containing 72 turbines giving an output of up to 804 MW and incorporating a road and rail crossing, locks and fish passes. A high level pumped storage scheme about 20 km away was also proposed to smooth the ebb-generated electricity. In 1944, the project was again the subject of a report to government. Some modifications were suggested including reducing the number of turbines from 72 to 32 units of 25 MW capacity. It was also appreciated that the pumped storage scheme was not an essential part of the proposal and might be omitted. The amenity value of a road and rail crossing, it was suggested, might be evaluated separately from the economics of electricity generation. The scheme, which was worked up in some engineering detail, was not pursued. An interesting and detailed discussion of plans for a Severn Barrage was given in 1944 by Headland [23].

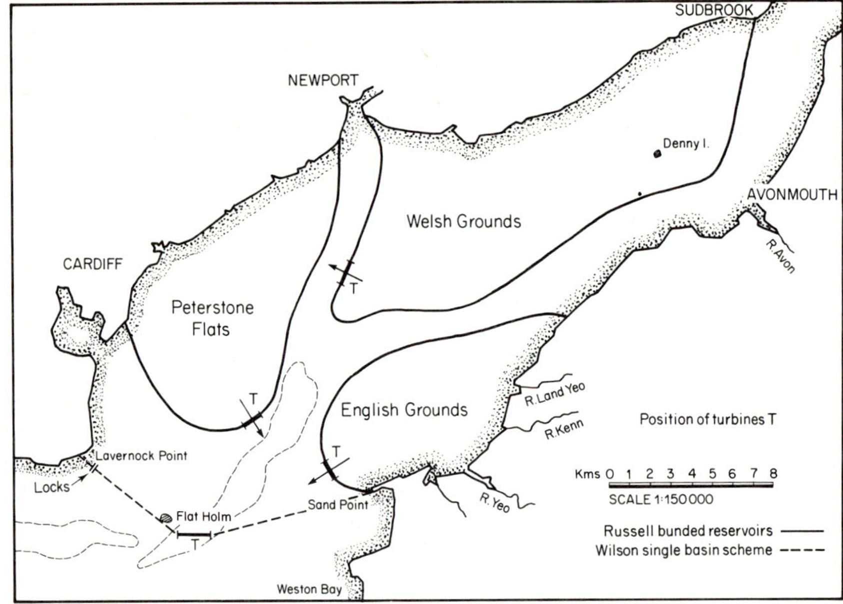

Figure 4.24 Various Severn Barrage proposals.

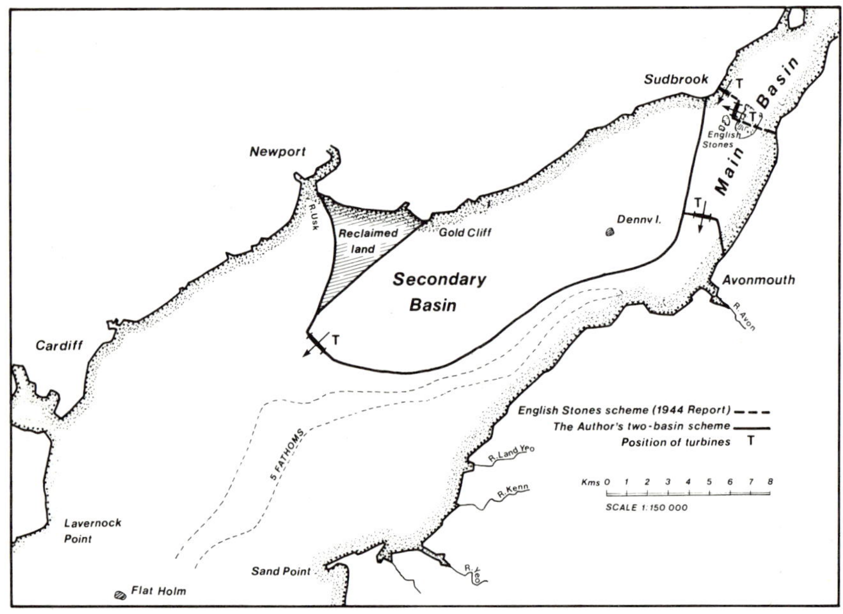

Figure 4.25 Two-stage Severn Barrage scheme (Hooker). (Courtesy Mr A V Hooker.)

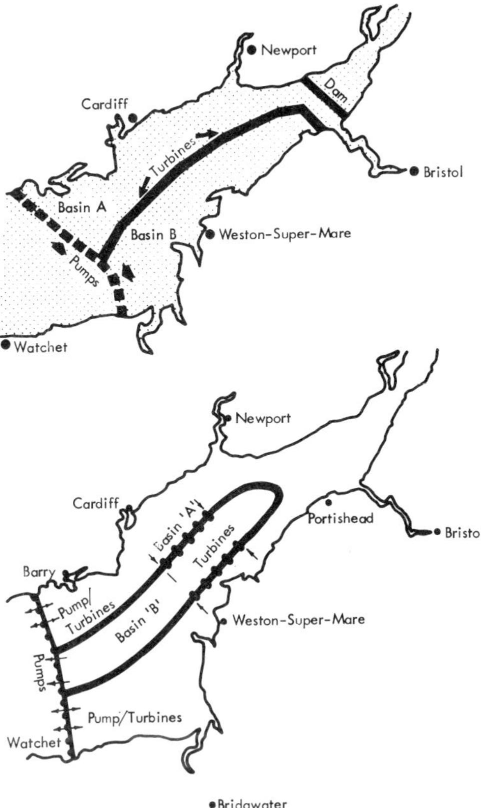

Figure 4.26 MEL schemes: the Severn Estuary with possible two-basin schemes.

In the late 1960s and early 1970s, several further and more ambitious schemes became the subject of discussion. These included a single-basin ebb generation scheme suggested by E M Wilson of the University of Salford [24], [25] (figure 4.24), two-basin schemes proposed by T Shaw (then of the University of Bristol) [3] (figure 4.6) and A Hooker (figure 4.25), an ambitious two-basin scheme investigated by the CEGB [26] (figure 4.26) and a system of bunded reservoirs proposed by R C Russell, Director of the Hydraulics Research Station [26] (figure 4.24). These schemes will now be discussed in more detail.

The Wilson scheme. This single-basin, ebb generation scheme is very similar to the stage one scheme envisaged in the 1981 study reported below. It has a barrage between Lavernock Point near Cardiff and Sand Point, north of Weston-super-Mare. Wilson suggested that 120 turbines of a little over 9 m in diameter could be housed in a 2060 m long barrage running along the line shown in figure 4.25. The sluices would be housed in an 1875 km length of dam on the English side of the estuary. Two locks would be provided in a deep channel on the Welsh side. The barrage, it is estimated, would generate 10.4 TWh p.a. with a power output ranging from 1.2 to 4.56 GW. In a report prepared for the DEn by NEDECO [27] (a firm of Dutch consultants involved in the Delta project of land reclamation), the cost of the Wilson scheme in 1977 was estimated to be £2200M, NEDECO looked closely at the civil engineering works required and the construction process. They proposed a slightly different line for the barrage terminating it at Brean Down rather than Sand Point.

The Shaw scheme. This two-basin scheme is illustrated in figure 4.6. In the version studied by NEDECO, Shaw proposed 300 turbines of 7 m diameter of which 180 would be located between the high level basin and the sea and 120 in that part of the barrage separating the high and low basins. Pumping from off-peak electricity provided by the grid was proposed in order to modify the levels in the two basins so that generation could take place during hours of peak demand. The high level basin would operate between high tide and mid-tide and the low level basin would operate from mid-tide to low tide. It has been estimated that the Shaw scheme could provide 13.6 TWh p.a., taking nearly 9 TWh p.a. from the grid for pumping.

The MEL (CEGB) study scheme. A very large scheme well downstream in the estuary was the subject of a computer simulation carried out by the Marchwood Engineering Laboratories (MEL) of the CEGB in 1974. Two possible double-basin schemes are shown in figure 4.26. The first of these was not studied in detail because it would leave Weston with a permanently low tide. Three options are available for operation. By filling basin A on the rising tide and emptying basin B on the falling tide, generation can take place by the flow of water from A to B with the option of achieving greater flexibility by employing pumping. In this case the power generated would be just over 3 GW at a spring tide falling with continuous output to about 2 GW at neaps (figure 4.27(a)). Another option is to restrain the generation to times when it is most needed, giving large blocks of power, the peak value of which falls away during the spring to neap cycle (figure 4.27(b)). Lastly, during neap tides, power can be taken from the grid to pump electricity into the main basin giving outputs as shown in (figure 4.27(c)). These modes of operation were also analysed for the much shorter barrage lines near Weston. This was called the reduced MEL scheme.

The Hooker scheme. This variant on the double-basin concept put forward by A Hooker of W S Atkins Ltd [28] involved a staged development with a main basin some distance up the estuary closer to the English Stones. This

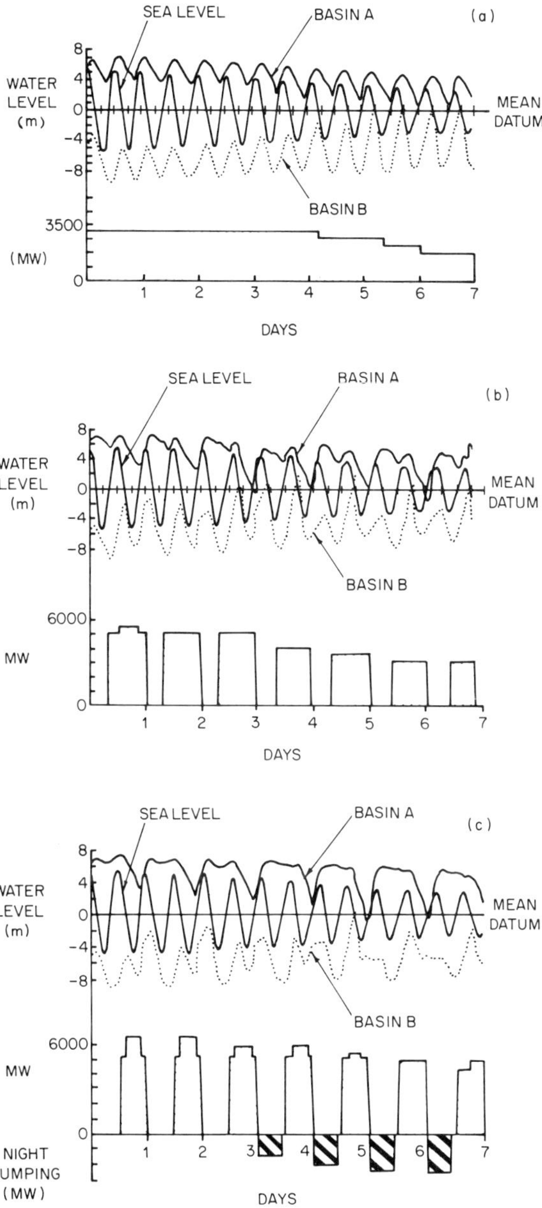

Figure 4.27 (a) Re-phasing of output from the MEL scheme. Basin A is filled by rising tide and basin B emptied on the falling tide. (b) Shows how output is retained to be available when it is most needed. (c) Showing how during neap tides power from the grid at night is used to pump water into the main basin to maintain high output the following day.

is illustrated in figure 4.25. Part of the project would involve land reclamation on the Welsh side of the estuary near Newport which could be used as a deep water port and for related industries. The main basin (fairly modest in size compared with other schemes) generating about 1 GW at peak output would be built first and the secondary basin, involving construction of a long embankment and giving about the same output, could be developed at a later date. The total scheme would generate over 5 TWh p.a., approximately equal amounts of energy coming from the two basins.

The Russell scheme. In this proposal [20], water would be impounded in a number of bunded reservoirs and discharged to the sea through turbines in the bund walls (figure 4.24). It has the advantage of leaving the estuary open for shipping and obviates the need for dredging to overcome any silting. The scheme could again be built in stages and as soon as two bunds were completed, continuous output could be generated by operating one barrage in low tide conditions and the other around high tide.

These various proposals and study schemes, stimulated by the rising cost of fossil fuels, led to the setting up in 1978 of a pre-feasibility committee under the chairmanship of the then Chief Scientist at the DEn—Sir Hermann Bondi. This reported in mid-1981 on the feasibility and economics of a Severn Barrage, the main findings and recommendations of which are summarised below.

Report of the Severn Barrage Committee. One of the main findings of the Severn Barrage Committee [2] was that it is 'technically feasible to enclose the estuary by a barrage located in any position east of a line drawn from Porlock due north to the Welsh coast'. Most of the schemes which have been discussed therefore would be capable of being built. The committee thus addressed itself to the economics of each of these and the environmental and amenity impacts.

Three schemes were considered the most attractive and were considered in detail. These are shown in figure 4.28. One of these was a barrage running from Minehead to Aberthaw. It was concluded that this would generate 20 TWh p.a. of electricity (about 9% of the electricity demand met by the CEGB) with an installed capacity of 12 GW. The cost of this scheme, which would tap most of the energy available from tides in the estuary, would be about £8900M at late 1980 costs. The most cost-effective option, however, was a single-basin ebb generation scheme running from Brean Down to Lavernock Point. This would produce about 13 TWh p.a. with a peak output of 7.2 GW and would cost £5600M. The third scheme, which it was felt could be an extension of the second at some future date when it might be economic, would involve a new barrage being built enclosing Bridgwater Bay. The staged scheme would generate about as

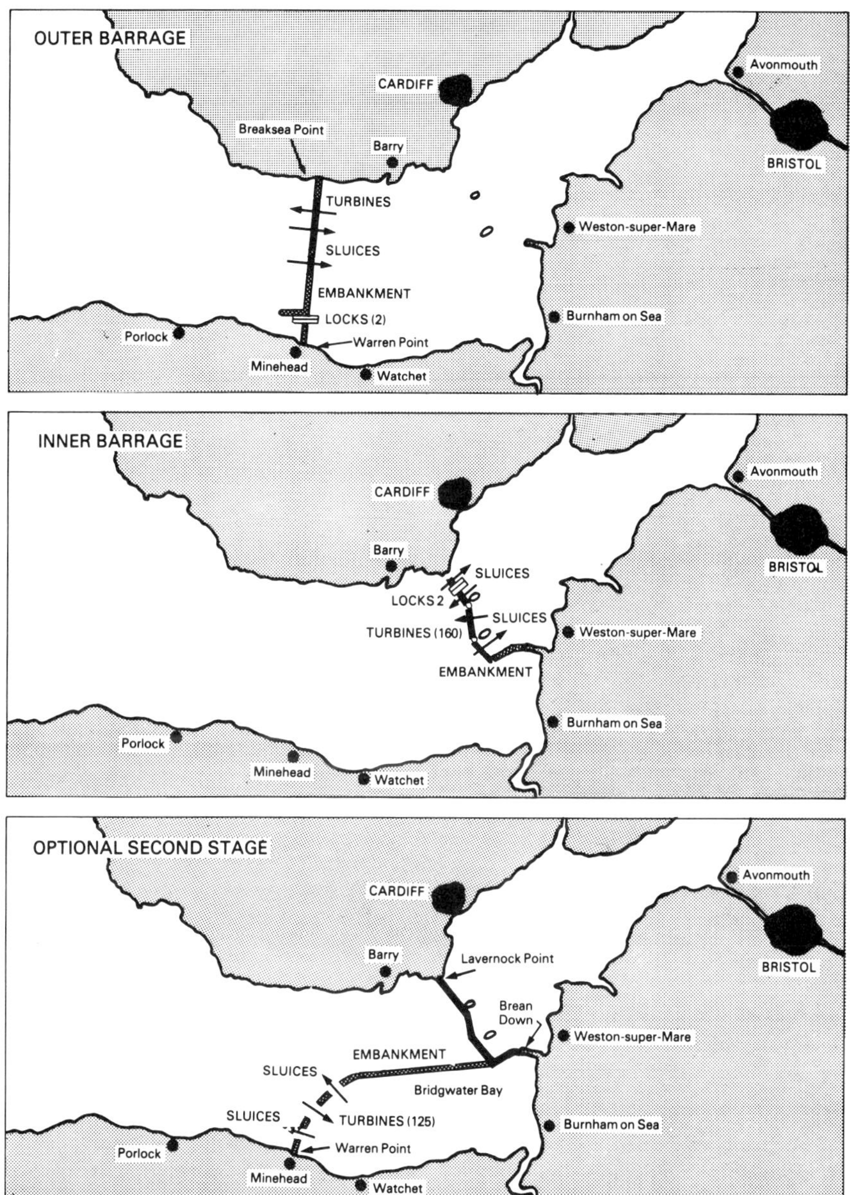

Figure 4.28 Leading Severn barrage schemes. (Reproduced from *Energy Paper* 46, courtesy HMSO.)

much electricity as the outer barrage and would show roughly the same benefit to cost ratio. If the second stage were operated in a flood generation mode, power would be generated for 20 h per day. The idea of this staged scheme, which in concept owes something to the Hooker proposal, emerged during the course of the study.

Double-basin schemes of the conventional type such as that proposed by Shaw and the more ambitious schemes analysed by the CEGB, were found, by a considerable margin, not to be cost effective. The storage provided by these schemes could be offered, it was felt, more cheaply by other forms of storage on the electricity supply system as a whole.

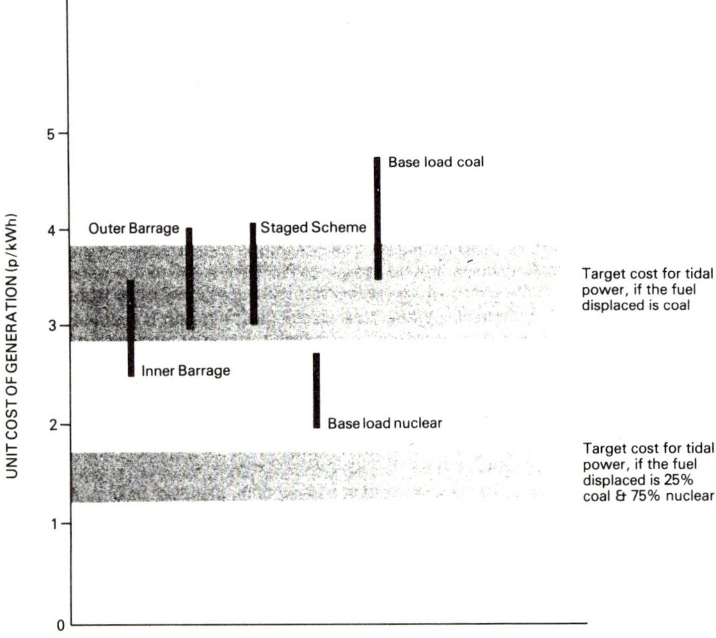

Figure 4.29 Costs of electricity generation in the year 2000 (a 15% range about the central value is shown in each case). Costs are given in December 1980 values. (Reproduced from *Energy Paper* 46, courtesy HMSO.)

The economics of the three most attractive schemes compared with other possible methods of electricity generation are summarised in figure 4.29. There was a significant increase in the estimate of the amount of electricity which could be generated by the simple ebb generation scheme during the course of the study, because it was found possible to increase the output by fitting more turbines into the barrage. Furthermore, it was concluded that generation could begin only nine years from the start of construction. The cost of electricity generated from the cheapest scheme was estimated to be 3.1 p/kWh (1980 prices). This, it was concluded, is likely to be cheaper than the cost of electricity from fossil fuel plant but more expensive than nuclear. The capacity credit of an ebb generation scheme would not exceed about 1 GW, so the main role of such a barrage would be to conserve expensive fossil fuels. As discussed in §4.8, the extent that this would be an attractive option depends upon the mix of generating plant which might be on the CEGB system during the first half of the 21st century. The more nuclear plant which can be built, the less attractive electricity at 3.1 p/kWh would appear to be.

It was accepted by the Severn Barrage Committee that the environmental and other impacts are still imperfectly understood and further work has been recommended in this area. Turning to the list of possible impacts given in §4.10, the following points were made as a result of the preliminary assessment of effects of the inner barrage scheme.

(1) Ports and navigation. Large ships entering ports in the estuary may be affected by a slightly reduced maximum water depth, but time spent traversing the lock system would be offset by the ease of passage through deeper water in the basin. Sedimentation would increase but the local importance of this was not assessed.

(2) Recreation. It was believed that the lower tidal range and reduction in currents would benefit water-based recreational activities.

(3) Water quality. Reduction in the strong tidal currents in the estuary would mean that pollutants would not be readily dispersed. They would, however, be diluted more effectively as a result of higher average water levels. Salinity and turbidity would be reduced. Problems of maintaining water quality could be overcome in the Severn Estuary but at a significant cost.

(4) Sea defences. The risk of flooding behind the barrage at extreme spring tides would be reduced but extra protection may need to be given to sea-defence banks because of the longer periods of wave attack at high water.

(5) Agriculture and land drainage. Extra pumping to clear surface water run-off behind the barrage would be required. A second stage may improve drainage but at the cost of habitat destruction.

(6) Visual amenity. This is a subjective matter and the effects of any scheme are thus difficult to quantify. An artist's impression of one view of a Severn Barrage is shown in figure 4.30.

(7) Industry and employment. It was estimated that more than 20 000 jobs would be created for up to 10 years. Provision of rock-fill for embankments and supply of the turbines at a fast enough rate might present difficulties. Social and industrial impacts were judged to be particularly difficult to assess.

(8) Estuary crossing. Some benefit would be obtained by providing a road crossing.

(9) Birds. Some species (bottom feeding and suspension feeding) would benefit but reduction of intertidal banks and of salinity may reduce the population of other species. The overall effect was considered difficult to evaluate.

(10) Migratory fish. It was felt that migrating salmon would be at some risk but fish passes might alleviate any problems.

Figure 4.30 Artist's impression of the stage 1 Severn Barrage. (Courtesy Department of Energy.)

The Barrage Committee recommended that an acceptability and preliminary design study should be undertaken at a cost of about £20M over four years. This study would include deeper analysis of environmental, social and industrial factors, further economic studies and preliminary design work. It was also felt that handling and foundation trials of a prototype caisson in the estuary should be considered halfway into the further phase. This would cost an estimated £25M.

A decision is now awaited from the UK government as to whether to accept these recommendations and commit itself to this next phase which might precede a decision on construction in the late 1980s.

4.11.3 Bay of Fundy schemes

The Bay of Fundy, straddling the border between the USA and Canada, north of Boston, has the highest known tidal range in the world (up to 17m) and is believed to offer a potential tidal resource of about 400 TWh p.a. Many studies have now been carried out of various possible schemes including Passamaquoddy Bay (partly in the USA), Shepody Bay, the Cumberland Basin, Minas Basin and the Annapolis Basin. A map of the Bay of Fundy is shown in figure 4.31.

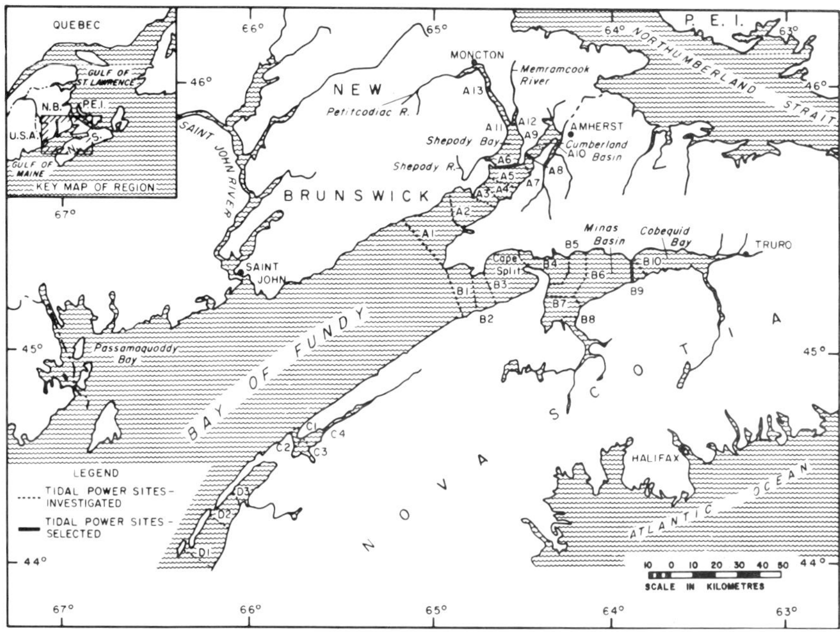

Figure 4.31 Bay of Fundy tidal power scheme. (Courtesy Mr R H Clark and the IEE.)

Various studies on the potential for tidal generation from this area have been carried out by Canada and the USA (including joint studies) since early in the century. Several companies have been set up to explore or develop tidal schemes and in 1934, the US government spent $7M acquiring rights and carrying out initial construction work but the project was then abandoned. In the late 1950s, the International Passamaquoddy Engineering Board was set up and looked at the design and costings for a 300 MW scheme using Passamaquoddy Bay as the high pool and Cobscook Bay in Maine and Friar Roads in New Brunswick as the low pool. Various possibilities were considered including systems combining this tidal output with steam plant and nearby hydroelectric plant. However, all schemes were deemed uneconomic.

A Canadian study in 1969 *Feasibility of Tidal Power Development in the Bay of Fundy* [29] which was a comprehensive joint federal and provincial investigation carried out over two years also concluded that the development of tidal power in the Bay was not economically justified at that time. They recommended that the position should be kept under review and should be re-examined if substantive changes in the cost of energy took place. In 1974, the Bay of Fundy Tidal Power Review Board [30] concluded that such changes had occurred and at the end of 1975 a further detailed appraisal was carried out and a final report issued in November 1977 [31].

Thirty possible sites in the Bay of Fundy were assessed and three—the Cumberland Basin (A8), Cobequid Bay (B9) and Shepody Bay (A6) in figure 4.31—were thought to provide the best prospects. An analysis of the costs using caissons 'floated-in' rather than temporary coffer-dams was as in table 4.4.

Table 4.4.

	Capacity (MW)	Output (TWh p.a.)	Total cost (1976 $M (approx.)) Estimates for 1981$ in brackets
Cobequid Bay	3800	12.7	4000 (5400)
Shepody Bay	1550	4.5	2200 (3000)
Cumberland Basin	1085	3.4	1230 (1700)

Bulb-type generators were considered in the study, but it was believed that Straflo turbines may have advantages. It was not thought likely from computer modelling of the tidal regimes that building a barrage at any of the three sites specified would change the tidal range appreciably. However, it was pointed out that reduction in the tidal range by altering the resonant response of the Bay may ultimately limit the tidal energy available from the region.

It was concluded that single-basin ebb generation schemes operated to give maximum energy output to displace fossil fuel plant would be most economic. On various scenarios, the integration of this output into the generation system of the Maritime Provinces was shown to be feasible but it was concluded that it would 'neither eliminate nor reduce the need for nuclear generation'.

A preliminary study suggested that there would not be insuperable environmental or social problems resulting from the development of any of the sites. Agricultural effects, drainage and flooding were thought to require considerable further investigation. Bird habitat, fisheries, recreation, minerals, navigation and land erosion were also recognised as important.

The Cumberland Basin site (A8) (figure 4.31) was preferred because it would enable benefits to be equalised between Nova Scotia and New Brunswick, 'socio-economic considerations' suggested that this would be the best site, environmental effects were thought to be least and as the smallest scheme, the technical problems and the capital investment requirements would be minimised. The Board recommended that a three-year study to develop detailed designs and specifications and further investigate environmental questions should be set up at a cost of $33M.

This, however, did not occur because of the high capital cost. A recent report of a Special Committee on Alternative Energy and Oil Substitution to the Parliament of Canada [32] has recommended that a $300 000 review should be initiated to try to confirm the 1977 study findings and if they can be verified $50M should then be made available for the detailed study.

The most recent efforts to harness tidal energy in the Bay of Fundy have been more practical, although at a somewhat smaller scale. As mentioned in §4.6, to test the promising Straflo turbine with rim-mounted generator, the Nova Scotia Tidal Power Corporation recently decided to instal a 7.6 m diameter, 17.8 MW turbine in an existing barrage across the Annapolis River at Annapolis Royal, Nova Scotia. It is hoped that demonstration of this unit will be achieved in 1983 [11].

4.12 Other possible tidal sites

Some other tidal sites around the world have been reviewed by Charlier [1]. A map of the world showing the main areas of interest is given in figure 4.32.

4.12.1 United Kingdom

Several other sites apart from the Severn Estuary have been studied to evaluate their potential for tidal schemes, although none in the same detail. The results of the most recent studies have been given by Wishart [33].

Five estuaries were considered—the Solway Firth, Morecambe Bay, the Dee, Humber and the Wash—and energy outputs and construction costs for potential barrage lines were estimated. The results are shown in table 4.5 adapted from similar tables in [33].

Table 4.5.

	Spring tide range (m)	Annual output (TWh)	Estimated cost (£M)	Estimated cost of electricity (March 1979)[†]
Morecambe (inner scheme)	8.2	5.4	2350	3.0
Solway Firth	7.2‡	10.1	4870	3.5
Dee Estuary	7.7	1.25	800	4.2
Humber (outer line)	5.6	2.0	1390	4.5
Wash (outer line)	6.0‡	4.7	3165	4.7

† The cost of inner Severn Barrage electricity at the same cost datum would be about 2.5 p/KWh.
‡ Reduction in present tidal range assumed.

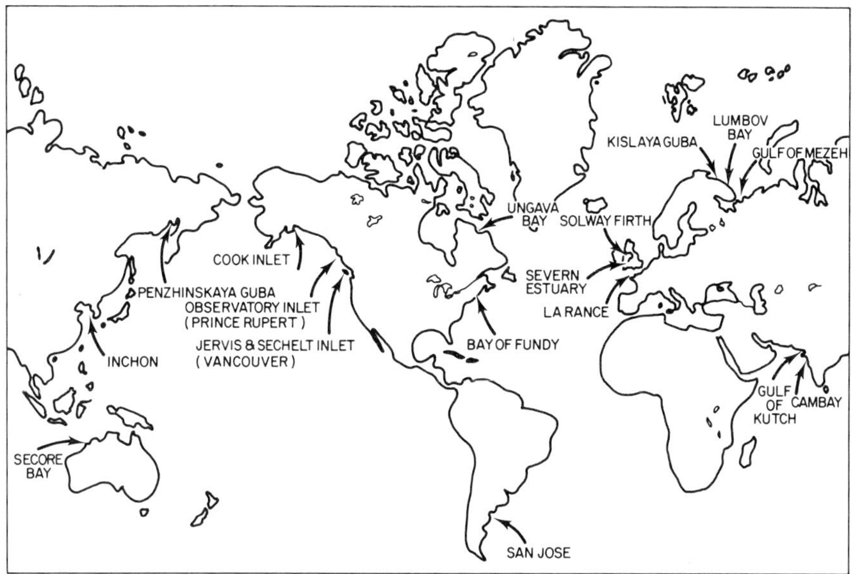

Figure 4.32 The location of world potential tidal power sites. (Courtesy Mr R H Clark.)

It is clear from this table that none of the schemes approach the estimated cost of electricity from the inner Severn Barrage except possibly Morecambe, although they do give a total tidal energy resource in the UK of over 42 TWh p.a. (about 20% of present UK electricity consumption) at less than about 6 p/kWh if the Severn Barrage first and second stages are included.

The time of high water at the Wash and Humber is within an hour or so of that in the Severn, but the first three sites would give their peak output five or six hours later. The geographical diversity may be of some value in integrating the output into the system and would add to its firm power contribution. There is also a possibility that Strangford Lough in Northern Ireland could be used for tidal generation. This is being explored at present.

4.12.2 Rest of Europe

Mention has already been made of the Iles de Chaussey plan in France. Perhaps because of the ambitious nuclear power programme in France this has not been actively considered for some years (although there may now be some renewed interest). Some thought has been given in the past to the possibility of tidal generation in the Somme Estuary.

Some other areas in Europe have been considered. Tidal generation was excluded from the Dutch Delta Plan in 1957 because of the effect on the shellfish industry [1]. One site near the Spanish–Portuguese border (Vigo) has previously been considered.

4.12.3 Soviet Union

A 400 kW tidal power station was completed in the Bay of Kislaya about 100 km from Murmansk in 1968 and this has been successful in operation. The USSR appears to have several sites which show great potential for tidal generation in the White Sea between Murmansk and Archangel [34]. The first site likely to be developed (Lumborskaya Bay) would generate up to 400 MW and the second (Mezen) would have a capacity of up to 10 GW. In the sea of Okhostsk, one site may have a potential capacity of 10 GW whilst a design for a 35 GW scheme (Penshinsky Bay) has been considered. For the large schemes, it is difficult to see how such large amounts of electricity could be used locally and transmission costs could thus be prohibitively high.

4.12.4 Australia

The tidal range of 9–12 m in the Kimberley Region of Western Australia appears to offer considerable promise for the generation of electricity from the tides [35]. Again, however, the absence of a nearby large demand for power may limit the potential. The major sites—Secure Bay, Walcott Inlet, George Water and St George's Basin—offer potential installed capacities of 570, 1250, 800 and 1000 MW respectively. The total energy which might be produced from these sites has been estimated to be as much as 12 TWh p.a.

4.12.5 South America

Preliminary studies were carried out in 1959 into constructing a canal through the Valdez Peninsula in Argentina (the San Jose proposal) to generate tidal energy. The Gulf of San Jose offers very great potential with a mean tidal range of nearly 6 m. Other possible sites in Argentina include [36], the Gulf of San Jorge, San Julian and the Deseado, Gallegos and Santa Cruz estuaries. At Gallegos, the maximum tidal range is estimated to be 14 m. Other sites with high tidal ranges are found in Chile, Panama and Baja California.

4.12.6 Asia

There are several locations in India where a large tidal power potential exists including the Gulfs of Kutch and Cambay in Gujarat and in West

Bengal. In the Gulfs of Kutch and Cambay, the mean tidal ranges are 5.3 and 6.8 m. Full use of the available energy would not be possible at present in the region but medium-scale developments [37] may be economically justified.

China has a 500 kW two-way tidal power station in operation at Jangxia Creek in the East China Sea. It has been estimated that there are 500 or more bays along the Chinese coast which could, in principle, be developed. A total potential of installed capacity in excess of 110 GW has been estimated.

Korea [38] has a particularly favourable tidal regime along its extensive and indented coast line. Feasibility studies are currently being carried out for three of the most promising sites—Asan Bay, Gargorim Bay and Cheonsu Bay on the Yellow Sea—where the tidal range is between 7 and 10 m. It is hoped that the first tidal scheme with an output of a few hundred megawatts will be completed within the decade.

4.12.7 North America

Several sites (apart from the Bay of Fundy) offer high tidal ranges including sites in Alaska. At Anchorage, the tidal amplitude reaches nearly 8 m.

In summary tidal schemes look promising in competition with fossil fuels in a number of countries and it is likely that schemes will be built towards the end of the century. If it were not for the massive capital costs frequently associated with tidal schemes, it is probable that several more would already be in operation.

References

This chapter tends to concentrate on the results of the Severn Estuary pre-feasibility study and, as an example of an in-depth analysis of tidal energy at one location, reference [2] provides a comprehensive survey. The article by Cotillon [16] likewise presents a good general review of the Rance scheme. References [29], [30] and [31] review the Bay of Fundy feasibility studies.

1   Charlier R H 1979 Tidal power plants: sites, history and geographical distribution *Proc. 1st Int. Symp. on Wave and Tidal Energy, Canterbury* paper A1–9 (Bedford: BHRA)
2   Tidal power from the Severn Estuary 1981 *Energy Paper* 46 vols 1 and 2 (London: HMSO)
3   Shaw T L and Thorpe G R 1971 *Proc. Am. Soc. Civil Eng., J. Power Div.* **98** 159–80

4 Count B M 1981 *CEGB Report* MM/MECH/TF 257

5 Robinson I S 1981 Surges in tidal power basins—can they increase power output *Proc. 2nd Int. Symp. on Wave and Tidal Energy, Cambridge* paper B3 (Bedford: BHRA) pp53–68

6 Jefferys E R 1981 Dynamic models of tidal estuaries *Proc. 2nd Int. Symp. on Wave and Tidal Energy, Cambridge* paper B4 (Bedford: BHRA) pp69–84

7 Wheeler S J 1981 Optimisation of tidal power schemes *Proc. 2nd Int. Symp. on Wave and Tidal Energy, Cambridge* paper H1 (Bedford: BHRA) pp237–48

8 Heaps N S 1981 Prediction of tidal elevations: model studies for the Severn Barrage *Proc. Conf. on the Severn Barrage* paper 6 (London: Institution of Civil Engineers) pp35–42

9 Carson J L and Samuelson R S 1978 *Power* March issue 54–7

10 Lewis A R 1981 Turbines for the barrage *Proc. Conf. on the Severn Barrage* paper 11 (London: Institution of Civil Engineers) pp75–82

11 Focas D 1981 *Proc. 2nd Int. Symp. on Wave and Tidal Energy, Cambridge* (unpublished)

12 Miller H, Strub W and Wilson E M 1981 Optimising Straflo turbines for tidal applications *Proc. 2nd Int. Symp. on Wave and Tidal Energy, Cambridge* (available from authors or BHRA)

13 Pearsall I S 1977 Low head hydraulic machinery *Proc. Conf. on Tidal Power and Estuary Management* ed R T Severn, D L Dinely and L E Hawker (Bristol: Scientechnica)

14 Gardner G E and Jervis W B 1981 Low head hydraulic machinery *Proc. Conf. on the Severn Barrage* paper 10 (London: Institution of Civil Engineers) pp65–74

15 Arnold P J 1981 Transmission aspects *Proc. Conf. on the Severn Barrage* paper 12 (London: Institution of Civil Engineers) pp83–90

16 Cotillon J 1978 La Rance tidal power station—review and comments *Proc. Colston Symp. on Tidal Energy* (Bristol: Scientechnica) pp49–66

17 Stokes C J and Street R D J 1981 Turbine caissons for the Severn Barrage *Proc. 2nd Int. Symp. on Wave and Tidal Energy, Cambridge* paper F1 (Bedford: BHRA) pp167–76

18 Murray W T 1981 Caisson design and construction methods *Proc. Conf. on the Severn Barrage* paper 15 (London: Institution of Civil Engineers) pp111–20

19 Clare R and Oakley A J 1981 The towing and positioning of caissons in a tidal barrage *Proc. 2nd Int. Symp. on Wave and Tidal Energy, Cambridge* paper F2 (Bedford: BHRA) p177

20 Campbell R O 1981 Preparation of caisson foundations *Proc. Conf. on the Severn Barrage* paper 14 (London: Institution of Civil Engineers) pp103–10

21 Brown C J C, Bush P R and Norgett M J 1981 The economics of electricity generation by tidal barrages *Proc. Conf. on the Severn Barrage* paper 22 (London: Institution of Civil Engineers) p157

22 Banal M and Bichon A 1981 Tidal energy in France. The Rance tidal power station—some results after 15 years of operation *Proc. 2nd Int. Symp. on Wave and Tidal Energy, Cambridge* paper K3 (Bedford: BHRA)

23 Headland H 1949 *Proc. IEE* **96** 428–51

24 Wilson E M, Severn B, Swales M C and Henery D 1968 The Bristol Channel barrage project *Proc. 11th Conf. on Coastal Engineering* paper 103 pp1304–25

25 Wilson E M 1966 *Water Power* April issue 135–42

26 Severn Barrage seminar 1977 *Energy Paper* 27 (London: HMSO)

27 Tidal power barrages in the Severn 1977 *Energy Paper* 23 (London: HMSO)

28 Hooker A V 1970 Severnside of the future *Proc. ICE* **47** 1–8

29 Atlantic Tidal Power Programming Board 1969 *Feasibility of Tidal Power Development in the Bay of Fundy* (Ottawa: Department of Energy, Mines and Resources)

30 Review of the Bay of Fundy Tidal Power Review Board 1977 *Reassessment of Fundy Tidal Power*

31 Clark R H and Karas A N 1979 Studies of tidal power from the Bay of Fundy *Proc. IEE Future Energy Concepts Conf.* (London: IEE) p143

32 Energy Alternatives 1981 *Report of the Special Committee on Alternative Energy and Oil Substitution to the Parliament of Canada*

33 Wishart S J 1981 A preliminary survey of tidal energy from five UK estuaries *Proc. 2nd Int. Symp. on Wave and Tidal Energy, Cambridge* paper K1 (Bedford: BHRA) pp299–314

34 Bernstein L B 1974 *Water Power* May issue 172–7

35 Scott W E 1976 *Energy Int.* 41–3

36 Subrahmanyam K S 1978 *Water Power and Dam Constr.* June issue 42–4

37 Song W and Chun Y 1978 Korea tidal power project *Proc. Int. Symp. on Korea Tidal Power* (Seoul: Korea Ocean Research and Development Institute) pp1–18

5

Geothermal energy

5.1 Introduction

Geothermal energy presents a spectrum of resources and methods of use which are at various stages of engineering and economic development. Broadly, the resources occur naturally in the forms of steam, hot water (aquifers) and hot rocks and the stage of development has been dictated by natural availability and the cost of extraction. Steam and hot water (hydrothermal) resources have existing technologies for extraction and are extensively used in many parts of the world. The temperature of the hot water can be greater than 200 °C. The 'mining' of heat from hot dry rocks by pumping water round a closed cycle to extract the heat, is still in the research phase, but possibly offers the best prospect for large-scale electricity generation.

All geothermal resources are strictly non-renewable because the average heat flow from the centre of the earth is so small (0.04 to 0.06 W/m^2) compared with the rate of extraction required for economic operation. Even in exceptional areas where the heat flow can be hundreds of times this value, the extraction rate required to support a power plant of a few hundred kilowatts will lead to a gradual run down of the field. The life of a geothermal field is therefore normally a few decades while replenishment could take centuries. However, geothermal fields can be extensive and provide steady power for many years. In general, capital costs are high and running costs are low so that, where it is available, the energy is used for base load applications.

In the last thirty years, the science of geophysics has advanced rapidly and our knowledge of the structure of the planet has increased enormously [1]. In particular, the theory of plate tectonics has allowed an understanding of why certain regions have greater seismic and volcanic activity than others. Techniques have also improved. Although the deepest mines are only a few kilometres deep and boreholes have generally been drilled only to a depth of up to 10 km (and usually much less), seismological techniques, together with indirect evidence have allowed a generally accepted picture of the earth's structure to be pieced together. The uppermost layer—the crust—is on average about 30 km in thickness over the continents but thinner over the oceans. The upper mantle extends to a depth of more than 500 km, the lower mantle to nearly 3000 km and the outer core to over 5000 km (figure 5.1). As the earth's centre (1000 km radius inner core) is approached, the density of material is believed to increase progressively from an average of about 3 g/cm^3 at the surface to more than three times this value in the liquid metal outer core and more than four times at the centre. Thus, although much can be inferred from surface measurements, it is clear that man's activities, including the possible exploitation of geothermal energy, are of insignificant importance in comparison with total earth geology and structure.

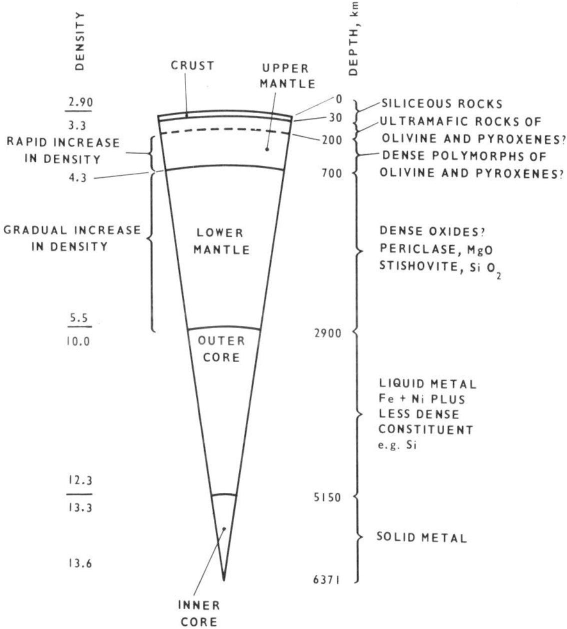

Figure 5.1 Schematic section through the earth.

Temperature gradients vary widely over the earth's surface and this shows that whilst generally the picture of a substantial crust covering an upper mantle is good, non-uniformity leads to the existence, particularly at plate margins, of anomalously high thermal gradients. This results from local melting due to pressure and friction as the plate boundaries move against each other. At actual rifts, an outflow of magma from below may occur. The location of the plate boundaries also corresponds to regions where active volcanoes are found and their global distribution is illustrated in figure 5.2.

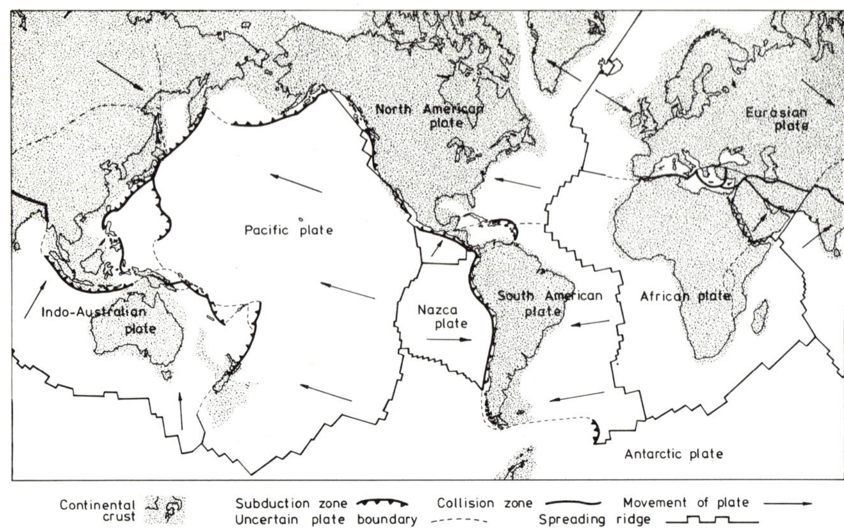

Figure 5.2 Plate boundaries and active zones of the earth's crust. (Reproduced from *Energy Paper* 9, courtesy HMSO.)

The heat measured close to the surface thus frequently arises from magma which has been intruded to shallow depths but other factors may also affect the heat flow and thermal gradient. In some cases, convection of natural ground water disturbs the pattern of heat flow and in other cases it is thought that the release of hot gases from deep rock may increase the flow. Another important mechanism is heat generation from radioactive isotopes of elements such as uranium, thorium and potassium. This mechanism is not fully understood, but certain areas of the crust have undergone successive melting and recrystallisation with time and this could have led to a concentration of these elements at certain levels in the crust. To a lesser extent, exothermic chemical reactions may also make a contribution to local heating. These mechanisms are possibly of more significance in anomalous areas of high heat flow distant from plate margins.

Armstead [2] has made a useful classification of geothermal sources. Areas classified as hyperthermal exhibit very high gradients (many times as great as non-thermal areas) and are normally close to plate boundaries and are enhanced by geothermal ground water. Semi-thermal areas with gradients from 40–70 °C/km may be due to anomalies in crust thickness in otherwise stable regions or due to local effects such as radioactivity. Finally, and by far the greatest classification in terms of surface area, there are the normal regions with gradients below 40 °C/km. Because it is distant from a plate boundary, the UK has no hyperthermal and probably no semi-thermal areas but it is likely that several deep aquifers in normal areas exist which may be usefully exploited.

Finally, before discussing the technology associated with electricity generation from hyperthermal aquifers, it is important to draw the distinction between thermal gradient and heat flow (which have so far been rather loosely interchanged). If heat flow is only considered in the vertical plane, the two are linked by the expression

$$q = K\frac{\mathrm{d}T}{\mathrm{d}z}$$

where q is the heat flow usually expressed in $\mu\mathrm{cal/cm^2/s}$ (usually abbreviated to heat flow units or hfu), K is the thermal conductivity of the rock and $\mathrm{d}T/\mathrm{d}z$, the gradient (usually quoted per kilometre). Clearly high thermal gradients will be found in rocks of low thermal conductivity and vice versa, and thus a high thermal gradient is not always synonymous with high heat flow. In reality, flow will occur through many strata of rock of varying conductivity and simple analysis will be confounded by effects due to ground water convection, radioactivity and any exothermic chemical reactions which are taking place. Average heat flows throughout the world are about 1.4 hfu but a very few reach 10–20 hfu where aquifers draw heat from a large body of rock.

5.2 Geothermal energy from aquifers

The word aquifer has been taken to mean any source involving ground water, either as a deep hot water system or steam field or as a geyser or fumarole which breaks the surface. Before discussing in some detail the technology which has been developed, a brief description will be given of the major developments which have already taken place to exploit this resource.

The first attempt to generate electricity from geothermal sources took place in 1904 at Lardarello in Tuscany. However, attempts to drive a steam engine from underground sources were unsuccessful because the chemical content of the steam led to chemical attack of the plant. Progress was then made both in designing heat exchangers and in choosing materials able to withstand the corrosive environment in which they operated in the natural steam. In 1913, a 250 kW power station was successfully commissioned and by the Second World War, over 100 MW of plant was being operated from the Lardarello steam field. This was destroyed in the war but new plant has been built up since then, and a capacity of over 400 MW now exists. An account of the development of geothermal power in Italy has been given by Villa [4].

In New Zealand the Wairakei steam field in the North Island was developed in the late 1950s and early 1960s and by 1964, 192 MW of plant was in operation [4]. This field is now declining after 23 years of generation but generation of 160 MW by 1990 is planned at a new field at Broadlands. A view of the Wairakei steam field is shown in figure 5.3.

Figure 5.3 Aerial view of Wairakei geothermal field. (Courtesy Electricity Division, New Zealand Ministry of Energy.)

The Geysers field in California (see figure 5.4) was producing over 500 MW of electricity in the mid-1970s and a comparable amount of plant has been commissioned. The growth in generation from this field has been dramatic, since in 1960 only 12 MW of plant had been installed and only 25 MW by 1963 [5]. The prospects for further growth in exploitation of geothermal electricity from the west coast of the USA are very encouraging—some estimates have put the exploitable power as high as nearly 400 GW. The tightening of environmental legislation in the 1970s may, however, restrict growth to some extent.

Figure 5.4 The Geysers field, California.

Mexico, Japan, the Philippines, Kenya and Iceland are among a number of countries who have recently been expanding their output of geothermal electricity. This will be discussed further in §5.2.9. Use of low enthalpy sources for district heating, industrial and horticultural applications, whilst not within the scope of this book, has expanded considerably. Hungary, Iceland, France (at Creil near Paris) and parts of the eastern USSR have developed sizeable schemes. In the United Kingdom there are several possible aquifers (figure 5.5) which are expected to yield water at a temperature in excess of 70 °C and these may also be developed for heating. The aquifer below Southampton confirmed by exploratory drilling, is an example of a geothermal resource which might be exploited in this way; others are expected to be found in Larne (Northern Ireland), Cheshire, Humberside and central Scotland but information is not yet available on their temperature, depth, flow rate and extent. An exploratory borehole drilled at Larne successfully located the expected formations, but these were found to be so impermeable that the rock was virtually dry. This may not necessarily apply over the whole basin. It has recently been suggested [6] that even low enthalpy sources such as those expected in the UK, could be used for electricity generation either by using advanced cycle plants (similar to those proposed for OTEC) or indirectly by preheating power station feedwater.

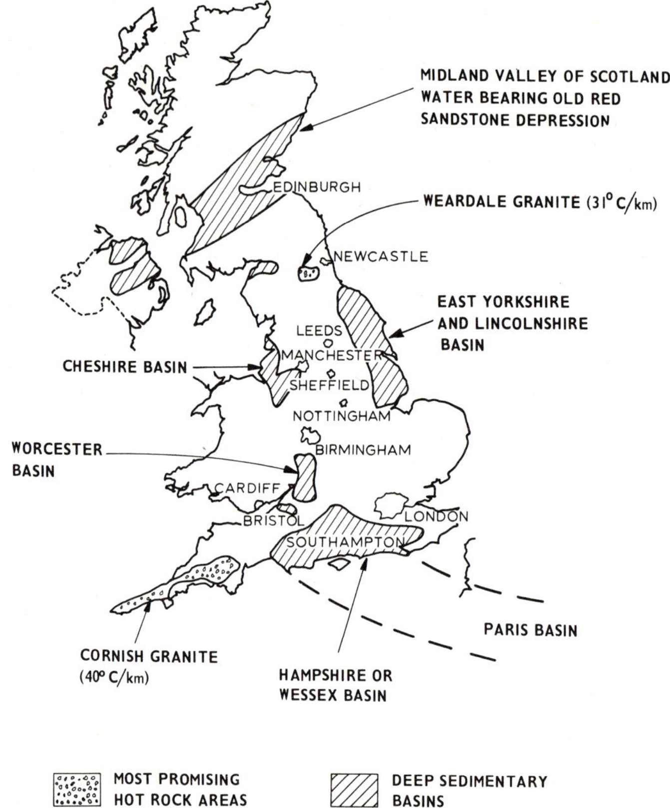

Figure 5.5 Areas of geothermal interest in the UK.

5.2.1 *The structure of typical aquifers*

The existence of a high temperature gradient is not the only criterion by which the usefulness of a geothermal area must be judged. The presence of a source of hot water or steam is also required if existing technology is to be used and the association of water with the high gradient depends on the detailed geological conditions. In addition, a geothermal field, like an oil field, must contain sufficient working fluid or a source of replenishment, this fluid must be of suitable quality (see §5.2.4) and the permeability of the fluid-bearing rock must be high enough to allow adequate extraction rates. Furthermore, the extent to which fields are water dominated or steam dominated has an important bearing on the plant required for exploiting the field. In general, semi-thermal fields contain water up to about 100 °C in temperature and no steam, but hyperthermal fields vary widely in their steam to water ratio. Wet hyperthermal fields contain pressurised water above 100 °C and when de-pressurised at the surface, the working fluid contains a mixture of flashed steam and boiling water. Dry hyperthermal fields contain dry saturated or slightly superheated steam at pressure.

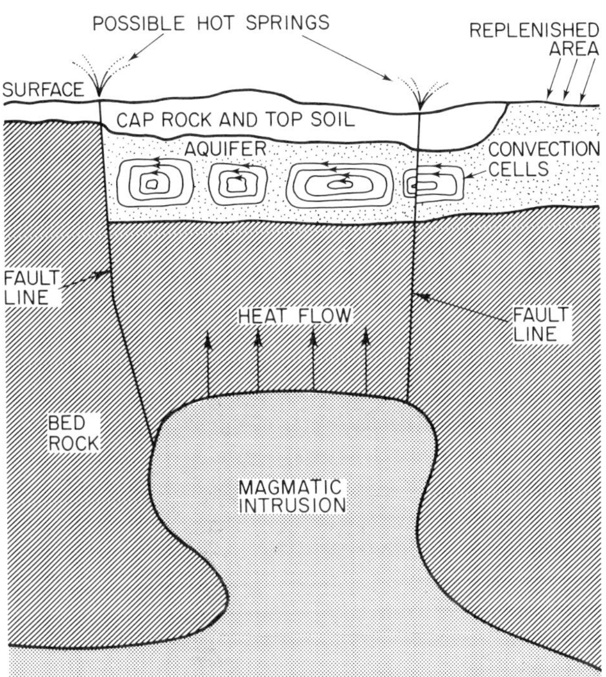

Figure 5.6 Main features of a hyperthermal field.

Figure 5.6 shows a schematic illustration of the geology of a typical hyperthermal field. Semi-thermal fields are similar. The structure of a number of real fields has been discussed by Facca [7]. A magmatic intrusion is shown feeding heat through a layer of bedrock to an aquifer which typically comprises a highly permeable rock which retains water or steam. This water may have arrived in the rock by a variety of routes, the most frequent being natural ground water (sometimes referred to as meteoric water). In some aquifers, the water originates from vapours released by magmatic materials on solidification but this is less common. Such water can be distinguished from meteoric water because of a lower ratio of hydrogen to deuterium. A third source of water is of 'connate' origin having been entrained between rocks during their formation. In the case of connate and most sources of magmatic waters, these are not able to replenish the aquifer and may be rapidly exhausted. Figure 5.6 thus shows the existence of a surface replenishment area as being typical of a hyperthermal system. The existence of an impermeable bedrock is normally found in the form of basaltic material at a depth of a few kilometres in the crust where the pressures will tend to lead to impermeability by closing fissure systems. Finally, a layer of cap rock is normally associated with hyperthermal fields and acts as an impermeable shield preserving heat in the water or steam.

Whether the field is wet or dry depends to some extent on the shape of the aquifer. At Lardarello, for example, the aquifer is thought to be convex upwards in shape and steam tends to collect at the highest point, having separated from the boiling water away from the 'summit'. In other cases, the boiling water may tend to lie in a central depression and steam rises along the bottom sides of the bedrock. The system will almost certainly contain strong convection currents whatever its configuration and these frequently occur in a number of distinct cells of interconnected low permeability fissure systems.

5.2.2 *Extraction of the working fluid*

The plant associated with extracting the fluid from a borehole will be discussed in §5.2.4 below, but it is interesting to examine under what conditions an aquifer will spontaneously 'blow' if connected to the surface either naturally or artificially via a borehole. The problem has been discussed for a wide variety of conditions by Armstead [2].

If a geothermal borehole or natural fissure connecting the field to the surface and filled with water is idealised in the form of a long tube with insulating sides, heated at the bottom by boiling water, it is possible to see why wet hyperthermal fields sometimes 'blow' spontaneously with the emission of steam. If the lowest layer in the tube is boiling, some steam will rise to a higher and cooler level in the liquid column. On condensing, the latent heat given out will raise the temperature of this layer of liquid but because the hydrostatic pressure at this layer is a little lower, it will boil at a lower temperature. In principle, therefore, this process will occur progressively up the water column leading to the emission of steam at the surface. If the steam at the top of the liquid column can only be ejected through narrow surface fissures, it may be propelled into the atmosphere with some force.

In a real situation, convection would occur in the column, heat transfer at the walls would take place and the heating at the bottom of the column may not be maintained if the permeability of the aquifer is insufficient to maintain the supply of water. Circulating ground water near the surface may also disturb the pattern. Nonetheless, it is believed that for some systems there exists a good approximation to the profile shown in figure 5.7. The curve is useful since it indicates that, in principle, a non-discharging bore with a temperature below the critical value given by the curve will be incapable of spontaneous blowing. To the right of the line, however, it should be possible for blowing to occur. A bore with a profile to the left of that shown could be stimulated into blowing by, for example, drawing off some of the water at the top to reduce the hydraulic pressure. Once a bore has blown, it should be capable of maintaining its output assuming that the permeability and temperature of the aquifer are sufficient and that the bore is not too wide. If its diameter exceeds a critical value, the steam volume will be insufficient to maintain buoyancy and the output at the surface will be quenched.

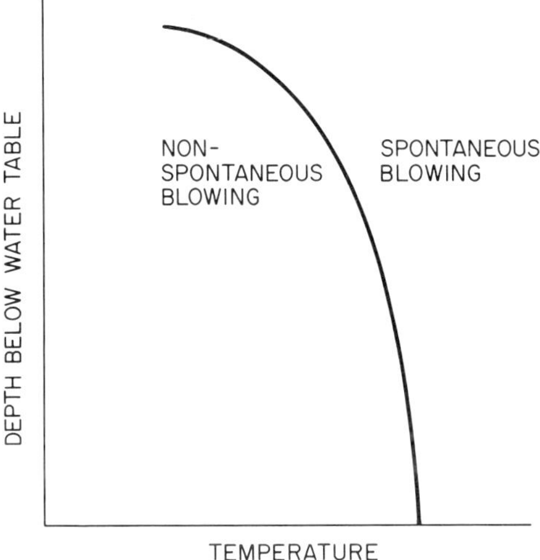

Figure 5.7 Variation of blowing temperature with depth below the water table.

Dry geothermal fields will generally blow without stimulation because of the pressure of trapped steam adding to the hydrostatic pressure of the water. It has been observed both in the California Geysers and at Lardarello that the steam emitted is superheated (by about 10 °C at the Geysers and as much as 70 °C at Lardarello). It is now believed that this arises because the water level in the aquifer has dropped [8], leaving rock at a higher temperature than the boiling water below. Heat from this rock is passed to the dry saturated steam as it escapes. Semi-thermal fields will not blow spontaneously except where the field is geo-pressurised or is part of an artesian system. Often pumping is required.

The rate of extraction from a geothermal field clearly influences the lifetime over which electricity may be generated. Fields are normally exploited cautiously at first, since if the resource has been underestimated for some reason, wells will not suffer a rapid drop of temperature or reduction in flow. Inevitably, however, extraction of heat and water or steam tends to lead to a decline in the extractable resource in a field but if it is large, drilling new bores or deepening existing ones allows a fairly constant output to be maintained. Output from individual bores normally declines fairly rapidly for two reasons. Firstly, chemical deposition tends to close fissures and pores and this reduces flow. Secondly, in dry fields, steam has to be drawn from a greater distance leading to cooling. In most fields, bore lifetimes of about ten years are assumed. Figure 5.8 shows a plot of output against time for a bore at Lardarello. Theoretical methods of estimating the output of wet fields [9] have been proposed but to arrive at dependable figures, a range of parameters such as area, depth, porosity and specific heat are needed and these are usually not well known.

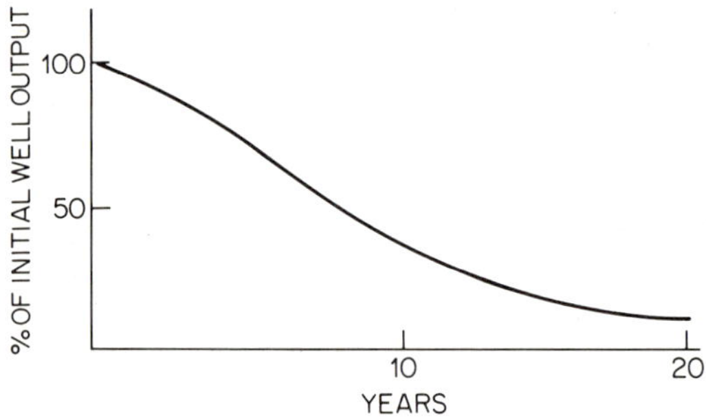

Figure 5.8 Typical bore aging curve for Lardarello.

One method that in principle ensures that the water in a field will not be depleted rapidly is to re-inject the extracted water into the aquifer. This is more costly than direct disposal, and in some circumstances could lead to a temperature drop in the aquifer. On the other hand, it has the advantage of minimising some of the environmental consequences of disposing of the brine (see §5.2.8). Furthermore, re-injection allows heat to be extracted from the body of rock rather than just the pore fluid initially trapped in the rock. Typically, this can increase the resource from a particular field by a factor of two to three.

Until recently there have been technical doubts about the possibility of re-injection, particularly into pressurised aquifers. Doubts were expressed as to whether re-injection could be achieved without incurring high pumping costs, whether unacceptable aquifer cooling would result or whether the polluted water would find an outlet elsewhere into the water supply. These problems depend very much on local conditions but there is now growing evidence that in most aquifers, re-injection can be achieved by the effects of gravity alone with few deleterious effects. Successful re-injection has been achieved in California, France, Japan and El Salvador. Experiments with radioactive tracers in this latter country [10] have supplied interesting data on dispersion of the re-injected fluid.

5.2.3 Drilling

Although geological information [11], measurement of heat flow, and use of a variety of geophysical tools [12] as well as estimates of the geochemistry [13] are important in exploring areas of geothermal potential, bores must always be drilled to confirm the indirect evidence and to begin production.

The depth to which bores are drilled clearly depends upon the nature of the geothermal field. In hyperthermal fields, it is seldom necessary to drill to a depth of more than 500–2000 m, whilst the average depth for a semi-thermal field is at the deeper end of this range. The costs of the drilling, including bringing the rig to site, preparing a suitable surface to support the rig (a 'cellar') and carrying out the drilling itself (which typically lasts for some 60–80 days for a depth of 2 km in sedimentary rock) are a major determinant in the costs of winning energy from a geothermal field. Costs also depend upon the width of the boreholes. The optimum width is difficult to assess in advance since if it is too narrow it will restrict the flow, whilst if too wide the cost escalates rapidly. Boreholes narrow with depth, thus the surface diameter depends on how deep the geothermal field is likely to be. The diameters of production bores generally lie in the range 0.25–0.6 m.

The techniques [14] employed are similar to those used in winning oil and natural gas. A typical rotary drilling rig of the type normally used is sketched in figure 5.9. High temperatures and the occasional presence of corrosive fluids, require special components in some cases. The drilling bit is conventionally of the serrated tricone type [14] with the depth of the serrations determined by the type of rock being drilled. The bits are normally made from very tough steels. The bit is connected to the surface by a hollow drill stem, down which special cooling muds are passed. These are then forced back to the surface up the annular space outside the stem, carrying with them the debris from the drilling. After filtering, the muds are recirculated. The pressure of the mud must be carefully adjusted so that its hydrostatic pressure is greater than the pore pressure of the formation being drilled otherwise a blow-out is likely to occur. Conversely, an overpressure may lead to damage of weaker formations. It is particularly important to avoid such damage near the production zone and compressed air cooling is sometimes used in such circumstances. Each production hole must be cased during drilling with several concentric corrosion-resistant steel tubes which are cemented (using special high-temperature cements) to the surrounding rock. Carrying out this task, together with periodic replacement of the drilling bit and 'fishing' for lost components is the main reason why the total drilling time is frequently more than twice the time that it takes to cut the rock.

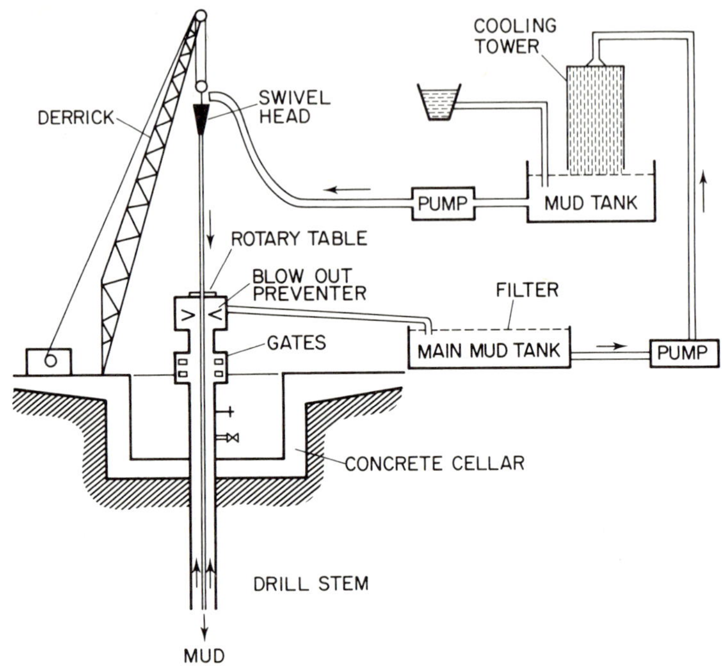

Figure 5.9 Schematic diagram of a rotary drilling rig and mud circulation system.

As mentioned above, drilling costs have a major influence on the success or otherwise of a geothermal project. It is considered technically possible to drill to depths of 15 km but the costs would be astronomical at present. Garnish [16] has shown that there is an exponential increase in time and cost with depth drilled and it is not often found to pay to drill deeper than 2 km or so. Deeper holes need a wider entrance, a larger and heavier drilling rig and employ commensurately greater quantities of cement and steel. There is a greater chance of error, more probability of drilling through igneous rocks where rates as low as 10 m/day can be obtained† due partly to slower drilling and partly to the greater number of deep round trips to change bits. Although costs are notoriously difficult to predict (not least because of the availability and pressure of competing projects for which rigs may be required), Garnish estimated that it may cost three times as much to drill to a depth of 8 km as it does to drill to 4 km. This factor

---

† This is not always the case. Recent drilling in granite at Camborne School of Mines has achieved rates of 3 m/h.

appears to be largely independent of rock type. It is hoped, however, that new techniques (see p. 245) may eventually reduce the costs.

The spacing of holes at the surface depends upon a number of factors. Close separations reduce the amount of interconnecting pipework and may minimise land use. On the other hand, if the production zones at depth are too close, there may be interference between them thus reducing yield. The exact relationship depends upon the details of the fissure system in the aquifer, but as a rule of thumb, separations of 100 m for depths of 500 m, rising to separations of 300 m for depths of 2000 m are often assumed. Interconnection costs can often be reduced by the useful technique of directional drilling as illustrated in figure 5.10. In addition, the technique can be used for sealing 'rogue' boreholes where blow-outs or other accidents occur. At Wairakei in 1960 [14] (and subsequently in Mexico) a directional bore was drilled and cement injected to seal a rogue hole from which steam and debris were emerging.

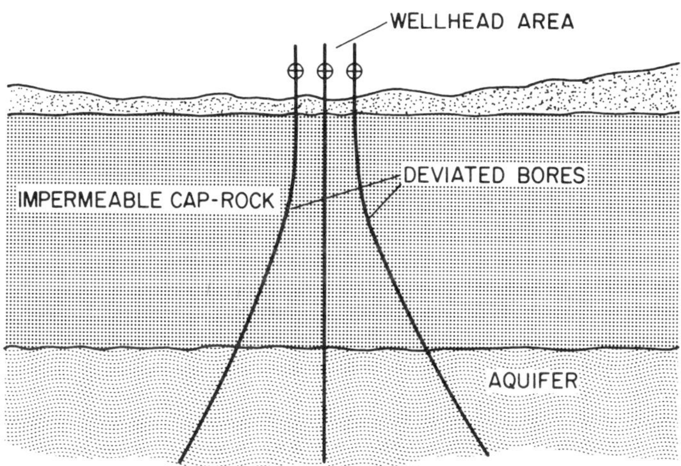

Figure 5.10 Illustration of the use of deviated bores to extract geothermal fluid from a wide area.

This summary of drilling methods cannot do justice to a major technology. The interested reader is referred to the review by Patterson *et al* [17]. Some innovative methods are discussed in §5.4.

When successfully drilled the output from a bore must be assessed. The techniques and pitfalls of carrying out the necessary measurements have been described by Armstead [2]. The major measurements required are as follows.

(1) The well-head temperature and pressure as a function of flow rate.

(2) The fluid enthalpy.

(3) Chemical composition.

5.2.4 *Surface plant*

Surface plant for fluid collection and transmission at various types of aquifers are described in some detail by Armstead [2] (a useful list of references is given in chapter 9 of his book). In this section, no mention will be made of detailed engineering such as pipe connections, silencers, separators and other parts of a typical geothermal power station but a description will be given of the main options for the power plant.

Before examining this, it is instructive to consider how the output of an aquifer should be optimised to give peak electrical power. Using aquifers for direct heating purposes implies using a pressure at the well head at which maximum heat can be extracted. Generating electricity, however, leads to a different optimum pressure [2]. An illustration of the type of curve of power against well-head pressure for a wet bore is shown in figure 5.11. The maximum reflects the fact that increasing the pressure reduces the amount of steam available but clearly raises the heat available in a given mass of steam.

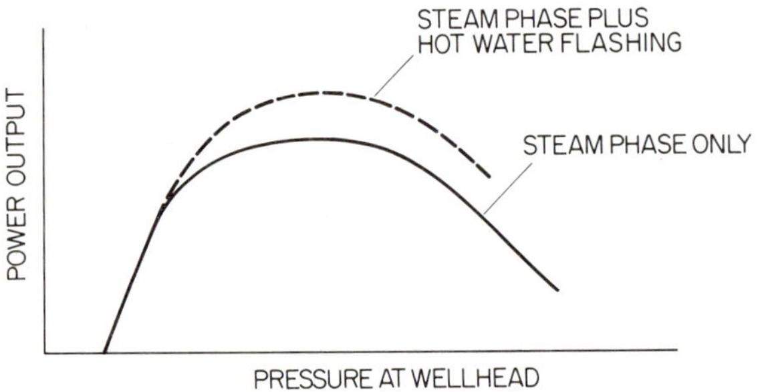

Figure 5.11 Typical power output curve from a 'wet' bore as a function of wellhead pressure.

The pressure which optimises power from the borehole is not necessarily overall the most economic pressure which can be used. Although maximising power output will reduce the number of bores required, other factors are also important. For example, low pressures tend to require bulky pipework for steam transport and high pressures require more costly fittings such as valves, thicker insulation and tend to lead to greater chemical build-up on turbine blades.

The design of the power system depends on many factors. Clearly the pressure and temperature of the aquifer are two of the most important, but the corrosiveness of the working fluid and matters of environmental concern such as reject heat disposal are of significance. A number of generation cycles are possible and are illustrated in figures 5.12 and 5.13.

Milora and Tester [18] have discussed in detail possible cycles for geo-thermal plant and have also examined their thermodynamics. A brief description of several systems will now be given.

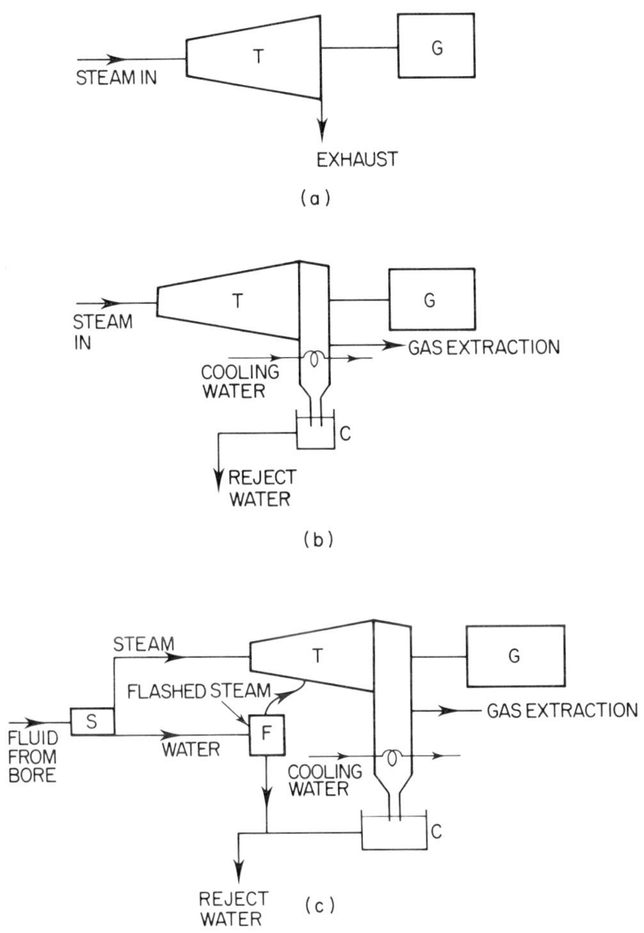

Figure 5.12 Main types of geothermal steam cycles: (a) non-condensing, direct cycle; (b) simple condensing cycle; (c) flash cycle (only one flash shown). Key: T, turbine; G, generator; C, condenser; D, degasifier; H, heat exchanger; P, pump; F, flash vessel; S, separator.

Non-condensing, direct cycle. Such cycles (figure 5.12(a)) have largely been superceded by more sophisticated plant but are still useful for pilot schemes, providing standby power or rapidly meeting peak loads. The method is very simple. Steam is separated from the water at the borehole and passed directly through a turbine, the exhaust going into the atmosphere. Their simplicity minimises capital cost but the efficiency is only about half that of condensing cycles and large quantities of steam and gases are released into the atmosphere, which could lead to environmental problems.

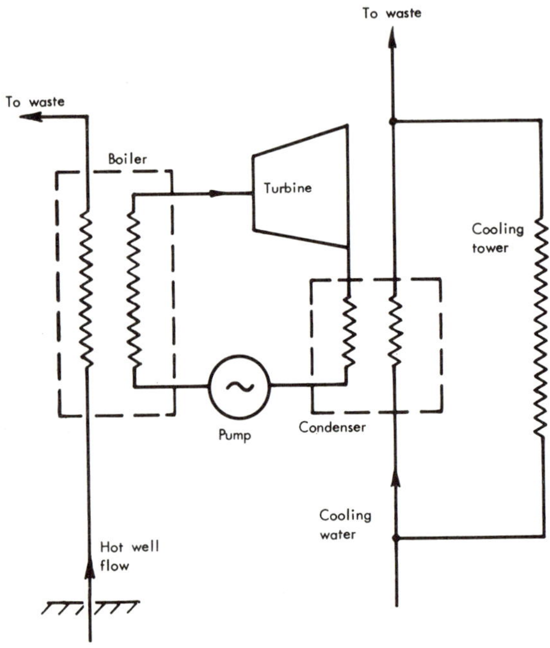

Figure 5.13 Schematic diagram of a binary cycle system.

Simple condensing cycle. This differs from the previous cycle only in that cold water is used to condense the exhaust gases before rejection or re-injection into the aquifer (figure 5.12(b)). Since no boiler feed is used in a geothermal cycle, the cost of condensing plant is much less than in conventional power stations because simple jet condensers may be used rather than costly surface condensers. Again, pollution problems may arise.

Flash cycles. In a bore which produces hot water as well as steam, some of the energy can be extracted from the hot water phase by passing it into a lower pressure vessel and using the steam produced to drive the lower pressure stages of a turbine. Some cycles operate with one such flash vessel operating at a temperature about half way between that of the hot water and condenser. In principle, it is possible to have a cascade of flashing vessels but clearly there is a trade-off between the increased capital cost of this and the increased energy yield. In practice, two stages about equally spaced in temperature are likely to be the most which would prove economic. A single flash cycle is illustrated in figure 5.12(c). A new device being explored at hydrothermal facilities in the Imperial Valley, California, is the rotary-separator turbine. This recovers kinetic energy from the high velocity water phase which has been flashed.

Binary cycles. Much interest has been expressed recently in using working fluids of low boiling point such as freons or hydrocarbons in a closed Rankine cycle such as that illustrated in figure 5.13. Such systems could cost more than the other cycles described above because heat exchangers, surface (rather than jet) condensers and feed pumps are required. Furthermore, it is necessary to guard against the release of toxic or inflammable materials and the use of heat exchangers and feed pumps reduces overall efficiency. Set against this, however, there are several important compensations [18], [19]. Lower temperature sources can be used for power generation and lower rejection temperatures enable more heat to be extracted. In addition, more corrosive fluids can be used (or those containing high proportions of trapped gases) with chemical attack likely to occur only in the heat exchangers. The higher vapour pressures which are achievable allow more compact turbines to be used. Noise problems are also reduced. Despite this considerable promise, few such systems have so far been operated in geothermal plant and these advanced cycles are not fully commercially proven. In the USA, however, a 65 MW demonstration plant is being designed for evaluation at Heber in the Imperial Valley. Milora and Tester [18] have shown that, in principle, organic fluid Rankine cycles will be cheaper than flash cycles for electricity generation from sources with temperatures of less than 200 °C.

Because working temperatures are frequently low, the efficiency of geothermal power stations (typically about 10%) is much lower than conventional plant. This will be further discussed below. One implication of this is that the heat rejection requirements of such plant are very considerable, 9 kWh of thermal energy being rejected for every 1 kWh of electricity produced. Thus cooling towers, if used, are much larger than would be the case in a conventional power station of the same size. This may present siting difficulties.

Finally, whatever form of power plant is used, it is important to note that operation may be affected by the corrosive nature of the aquifer—particularly in wet fields [20]. In the worst cases of highly acidic water or strong brines, it has not proved possible to exploit the resource. More typically, plant must be designed to withstand the corrosive environment and in some cases, maintenance has to be carried out fairly frequently (thus significantly reducing plant load factor). Chemical attack often occurs as a result of gases in the water (CO_2 and H_2S are the most prevalent) and the build-up of deposits in various parts of the plant. Dissolved calcium salts and silica are particularly troublesome in this respect, leading to the partial blockage of boreholes, surface plant or even pores in the aquifer itself. Deposition occurs mainly in places where hot water is flashed to steam. It is clearly most important that chemical attack or deposition does not occur in the turbine. Corrosion in the condenser is frequently quite severe because of the presence of oxygen.

5.2.5 Plant efficiency and size

The thermal energy available from a reservoir clearly depends upon the flow rate, temperature of the aquifer and the specific and latent heat of the working fluid. As mentioned above, however, the electrical output is only a small fraction of this resource.

It has been shown [21] that the ideal thermodynamic efficiency of a cycle accepting heat at a temperature T_2, rejecting heat at T_1 and operating with a condenser temperature of the working fluid, T_0, is given by:

$$\eta_i = 1 - \frac{T_0}{T_2 - T_1} \ln \frac{T_2}{T_1} .$$

In practice, however, Milora and Tester [18] have shown that for sources with temperatures in the range 150–200 °C, practical efficiencies of about 60% of η_i might be obtained. Thus, for a geothermal source with a well-head temperature of 200 °C, a rejection temperature of 35 °C and a condenser temperature of 25 °C, the practical efficiency might be about 12%. This does not take into account the power required for pumping or running auxiliaries.

The size of plant is thus limited by the thermal output and the plant efficiency. In practice, sizes range from a few megawatts up to the largest of 130 MW in the Geysers field. The number of individual boreholes connected to the generator depends upon the size and configuration of the field, the possibility of directional drilling or the economics of connecting together steam pipes from separate wells. By comparison with major established fields, the size of individual plant from the low enthalpy sources in the UK would be very small. Tozer *et al* [22] have estimated that for the Lincolnshire basin, assuming estimated figures for the permeability of the rock, the aquifer thickness and the density and viscosity of the brine and taking a temperature of 100 °C, a gross electrical output of under 1 MW would be obtained. Thus for one complex, consisting of several boreholes, about 5 MW or so might be generated.

5.2.6 Use of the heat

This book restricts its attention to electricity production, but it is interesting to ask under what conditions the geothermal heat should be used for district heating, direct generation or for improving the efficiency of conventional power stations by preheating feedwater. No firm answers are available and the subject is still controversial.

Richardson and White [6] have analysed the return on capital achieved by generation of electricity using a binary cycle compared with a district heating scheme for geothermal hot water at 100 °C and have concluded that generation of electricity is likely to be at least as attractive as district heating for the assumed conditions. The result is contrary to the general view that electricity generation would only be economic *vis-à-vis* district heating for temperatures in excess of about 150 °C. Garnish [23] considers that only with combinations of high temperature and/or flow rate (e.g. 100 °C at a flow rate of 50 kg/s from a doublet) would electricity production be favoured.

The analysis of Richardson and White has recently been extended by Tozer *et al* [22] for the conditions which might be expected to prevail in UK aquifers. The study included an analysis of the economic performance for both electricity generation and district heating as a function of borehole depth, thermal gradient and the number of dwellings per hectare. They have assumed in their study that the geothermal field is independently owned and electricity must be produced for less than 2.2 p/kWh or space heating achieved for less than the cost of gas used for the same purpose. Although some of the assumptions and input data might be revised or improved, the results appear to indicate that electricity generation is economically more attractive than space heating, except at very high heat load densities (apparently supporting White's previous assertion) and that electricity generation might be economic on the most favourable assumptions. For the conditions found in the first UK geothermal borehole at Marchwood near Southampton (well-head temperature 68 °C, flow rate about 30 kg/s, pumping head 270 m) it is likely that electricity generation costs would be several times those of a fossil-fuel fired station even assuming that the aquifer has an acceptable life.

Harrowell [24], Khalifa *et al* [25] and White [26] have also examined the case for low enthalpy sources to be used for pre-heating power station feed water. For such a scheme to be successful, it is obvious that a suitable power station must exist or be built close to a geothermal basin. It is also important that the power station operates with a high load factor in order to repay the capital investment. The principle is illustrated in figure 5.14. The hot water heats the feed water of the generating plant through a similar temperature rise to that existing across the low pressure heaters of a conventional generator. This allows the steam which is normally bled to these heaters to be retained within the turbine and expanded to condenser pressure, thus increasing the output of the alternator. It was suggested that this might lead to an increase in efficiency of up to 3% for a 60 MW set at

Marchwood Power Station. This estimate was based on fairly favourable assumptions for the aquifer on which this power station is built (100 °C, 55 kg/s) which have not been achieved in practice. A pilot experiment planned to test the technique was not carried out at Marchwood due to oil price rises and a subsequent reduction in station load factor. In a practical scheme, it is not clear to what extent the increase in flow would increase the blade loadings on the turbine.

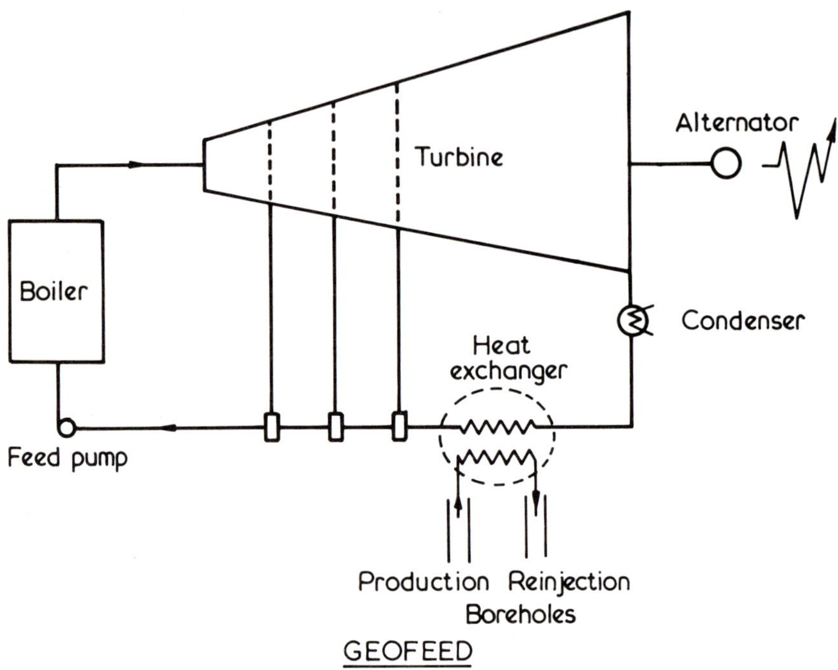

Figure 5.14 Principle of using geothermal hot water for power station feed-water pre-heating.

5.2.7 Economics of electricity generation from hydrothermal sources

Generally speaking in developed countries the most economic sources of geothermal energy have probably been developed. In these cases because of high temperature and flow rate and close proximity to the surface (low drilling costs), the electricity generated can be expected to be cheap. An EPRI report [27] estimated that in 1976, the cost of electricity generated from the Geysers field was about one-quarter the cost of electricity from an oil-fired power station. This may now be closer to a figure of one-third because of the effects of regulating H_2S emissions. Since the capital cost of geothermal plant was estimated to be about the same as that of oil-fired power station plant, the savings can be attributed to the cost of the fuel saved.

It is very difficult to generalise these results. The economics of exploiting a particular field depend upon a long list of parameters including the costs of exploration, reservoir development, purity and temperature of the water, flow rate (transmissivity of the aquifer), the type of power plant, disposal of the brine, the degree of environmental protection required, the operating and maintenance costs including reservoir management (re-boring, etc), taxes and assumed discount rates. It is also clearly depends upon who owns the system, the capital cost of plant with which it must compete and the future cost of fuel for that plant.

5.2.8 Environmental considerations

There appear to be six primary environmental considerations which may affect the development of aquifers. The severity of these constraints clearly depends upon the characteristics of each field and the importance attached to environmental matters in each country. Several authors have reviewed the problems in general terms [28], [29] and for particular plants such as the Californian Geysers [30].

Visual. Geothermal fields can be made to blend to some degree with the landscape, but for large fields, the plant and pipework may be extensive. An impression of this may be obtained from figures 5.3 and 5.4.

Water disposal. Most geothermal waters are brines, and some contain toxic material such as arsenic and mercury. The rejection of wastes may be overcome by re-injection into the aquifer as discussed above, but this requires further boreholes to be drilled, thus increasing costs. In a large field spent bores may sometimes be used for this purpose.

Non-condensible gases. Hydrogen sulphide is present in many steam fields and has given rise to environmental difficulties. In some cases (eg the Geysers) emissions can be reduced below an acceptable level, but in the worst cases this involves expensive technology.

Waste heat rejection. The large amount of waste heat to be disposed of in a geothermal plant compared with conventional plant of the same size has been discussed above. In some situations, the use of cooling towers will present a problem in terms of visual amenity. In arid regions, the requirement for cooling water may be hard to meet, so dry cooling towers would be needed at higher capital cost.

Subsidence. There are some cases [31] where subsidence has occurred particularly in cases where re-injection has not been carried out. For example, there has been considerable subsidence at Wairekei but not at the Geysers field. The amount of subsidence appears to depend largely on the prevailing geological conditions. There is little evidence for appreciable induced seismicity.

Noise. Drilling and venting of wells or steam lines can create high noise levels. Muffler systems are employed and these are normally adequate but clearly circumstances vary from field to field.

5.2.9 Future development of the resource

In §5.2, a short description was given of the major developed fields at Lardarello, Wairekei and the Geysers. Brief mention was made of some encouraging future developments. In this final section of the discussion on aquifers, reference will be made to the plans and prospects for electricity generation in a number of countries where a sizeable resource is believed to exist. Table 5.1 summarises these developments and gives estimates of the resource base. More details of the plant currently installed and plans for further development throughout the world are given in references [32] and [27]. A summary of some of the more promising areas will now be given.

Table 5.1 Electrical generating capacity for various geothermal aquifers (MW(e)).

Country	Installed 1980	Estimated 1986	2000
USA	938	2700	4000–20000
Italy	440	500–550	1000–1500
New Zealand	202	302	575
Mexico	180	700	1700–4000
Japan	168	350	3500
Philippines	446	1500	4000–4700
El Salvador	60	100	540
Nicaragua	—	35	100
Iceland	38	68	1500
Costa Rica	—	—	100–300
Guatemala	—	15	50
Honduras	—	100	
Panama	—	60	
Taiwan	0.4	3	500
Portugal (Azores)	3	13–18	400
Kenya	—	30	90
Guadeloupe	—	30	
Chile	2	20	100
USSR	6	10	310
Turkey	0.5	18	100
Canada	—	10	
Indonesia	0.25	30	1540
China	5	11	50
Djibouti	—	5–15	50
Ethiopia	—	15	50
Greece	—	3.5–10	100–200

Europe.

(1) Greece. There appears to be some potential for geothermal electricity in the Ionian islands. On the island of Milos, wells at temperatures in excess of 250 °C have been found but low flow rates were obtained.

(2) Iceland. The district heating schemes in Reykjavik are well known. The estimated geothermal electric power capacity in Iceland has been projected to be 70 MW by 1986 rising to 1500 MW by the year 2000.

(3) Italy. There are thought to be exploitable geothermal resources in a coastal band running almost continuously from Lardarello to Naples. It has been estimated that the installed capacity will double to nearly 1 GW by the latter part of the decade.

(4) Portugal. Islands in the Azores such as San Miguel show exceptional gradients and a long-term objective of developing 400 MW has been set.

(5) Soviet Union. As with several other eastern European countries, substantial amounts of district heating are presently obtained from geothermal sources. Two small power plants have been operating for some time but the area of the Kamchatka Peninsula is believed to be capable of providing as much as 500 MW. A 50 MW(e) plant is planned for the late eighties at Mutnovsky.

(6) Spain. Lanzarote in the Canary Islands has very high temperatures at shallow depths but it is not yet certain that this is associated with an aquifer. Nonetheless, a target of replacing 25 MW of diesel plant has been set for 1985 and as much as 200 MW by 2000.

The Americas.

(1) Canada. The western part of Canada is believed to have considerable geothermal potential. It has been estimated that 10 MW of geothermal plant will be operating by 1985. Exploration is still at a relatively early stage.

(2) Costa Rica. 100 MW is planned for the late 1980s.

(3) El Salvador. A 60 MW plant has been operating since 1976 and the output of this is currently being increased. It is hoped to achieve an output of 100 MW by 1985. Several promising areas for further exploitation have been identified.

(4) Honduras. There are two particularly promising areas and a target of 100 MW has been set for the mid-1980s.

(5) Guatemala. There are several promising areas which may shortly lead to a geothermal capacity of 15 MW.

(6) Mexico. There is now 180 MW of geothermal plant installed at Cerro Prieto. This is in the process of being expanded with a medium-term target of over 1000 MW. It is hoped to produce 700 MW from geothermal sources by 1986. Temperatures as high as 340 °C have been recorded in some bores.

(7) Nicaragua. A considerable programme of drilling has been initiated and it is hoped to achieve an output of 100 MW by 2000.

(8) South America. Like the central American countries mentioned above, Argentina, Brazil, Chile, Peru and Bolivia have excellent prospects for geothermal development. Surveys are presently being undertaken in all these countries and several promising areas have been identified. It is not yet possible to make an estimate of the likely rate of development except in Chile where 100 MW is planned by 2000.

(9) United States. Geothermal energy is being taken very seriously and the Geysers field alone is believed to be capable of being expanded to produce up to 4 GW. Imperial Valley in California (just north of the Cerro Prieto field) has been estimated to have a potential of anything from 2 to 20 GW. Much of this could be hot rock. Uncertainty stems from the high salinity of the geothermal fluids extracted in some parts of this area. The Valles Caldera in New Mexico may eventually be able to support an output of more than 100 MW and promising areas in Utah, Idaho, Virginia and New York State have been identified. The minimum target is to produce 2.7 GW by 1985 and 4–20 GW by 2000. Nearly 1 GW is already installed.

Asia.

(1) China. Several geothermal fields have been discovered in Tibet. The full potential is not known but 5 MW is currently thought to be installed.

(2) India. Very high gradients have been found in parts of the Himalayas and hot springs exist in the west of the country. Priority is being given to carrying out assessment of this resource.

(3) Turkey. A field with temperatures higher than 200 °C has been discovered but due to high levels of calcium carbonate this field has not been exploited. In western Turkey, and near Ankara, promising areas have been found. It is hoped to achieve an electric power capacity of up to 100 MW by 2000. One 500 kW unit is operating.

Africa. The rift valley of East Africa is a particularly promising geothermal area. Ethiopia, Kenya and Tanzania may benefit from development but transmission distances to demand centres may reduce the value of the resource at present.

(1) Kenya. Temperatures approaching 300 °C at 1.5 km depth have been found in low salinity sources about 50 km north of Nairobi. It is planned to build 30 MW of plant by 1985 and 15 MW should be installed by the end of 1982. The total potential has been estimated to be about 170 MW.

Far East and Australasia.

(1) Indonesia. There appears to be a major geothermal resource and exploration is underway in a number of areas. It is estimated that about

30 MW of geothermal electricity will be produced by 1985 and as much as 1.5 GW by the end of the century.

(2) Japan. Interest in geothermal energy has grown rapidly since the 1973 oil crisis. In the early 1970s there were 35 MW of installed capacity but this had doubled by 1976 and more than redoubled by the end of the decade. The objective is to achieve an installed capacity of about 350 MW by 1985 but the long-term goal is to achieve several GW by the end of the century. Low temperature binary plant systems are being developed.

(3) New Zealand. The total geothermal capacity in New Zealand is thought to be about 2 GW and it is hoped that this will be achieved in the next century. Intermediate targets of 300 MW by 1985 and 575 MW by 2000 have been set. The installed capacity at present is about 200 MW.

(4) Philippines. Extensive exploration is taking place and nearly 500 MW of power plant are now installed or operational. It is likely that several other large schemes will be commissioned by 1985 amounting to 1500 MW.

(5) Taiwan. Taiwan is thought to have a significant geothermal potential, although some bores have yielded very acidic fluids. Nonetheless 3 MW is projected for 1985 and 500 MW for 2000.

5.3 Hot dry rock (HDR) technology

Conventional geothermal power production relies upon the existence of permeable rock from which existing ground waters may be extracted. However, many parts of the earth's surface are covered with impermeable rock, and below a few kilometres the so-called basement rock is almost entirely impermeable.

The resources from geothermal energy would be vastly increased if the heat in such rock could be utilised. Electricity could be directly generated from 'heat mines'. There are some surface formations of magmatic origin which exhibit high thermal gradients which might be exploited. In the longer term it is possible that by drilling to great depth, power could be produced from deep basement rock. The main technological barrier which must be overcome before this technology could be exploited, is to establish reproducible methods of rock fracturing which will provide paths for the circulation of injected water at an acceptable impedance and with the maximum heat transfer.

The development of HDR geothermal power is clearly of great importance to the future of geothermal energy in the UK. Although some hot water exists this is unlikely to be economic for electricity generation (see §5.2). In addition to basement rocks which underly the whole country and offer an almost limitless resource, there are batholiths of granite in Cornwall and Northumberland which come close to the surface and where thermal gradients of 30–40 °C/km are expected [16], [33]. Thus, if drilling to a depth of 5–7 km in such batholiths could be followed by the establishment of a suitable fracture system at acceptable cost, there would be a considerable potential for electricity generation.

5.3.1 *Theoretical behaviour of HDR systems*

If hydraulic pressure at 70–100 bar is applied to a column of water in an impermeable homogeneous rock at a depth of 3 km, cracks are expected to form leading to the creation of a vertical disc-shaped fissure, such as that illustrated in figure 5.15. The crack system formed by hydro-fracturing in this way could be several hundred metres in diameter. If a second bore is sunk directionally so as to intercept the disc fracture, injected water could be circulated to extract heat. A successful system would need to achieve several objectives [34], [35].

(1) The contact area between water and rock would have to be maximised to achieve high heat transfer and the flow kept sufficiently slow for the same reason. The growth of a single low impedance channel would need to be avoided.

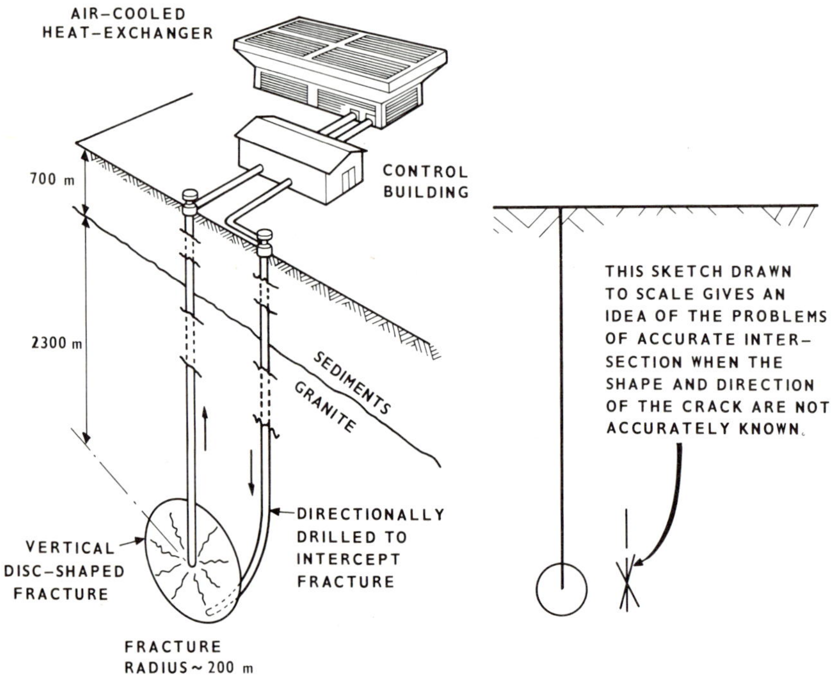

Figure 5.15 Schematic diagram of a hot rock geothermal system. (Courtesy Los Alamos Scientific Laboratory.)

(2) The system of fissures and voids created would have to be engineered to give an acceptable impedance.

(3) The lifetime of the system would need to be sufficiently long to be economic. Lifetime could be limited by cooling of the rocks local to the fracture or by closing of the fissures. Redrilling or stimulation by high pressure water might be used to achieve a sufficient life.

(4) Water loss from the system would need to be small (particularly where surface water is in short supply) and chemical leaching from the rocks into the continuously circulated water would need to be of an acceptable level, otherwise treatment of the water would be required.

For an HDR power station to be viable it is estimated that the parameters listed below must have the approximate values given:

heat transfer area	$1–2\,\mathrm{km^2}$
stimulated bulk volume	$0.2–0.3\,\mathrm{km^3}$
water loss	$<2\%$
circulating mass flow	$75–100\,\mathrm{kg/s}$
life (with re-drilling)	$20–30\,\mathrm{y}$
installed capacity	$\sim50\,\mathrm{MW(t)}$.

A range of different methods is being explored, particularly at the Los Alamos Scientific Laboratory (LASL) and the Camborne School of Mines, to achieve acceptable fracture systems [35]. The simplest is that described above, where in theory a disc-shaped fracture can be set up in a vertical plane by hydro-fracturing. The size of such a fracture must be 100 m or more in radius, with a cross section of only a few millimetres. It is important that the conductance of the fracture is such that buoyant circulation across the faces of the fracture can be maintained. If the original fracture becomes thermally exhausted, it is envisaged that further boreholes could be formed parallel to the first system (figure 5.16) and the inlet and outlet boreholes extended to exploit the new fracture. Another possibility, if large stable fracture areas cannot be produced, is to form multiple fractures grown from perforations in the casing of the lower borehole. It has been suggested that these perforations can also be used as flow regulators but this possibility now seems unlikely. Some of these techniques are illustrated in figure 5.16. The theory of such techniques has been analysed by Gringarten *et al* [36]. If other methods are not possible, the original boreholes could be diverted to exploit an entirely new region of hot rock.

It has been suggested that the lifetime of an HDR system might be much longer than the theoretical value because as the rock is cooled locally, new cracks will be opened up at right angles to the cooled surface (figures 5.17 and 5.16(b)). Such thermal stress fracturing or 'thermal cracking' would occur when the local thermal tensile stress exceeds the compressive stress in the rock.

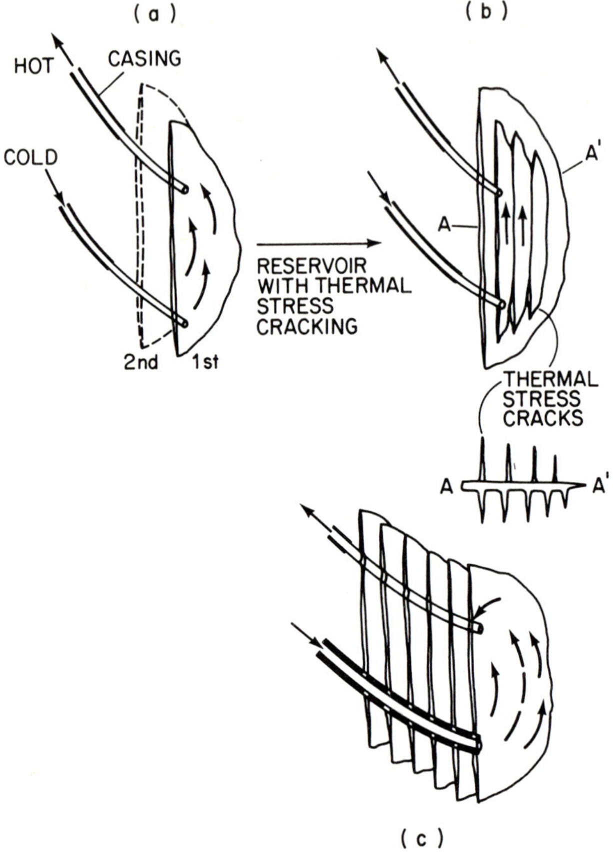

Figure 5.16 Several methods for ensuring a prolonged life for a hot rock geothermal system: (a) growth of a second fracture; (b) thermal cracking; (c) multiple fractures. (Courtesy Los Alamos Scientific Laboratory.)

Theoretical studies have shown that the subject of thermal stress cracking is very complex. Some analyses have suggested that secondary thermal cracks may grow such that closely spaced shallow cracks gradually coalesce to form deeper more widely spaced channels. Because these larger cracks will allow greater circulation of the fluid, it has been suggested that the heat extraction rate would increase as the cracks developed. These cracks themselves could form the nucleus of a further system and thus the reservoir would be self-sustaining without re-drilling. Whether or not this happens depends among other things on the heterogeneity and stresses in the rock at depth, and these are very uncertain. The only way that the effect can be tested is to carry out a full-scale experiment and this is planned at Los Alamos. Figure 5.18 shows schematically how the output of a fracture should vary with time, both with and without thermal stress fracturing. It is only after more than a year of continuous heat extraction that the increased extracted power due to thermal cracking might be expected to manifest itself according to theory.

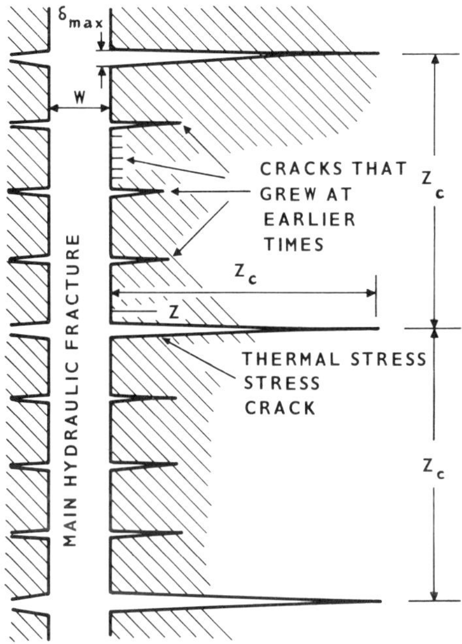

Figure 5.17 Thermal stress crack growth $(\sigma_\tau = E\alpha T/(1-\mu))$. (Courtesy Los Alamos Scientific Laboratory.)

McFarland and Murphy [37] have pointed out that even if thermal stress cracking does not occur, contraction of the rock around the fracture should lead to increased buoyancy and enhanced heat extraction. The type of fissure system discussed so far is essentially two dimensional. An ideal system, capturing heat from a much greater volume, would be three dimensional and it is possible that the use of explosive charges may enable three-dimensional systems to be developed. As discussed below, this possibility is being pursued at Camborne School of Mines and appears to be a more practical form of creating a suitable heat exchange surface than the hydro-fracture method alone which might lead to the two-dimensional disc structures discussed here.

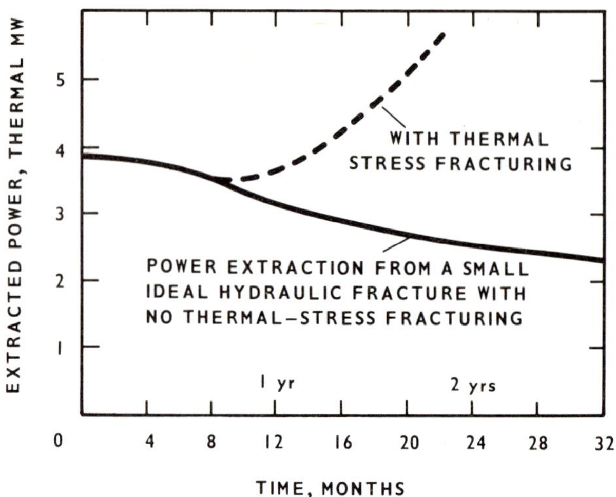

Figure 5.18 Thermal power output for an idealised 140 m radius fracture. (Courtesy Los Alamos Scientific Laboratory.)

Various studies [37], [38] have attempted to estimate the energy draw-down rate of HDR systems under various assumptions. This depends upon the density, thermal capacity and conductivity of the rock (ϱ_r, C_r and λ_r), the mass flow rate of the water ($\dot{m}_w$), the fracture surface area (A) and the distribution of fluid across this surface. For a single crack in rock of lower permeability and uniform flow, the recoverable power, $P(t)$, has been estimated to be

$$P(t) = \eta\, \dot{m}_w\, C_w\, (T_i - T_o)\, \text{erf}\, [(\lambda_r\varrho_rC_r/t)^{1/2}\, (A/\dot{m}_wC_w)]$$

where η is the fraction of recoverable power corresponding to uniform flow across an ideal plane fracture which could be recovered, C_w is the thermal capacity of the extraction fluid, T_i the initial rock temperature and T_o the fluid rejection temperature.

The parameter, η, is taken to lie in the range 0.4–0.9, the actual value depending upon the amount of buoyant circulation. Using this formula figure 5.19 shows plots of the thermal power versus time for $\eta = 0.9$ and various values of $\dot{m}_w/R^2$ where R is the fracture radius ($A = \pi R^2$).

In multiple-fractured systems, number and spacing are important [38] but as long as the spacing is greater than about 30 m, the low thermal conductivity of granite ensures that there is little interference between the fractures and the shape of the draw-down curve is similar to that for the single crack.

The two major projects in HDR technology are being carried out at Los Alamos in the USA and at the Camborne School of Mines in the UK. These will now be described.

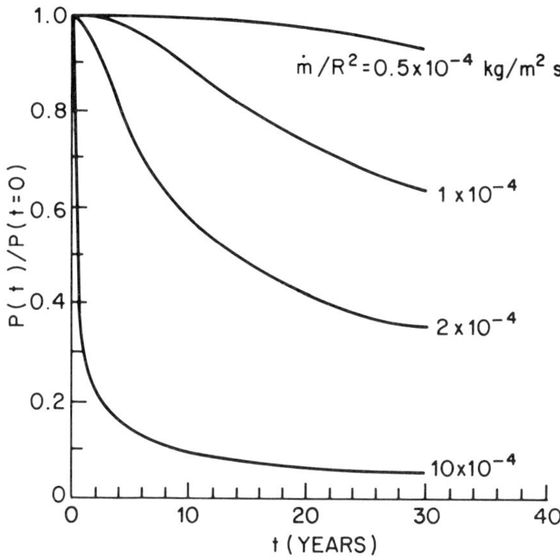

Figure 5.19 Parametric power drawdown curves for a single fracture with no thermal stress cracking, $\eta = 0.9$.

5.3.2 *The Los Alamos experiment*

The Los Alamos Scientific Laboratory is close to the Valles Caldera, a magmatic intrusion resulting from volcanic activity about one million years ago. In 1971 shallow bores (about 30 m deep) showed high temperatures, and heat flow measurements at 200–300 m depth indicated flows of 5 hfu. A preliminary test-bore drilled to a depth of 150 m into the granitic rock (over 750 m in total depth) achieved a temperature of 100 °C and verified that the rock was of very low permeability. As a result of these encouraging early tests, a site (the Fenton Hill site) was chosen for detailed study in 1973. An exploratory hole (GT-2) was first drilled to just over 2000 m and despite difficulties in drilling, cementing and logging the hole, hydrological experiments were carried out including attempts to hydro-fracture the rock. Subsequently GT-2 was deepened (GT-2B) to nearly 3000 m where a temperature of 197 °C was measured. It was found that a large fracture (an estimated 50 m in diameter) was created vertically by further hydro-fracturing.

A second bore (EE-1) was drilled in 1975 and reached a depth of over 3000 m where a temperature of 206 °C was recorded. From this position, directional drilling was carried out in order to intercept the fracture zone at the bottom of GT-2 and flow between the two holes was achieved. However, the connection was not a complete success, since the impedance was too high for large-scale heat extraction experiments. After several attempts GT-2 was plugged at 2500 m and side-tracked by directional drilling (GT-2C) to connect it with a fracture system generated at the base of EE-1. In a 96 h test run carried out in September 1977, 3.7 MW(t) was obtained with a water temperature of 132 °C [39]. Water losses were low. Since 1977, this system has been used to provide information on temperature draw-down, changes in impedance, water loss rate, geofluid chemistry and to serve as a test bed for reservoir enhancement techniques. Longer tests have confirmed the earlier work and output has been stable during test runs up to 23 days. A new fracture surface has been created which, with an area estimated at 55 000 m², is thought to be over six times larger than the older fracture system but is still far too small for a practical economic system. In May 1980, 60 kW(e) was generated using a freon turbine, in the first demonstration of electricity generation from an HDR system [40].

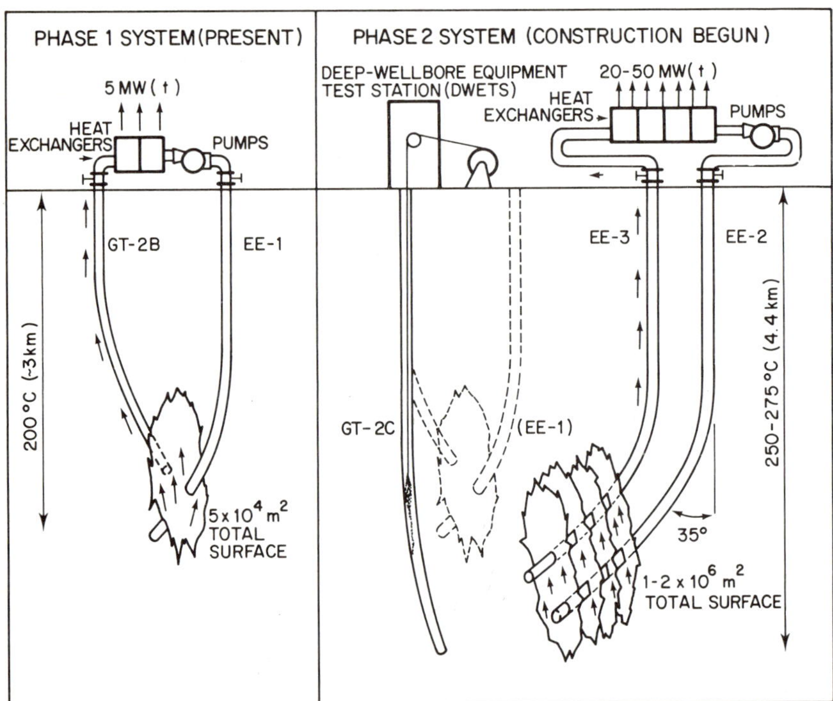

Figure 5.20 Arrangement of borehole systems at Fenton Hill, Los Alamos. (Courtesy Los Alamos Scientific Laboratory.)

In 1978, a second experiment (phase II) was initiated and is expected to provide 20–50 MW(th). Two boreholes are being drilled, separated vertically at depth by about 360 m and about 35 m laterally. The first of these

(EE-2) has been drilled to a depth of just under 3000 m and then drilled for a further 1600 m in vertical depth at an angle. The temperature at this depth has been estimated to be about 260 °C [40]. A second borehole (EE-3) has recently (late 1981) been completed and runs parallel and above EE-2. As many as fourteen fractures produced by hydro-fracture and explosive techniques are planned to connect the bores between the angled sections. The system will be a pilot plant for larger experiments and is expected to have a lifetime of at least 10 years. The layout of the various holes at Los Alamos is shown schematically in figure 5.20.

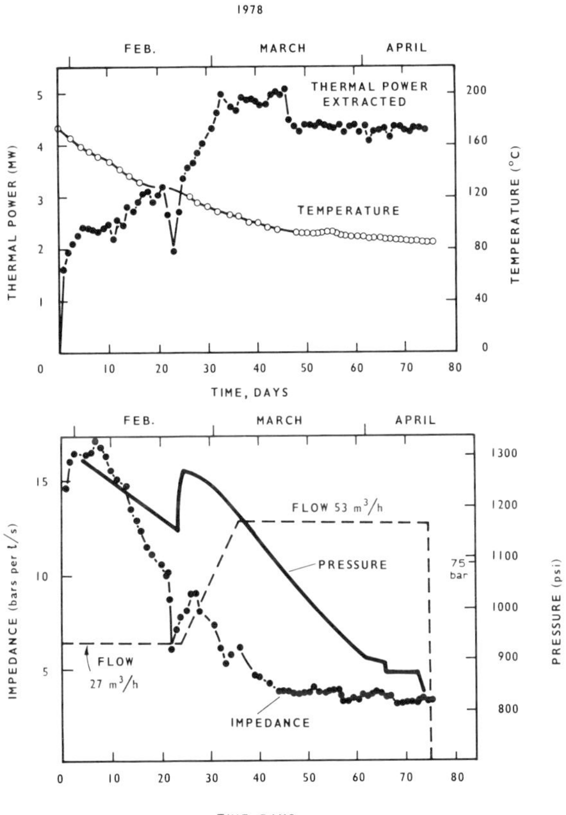

Figure 5.21 Los Alamos (Fenton Hill) hot dry rock field test—75 days operation of a prototype reservoir. (Courtesy Los Alamos Scientific Laboratory.)

The Los Alamos project has been of great significance in demonstrating the feasibility of HDR geothermal power. Many of the techniques developed for instrumentation of the holes (such as electromagnetic methods and high temperature TV) will be of continuing value [41]. Materials problems have been overcome in cementing the boreholes and testing underground valves. Injection pressures of up to $7200\,N/m^2$ have been used to generate fractures and information on the form of these (they are 'sinewy' rather than clean) has been obtained. There is, as yet, insufficient evidence to deduce whether thermal cracking has been taking place. Figure 5.21 shows the measured results for a 75-day test; this is clearly of insufficient duration to allow conclusions to be drawn about the development or otherwise of thermal stress cracks. It is now apparent that the simple disc-shaped fracture which has been extensively modelled is not, in fact, created. It is believed that the natural joint system in the rock determines the flow pattern at depth. Hydro-fracturing serves to link the boreholes to this complex system.

5.3.3 Experiments at Camborne School of Mines

The granite batholith in Cornwall appears to exhibit a thermal gradient of 30–40 °C/km and heat flow measurements have recorded values as high as 3.2–3.3 hfu. The batholith is thought to extend to a depth of about 12 km and the conjectural form is shown in figure 5.22. The reason why this batholith exhibits excess heat flow is thought to result either from exothermic kaolinisation of the granite at the base of a circulating water system, high levels of radiogenic heat below the weathered surface or because remnants of volcanic activity in tertiary times still manifest themselves.

Because of its central position in this batholith it is understandable that the Camborne School of Mines (which developed largely from the tin and china clay mining industries in the area) should take an active interest in geothermal energy from hot dry rocks. The effort has until recently been on a much smaller scale than the LASL work but the development of new techniques [42] by A S Batchelor and his co-workers at Camborne have established a world-wide reputation. Work has been carried out mainly at shallow depths (300 m) but particular interest has developed into the use of unfocused explosive charges detonated within a small space at the bottom of the bore. The explosion connects the borehole to the natural crack system in the granite through a multitude of cracks. Both these and the natural joint system are then developed by the use of hydro-fracturing. The technique is thought to avoid the creation of a single crack close to the borehole and the result is a lower impedance system, since a 'bottleneck' near the borehole is avoided. It is clearly important to ensure that the shock wave generated stresses remain below the plastic limit. The technique also seems to lead to self-propping of the cracks with debris from the explosion. The work carried out so far has allowed three holes to be linked with bottom distances from 10 to 40 m. It has been shown by tracer experiments that several connecting paths have been set up and these have been maintained for experiments of over three weeks at a flow rate of about 10 kg/s with only a 1% water loss.

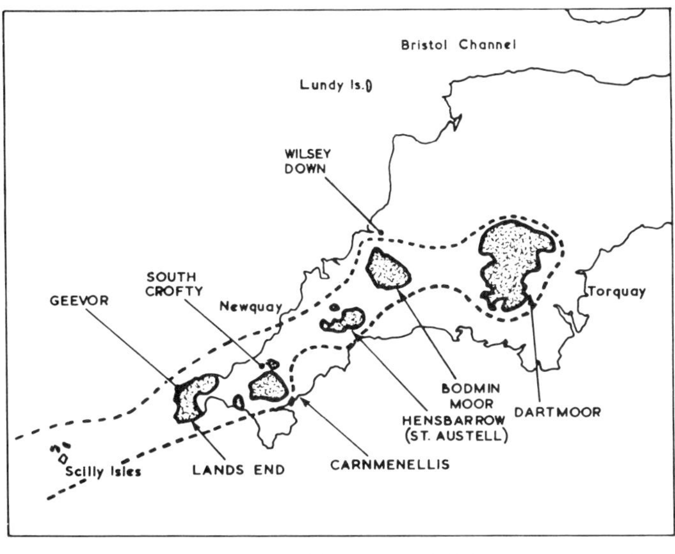

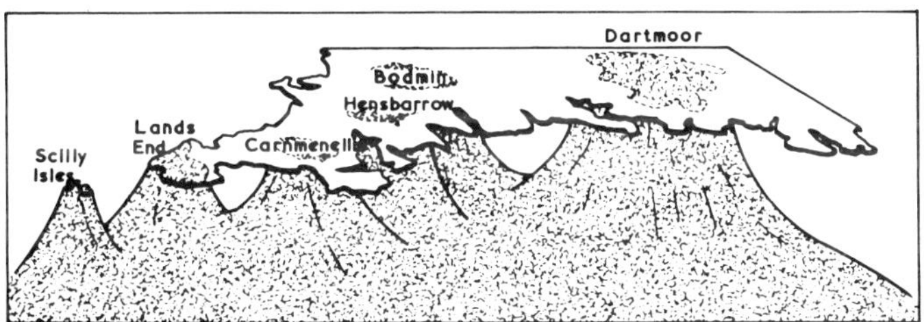

Figure 5.22 Conjectural form of the Cornish batholith. (Reproduced from *Energy Paper* 9, courtesy HMSO.)

The major advantages of the technique developed at Camborne appear to be as follows.

(1) Lower hydraulic pressures are required to extend the crack system.

(2) The larger specific acceptance (injection flow rate divided by overpressure) required for fluid in the injection well.

(3) Lower inter-well flow impedance.

(4) Larger fracture swept areas.

(5) Possible reduced water losses (it is not yet clear whether this is a consequence of the technique or reflects the type of rock at Camborne).

Further developmental work may need to be carried out to develop explosive systems which are stable at very high temperatures.

This work is now being extended to a depth of about 2000 m with funding of about £6M from the DEn and more than £1M from the EEC. At this depth the rock mechanics of fracture can be studied under fairly realistic conditions without the multitude of experimental complications that the higher temperatures at greater depths would bring. In late 1981, two deep bores had already been sunk twenty weeks ahead of schedule (due partly to the use of an advanced downhole drill system for directional drilling) and the attempt to create a suitable fracture surface was due to take place in late 1982. At 2000 m, bottom temperatures of 80 °C were achieved. The planned layout of the boreholes for this project is shown in figure 5.23. A view of the Camborne School of Mines site at Rosemanowes while drilling was being carried out is shown in figure 5.24.

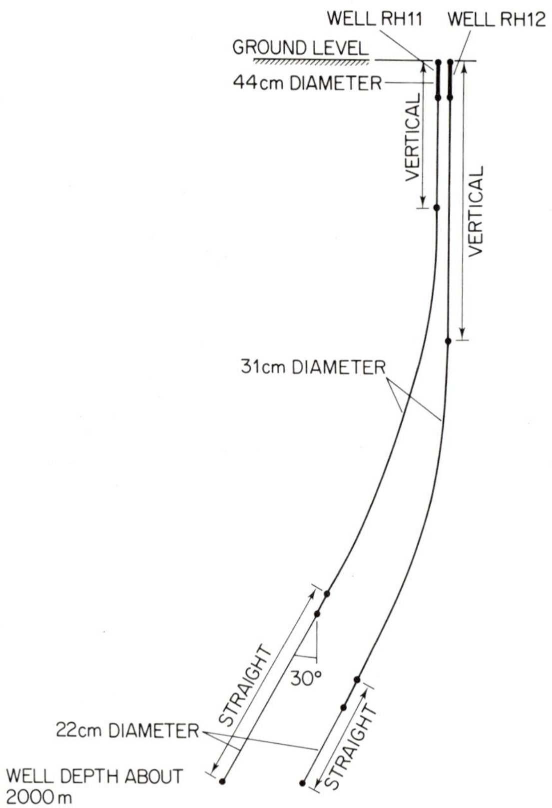

Figure 5.23 Planned layout of the boreholes in the Camborne School of Mines project. (Courtesy Camborne School of Mines.)

If this experimental study is successful, a pilot plant at full depth could be operational in the late 1980s or early 1990s.

Figure 5.24 Drilling the boreholes for the HDR experiment at Camborne. (Courtesy Camborne School of Mines.)

5.3.4 Other experiments [43]

At Mayet de Montagne in the Massif Central, France, experiments at 200 m have been carried out attempting to create a fracture between two bores by hydraulic means. Only 60% recovery of the injected water was achieved. Considerable interest has also been shown in West Germany. At

Falkenburg in Bavaria, excellent experimental work measuring stress fields in rock at 200 m has been carried out and attempts are being made to link bores to monitor heat transfer. At Urach, fracture was achieved at a depth of 300 m in crystalline basement rock and it is believed that two inclined fractures have been reopened and are intersecting by hydro-pressure in the same bore. Leaching experiments have also been carried out.

Several theoretical studies are underway. In particular the IEA is sponsoring a systems analysis of HDR with particular emphasis on hydro-mechanical modelling of 'ideal' rock. It is known as MAGES (man-made geothermal energy systems). Theoretical fracture studies have been carried out in the UK by the CEGB [44].

5.3.5 Hot dry rock plant

The technology associated with drilling and generating electricity is likely to be similar to that developed for exploiting aquifers.

There is less experience of drilling in granite but in some cases it has been slower and more expensive than drilling in permeable rock partly because of the rock hardness and because of the number of round trips to replace blunt drills. However, better tools are becoming available and recently (at Camborne, for example) much higher drilling rates have been achieved. The technology, however, is in principle little different from that discussed in §5.2.3. Some advanced drilling concepts which might further reduce the costs of HDR, if they prove to be successful, are discussed in §5.3.10.

Electricity generation will need to be achieved using closed cycle plant and thus the binary systems discussed in §5.2.4 will need to be developed commercially for a range of conditions if HDR geothermal energy is to be utilised for electricity generation [45]. Little work has so far been carried out in designing and costing surface plant requirements for this type of geothermal energy. This is an important part of the process of determining the optimum depth for dry rock systems.

5.3.6 Environmental consequences

The same general environmental effects would be likely to arise from an HDR plant development as those for aquifers except that the use of closed cycle plant should prevent the emission of toxic gases from the aquifer. Furthermore any subsidence would not in this case result from the removal of material (e.g. hot water) from below the surface but from contraction caused by cooling of a large mass of hot rock. Simple calculations suggest that ground subsidence could be as great as 3 m, although no such effects have so far been measured at Fenton Hill. It is possible that even in a full-scale system, the cracks which have opened will compensate for the effects of cooling.

At Fenton Hill, no significant induced seismicity has been recorded [40], although this may be site specific. The same problems of heat rejection discussed with reference to aquifers, will exist for HDR systems. Water loss may be difficult to overcome in arid regions.

5.3.7 Resource

It is difficult to begin to estimate the global resource for HDR but attempts have been made to evaluate the resource for the USA and for the Cornish granite. These, however, must be treated with extreme caution, although they do give an idea of the order of magnitude of the resource in these areas.

It has been estimated [27] that about 5% of the total US land area may have rocks with gradients of 40 °C/km and that 7.4×10^{22} J are potentially available to be tapped from rocks at temperatures in excess of 150 °C to a depth of 10 km. Of course, only a very small fraction would be technically recoverable even if the rock were suitable in terms of permeability, porosity and chemical reactivity with water.

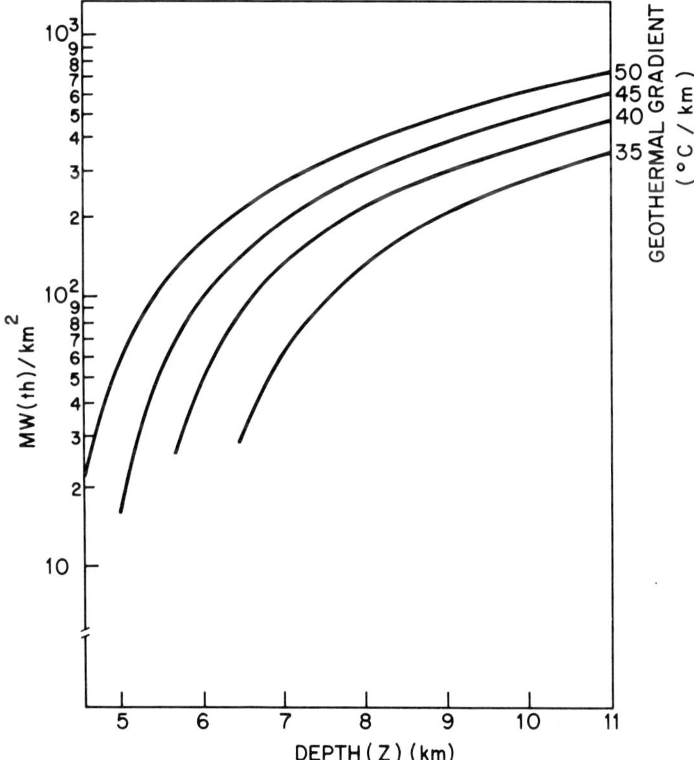

Figure 5.25 General form of thermal power output from dry rocks as a function of depth and gradient.

Garnish [16] showed that the total heat potentially available, Q_o, from a granite area of 1 km^2 between a maximum depth Z and a depth Z_o at which a minimum useful temperature is found is given by

$$Q_o = 1.1 \times 10^{15} G(Z - Z_o)^2$$

where G is the gradient in °C per km. This neglects any radiogenic heat generation within the rock and changes of thermal conductivity with temperature. These two effects, however roughly cancel each other. The curve of Q_o against Z for various values of gradient and minimum useful temperature resulting from this expression is shown in figure 5.25.

Cornish granite is thought to extend over an area of 6000 km^2 and Durham granite to extend over about a quarter of this area. Using Garnish's formula the gross resource from Cornish granite at temperatures above 200 °C would be about 6×10^{20} J. More realistically, if about 100 MW(t) could be extracted from the granite over each square kilometre, the total recoverable resource from Cornish granite over 20 years, assuming power station efficiencies of 10% and a load factor of 75%, would be about 8000 TWh (or 300 TWh p.a.). To give an idea of scale, current UK electricity demand is about 220 TWh p.a. This illustrates the vastness of the resource even in this one area of the UK. Being an area of great scenic value, even if all other conditions were acceptable, it is likely that only a very small fraction of this heat could be mined.

If the technique could be extended to basement rock, the total HDR potential in the UK would, in principle, be almost limitless.

5.3.8 Economics of hot dry rock electricity

Since the technology is in its infancy and the principle has not yet been reproducibly established even at pilot scale, the economics of HDR systems must remain speculative. However, several attempts have been made to examine the economics and these will be briefly reviewed.

It is not yet clear what density of power extraction can be achieved and this will clearly vary from site to site. Garnish [16] assumed six production bores and three re-injection bores per square kilometre. Costs were estimated for drilling in granite and for the well-head plant (in 1975 prices). The undiscounted unit cost of the heat as a function of depth for various gradients was calculated together with the cost of heat at the optimum depth as a function of minimum required temperature for different geothermal gradients. The curves produced were indicative of cost trends as various parameters are varied, but the actual costs calculated are now out of date.

A major study of the economics of electricity production from HDR sources was reported in a study by the University of New Mexico for LASL and EPRI [45]. The analysis was complex assuming detailed reservoir management scenarios. The study considered a multiple fracture hot dry rock system with electricity produced using an organic binary cycle system of the type discussed in §5.2.4. Electricity costs of 4.3 c/kWh (bus-bar† at $1978) were obtained for a reference case which assumed a geothermal gradient of 40 °C/km, a flow rate of 75 kg/s, a heat transfer area between pairs of boreholes of 1.7×10^6 m² and a plant design temperature of 160 °C. Some cost curves as a function of various parameters are shown in figure 5.26. Sensitivity analysis for a range of gradients, reservoir sizes, drilling costs and rates of return suggested that costs might realistically be expected to lie between −40% and +76% of the reference case estimate. The results cannot simply be translated to UK conditions because of the different economic parameters such as discount rates and tax assumptions. But bus-bar costs of electricity, should HDR technology prove successful in attaining the conditions for viability given in §5.3.1, have been estimated to lie in the range 3–5 p/kWh (at 1981 prices).

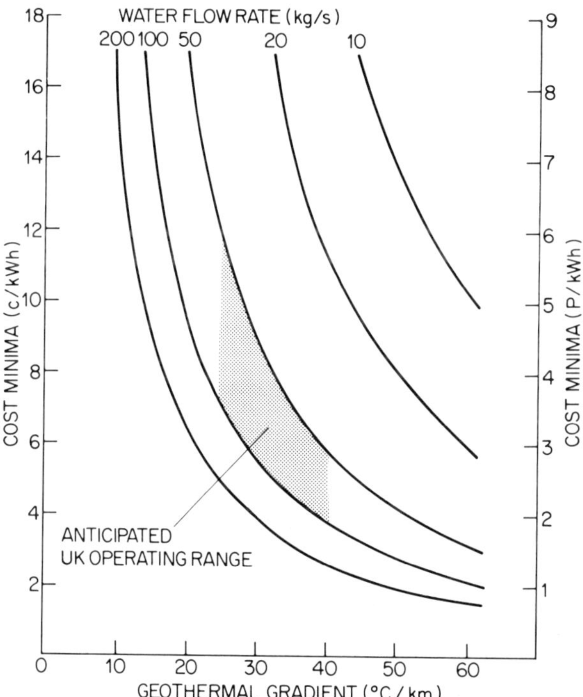

Figure 5.26 Predicted bus-bar cost minima for electricity generation from HDR systems (1981). (Courtesy Dr J Garnish, ETSU.)

A more recent study of costs of the HDR geothermal energy area which might be exploited in the Imperial Valley, California, has been carried out

† Bus-bar costs are the final costs on the supply network.

for the US DoE [46]. Costs were somewhat higher because of the lower gradient assumed, and the greater thickness of sedimentary rock overlying the crystalline material and higher state taxes. The results, however, were felt to 'provide some basis for cautious optimism concerning the future commercialisation of hot dry rock systems in the Imperial Valley once a proven technology is well established'.

5.3.9 *Remaining problems with* HDR *geothermal energy—a summary*

HDR techniques clearly offer exciting possibilities but there are still many problems which must be overcome. Some of these have been discussed above, but it may be helpful following Ashbee [34] to summarise these with special reference to the UK.

(1) It is yet to be proven that an adequate surface area can be formed reproducibly at depth.

(2) Actual temperature variations with depth in the more promising areas of the UK have yet to be established with certainty.

(3) Techniques for drilling and instrumentation at great depth in hot rock (particularly basement rocks where depths in excess of 5 km may be required) have still in many cases to be developed.

(4) Interception of a fracture with a second borehole may be difficult to reproduce.

(5) The nature of the crack system may vary from one area of rock to another using a given stimulation technique. Sufficient circulation may be difficult to achieve or, at the opposite extreme, channelling and thus short circuiting may occur.

(6) In some formations, water loss through natural faults may occur. This is likely to become less of a problem with increasing depth, however.

(7) Environmental effects such as visual amenity, subsidence and rejection of waste heat may limit the use of HDR geothermal power.

5.4 Possible future developments

Two interesting possibilities, one for increasing the resource base for geothermal energy and others for reducing the costs for deep drilling may enhance prospects and these will be considered in this last section of the chapter.

Andrews *et al* [47] have recently suggested that the techniques being developed to exploit HDR systems and conventional high-permeability systems may be brought together to exploit the large number of systems which may have only moderate permeability where very high pumping costs or low flow rates would make the economics of electricity production very unfavourable. The method suggested involves creating at least two vertical, parallel fractures at the bottom of boreholes. By maintaining the pressure in the bores, in one case above the hydrostatic pressure and in the other below, a pressure variation like the potential distribution between the plates of a capacitor could be set up. Heat could be convected from the system by flushing water between the fractures. The area of the fractures would offset the low permeability of the rock and give rise to an enhanced flow rate and thermal output. The theoretical analysis suggests that doublet systems engineered in this way, should be able to yield from three to several hundred times the thermal output of unfractured wells in aquifers in the same geothermal area. It was found that the re-injection pressure of such a doublet would need to be increased with time because of the change of fluid viscosity with temperature. For a system with a top temperature of 100 °C and a re-injection temperature of 30 °C it was estimated that, for constant output, a threefold increase might be required. If achieved in practice, the technique could be used to reduce the number of geothermal systems in a given location for a given power output.

The economics of geothermal energy clearly depend, to a large degree, upon the costs of drilling, particularly for HDR systems where boreholes must usually extend to depths of several kilometres and where drilling costs represent about 60% of the total costs of an HDR project. One suggestion made some time ago but which is technically very difficult is that melt-drilling might be employed. Most rocks melt below 1500–2000 °C and if a heated penetrator (sometimes referred to as a subterrene) [48] could be designed, it would avoid the problems of bringing large rigs to site, changing bits, supplying cooling muds, high levels of noise—and if the fused rock surrounding the device were suitable, it may even obviate the need for casing the boreholes. On the other hand, energy demand would be high (although this would reduce with depth in the hotter rock and uncased holes could easily be damaged). The rate of penetration in granites and other crystalline rocks might be much higher than is frequently achieved using conventional techniques, Electrically heated devices have been designed but it has been suggested that nuclear powered devices could be used. The major difficulty with melt drilling is that the granite, the volume of the melted rock is greater than for the solid and the low porosity of the surrounding material leads to blockage. Other ideas for melting and vaporisation include drills which would utilise lasers, electron beams or plasmas. Drills have also been designed which rely upon thermal spalling of the rock rather than melting. These include jet piercing drills which achieve a temperature of up to 2500 °C by burning oil and oxygen, rocket exhaust drills which burn more exotic fuels, microwave drills operating at about 1 GHz and others which employ induction heating. Chemical drills have been tried but the cost of chemicals has usually made these uneconomic. These concepts have been discussed in reference [15].

Two less revolutionary, but nonetheless very promising techniques, show more potential in the short term. The use at the Camborne School of Mines of a slow-speed, high-torque down-hole motor for directional drilling has allowed very rapid progress to be made at depth. The major development work which needs to be carried out on this device is to ensure survival at much higher bottom-hole temperatures. If this could be done, costs for drilling in granites might be significantly reduced. Secondly, the development at Sandia Laboratories of a shark's tooth retracting rock-cutting system offers great promise.

If any of these advanced drilling methods could be proven and could lead to major reductions in cost, they would further enhance the prospects for geothermal energy and HDR technology in particular.

References

For a full discussion of the history, uses, technology and prospects of geothermal energy from hot water and steam sources the reader is referred to Armstead's comprehensive book [2]. A new and updated edition of this volume is to be published shortly. There are few general reviews on HDR systems. To keep up to date with this rapidly developing field, the reader is advised to apply to Los Alamos Scientific Laboratories and the Camborne School of Mines to receive their regularly produced news sheets.

1	Harris P 1972 *Understanding the Earth* (Horsham: Open University/Artemis Press) ch 3 pp53–69
2	Armstead H C H 1978 *Geothermal Energy* (London: Wiley)
3	Villa F P 1975 Geothermal plants in Italy: their evolution and problems *Proc. UN Symp. on the Dev. and Use of Geothermal Resources, San Francisco* vol 3 p2061
4	*Wairakei: Power from the Earth* 1970 (New Zealand Government Brochure)
5	Finney J P, Miller F J and Mills D B 1972 Geothermal power project of Pacific Gas and Electric Company at the Geysers, California *Proc. IEEE Power Eng. Soc.* (New York: IEEE)
6	Richardson S W and White A A L 1980 *Nature* **286** 103–4
7	Facca G 1973 *Earth Sciences* (UNESCO) vol 12 p61
8	Chierici A 1961 Planning of a geothermoelectric power plant: technical and economic principles *Proc. UN Conf. on New Sources of Energy, Rome* paper G/62 (UN publication ref E/CONF 35/3/4)
9	Banwell C J 1961 Geothermal drillholes: physical investigations *Proc. UN Conf. on New Sources of Energy, Rome* paper G/53 (UN publication ref E/CONF 35/3/4)

10 Einarsson S S, Vides R A and Cuellar G 1975 Disposal of geothermal waste water by reinjection *Proc. UN Symp. on the Development and Use of Geothermal Resources, San Francisco* vol 2 p1349

11 McNitt J R 1973 *Earth Sciences* (UNESCO) vol 12 p33–41

12 Ward P L 1972 *Geothermics* **1** 1

13 Sigvaldason G E 1973 *Earth Sciences* (UNESCO) vol 12 p49

14 Craig S B 1961 Geothermal drilling practices at Wairakei, New Zealand *Proc. UN Conf. on New Sources of Energy, Rome* paper G/14 (UN publication ref E/CONF 35/3/4)

15 Collie M J 1978 *Geothermal Energy—Recent Developments* (New Jersey: Noyes Data Corporation)

16 Garnish J D 1976 Geothermal energy: the case for research in the UK *Energy Paper* 9 (London: HMSO)

17 Patterson S O, Sables B E and Kooharian A 1973 *NTIS Report* AD-774108

18 Milora S L and Tester J W 1976 *Geothermal Energy as a Source of Electric Power* (Cambridge, Mass.: MIT Press)

19 Cortez D H, Holt B and Hutchinson A J L 1973 *Energy Sources* **1** 74

20 Pruce L M 1980 Power generation *Power* July issue 84–7

21 White A A L 1980 *Proc. IEE* A No. 5 127

22 Tozer B A, Turtle J and White A A L 1981 The possibility of direct generation of electricity from geothermal resources in the UK *CEGB Report* No. RD/M/1165 N 81

23 Garnish J D 1980 *Nature* **286** 386

24 Harrowell R V 1977 *Nature* **266** 299–302

25 Khalifa H E, DiPippo R and Kestin J 1978 *J. Soc. Automot. Eng.* 1068–73

26 White A A L 1978 *CEGB Report* No. R/M/N 1031

27 EPRI 1978 Geothermal energy prospects for the next 50 years *EPRI Report* ER-611-SR

28 Bowen R G 1973 in *Geothermal Energy* ed P F Kruger and C Otte (Stanford: Stanford University Press)

29 Axtmann R C 1975 *Science* **187** 4179

30 Reed M J and Campbell G E 1975 Environmental impact of development in the Geysers geothermal field *Proc. 2nd UN Symp. on the Dev. and Uses of Geothermal Resources, San Francisco* (Lawrence Berkeley Laboratory, University of California) vol 2 p1399

31 Hatton J W 1970 Ground subsidence of a geothermal field during exploitation *Proc. UN Symp. on the Dev. and Use of Geothermal Resources, Pisa* vol 2 pt 2 p1294

32 DiPippo R 1980 *EPRI Report* TC-80-909, 7 pp63–7

33 Tammemagi H Y and Wheidon J 1974 *Geophys. J. R. Astron. Soc.* **38** 83–94

34 Ashbee R A 1979 *CEGB Report* No. RD/L/N 205/79
35 Tester J W, Morris G E, Cummings R G and Bivins R L 1979 *Los Alamos Scientific Laboratory Report* No. LA-7219-MS
36 Gringarten A C, Witherspoon P A and Ohnishi Y 1975 *J. Geophys. Res.* **80** 1120
37 McFarland R D and Murphy M D 1976 Extracting energy from hydraulically fractured geothermal reservoirs *Proc. 11th Intersoc. Energy Conv. Eng. Conf., State Line, Nevada* (New York: American Institute of Chemical Engineers) vol 1 pp828–35
38 Wunder R and Murphy H D 1978 *Los Alamos Scientific Laboratory Report* No. LA-7219-MS
39 Los Alamos Scientific Laboratory 1978 *Hot Dry Rock Geothermal Energy Development Program, Annual Report* LA-7807-HDR
40 Los Alamos Scientific Laboratory 1980 *Hot Dry Rock Geothermal Energy Development Program, Annual Report* LA-8855-HDR
41 Denis B R and Horton E H 1980 *ISA Trans.* **19** 49–58
42 Pearson C M 1982 *Int. J. Ambient Energ.* **3** 19–26
43 Commission of the European Communities 1980 *The European* HDR *Case—Where to Go Next?* (Commission of the European Communities)
44 Lewis D J 1979 *CEGB Report* No. RD/B/N4496
45 EPRI 1979 Economic modelling of electricity production from hot dry rock geothermal reservoirs: methodology and analysis *EPRI Report* EA-630
46 Cummings R G, Morris G E, Arundale C J and Erickson E L 1979 Power production from hot rock geothermal resources: a case study for the Imperial Valley, California *US DoE Report* DoE/ET/27017-T2
47 Andrews J G, Richardson S W and White A A L 1980 *CEGB Report* No. RD/M/N1130
48 Altseimer J H 1974 *Los Alamos Scientific Laboratory Report* LA-5689-MS

6

Solar energy

6.1 Introduction

Solar energy may be regarded as the driving force behind several of the renewable forms of energy discussed in previous chapters—wind, wave and OTEC. In its direct form alone, it is a vast resource. It has been calculated that the solar radiation falling on the earth's atmosphere every hour could, if fully exploited, meet the annual energy needs of the world. Even in the UK, where the lack of sunshine, particularly in winter, is notorious, the amount of solar energy received is eighty times the present total primary energy requirement.

It may be exploited in various ways including the direct production of heat for space or water heating, growing of crops which can be subsequently processed to form useful chemicals (the biomass route) and the production of electricity. Although biomass can be burnt and used to generate electrical energy, this is generally regarded as one of the least attractive options when food, oil or gas can be produced from some crops. Biomass will not therefore be discussed in this volume. The interested reader is referred to references [1], [2] and [3] where a full discussion of the various processes and their potential is given. In this chapter, the nature of solar radiation will be discussed and then production of electricity using photovoltaic cells (including the concept of the solar power satellite), photogalvanic systems which can also act as fuel cells, and the direct use of solar heat in a furnace to produce electricity will be reviewed. A short description will also be given of the concept known as the solar pond.

6.2 The characteristics of solar radiation

The rate at which radiant energy falls on a surface is known as the irradiance and the integral of this over time is the irradiation or insolation. Irradiance is measured in W/m^2, irradiation in J/m^2. The irradiance normal to the solar beam just outside the earth's atmosphere is on average

1350 W/m² with a variation of $\pm 3.5\%$ arising from the elliptical form of the earth's orbit about the sun.

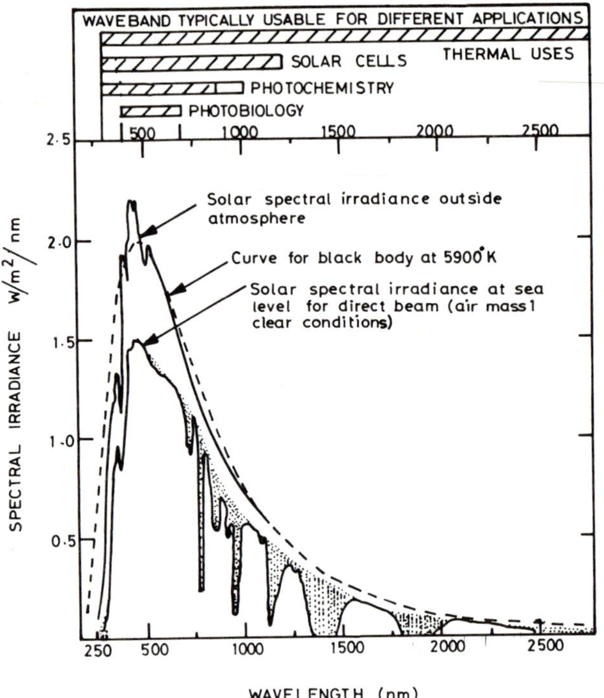

Figure 6.1 Spectral irradiance curves for direct sunlight extraterrestrially and at sea level. Shaded areas indicate absorption due to atmospheric constituents, mainly H_2O, CO_2 and O_3. Wavelengths potentially utilised in different solar energy applications are indicated at the top. (From reference [3] courtesy International Solar Energy Society.)

The solar spectrum is similar to that which would be obtained from a black body at about 5800 K and the spectral irradiance (incident energy per unit time and area within a particular range of wavelengths) is compared with such a black body in figure 6.1. The irradiance at the earth's surface, however, is strongly modified by the atmosphere even on clear days. The spectral irradiance at the surface depends upon the path length through the atmosphere and on the content of the atmosphere through which the radiation is passing. Vapour molecules and man-made aerosols such as smoke can have a significant effect, but even in perfectly cloud-free and unpolluted conditions, air molecules, naturally occurring gases and water vapour scatter or absorb up to 30% of the incident radiant energy. Carbon dioxide absorbs strongly in the mid-infra-red, ozone in the ultraviolet and water vapour in the near infra-red. Most of the scattering of incident energy occurs at the blue end of the spectrum (accounting for the colour of the sky) and this scattered component gives rise to a diffuse background which even on a clear day can account for up to 20% of the total irradiance at ground level at midday. It can be seen from figure 6.1 that a large part of the incident energy at the earth's surface lies between 300 and 2500 nm and about 50% of this lies between 400 and 700 nm (the range of the human eye). Also shown are the ranges of wavelength which are typically utilised in various applications.

The presence of clouds has a major effect on the amount of radiation reaching the earth's surface, since they scatter sunlight strongly in both upward and downward directions. The proportion of direct and diffuse energy is clearly strongly affected by cloud presence. In the UK at midday in summer on a clear day, the irradiance on a receiver placed normal to the sun's direction can be as high as 750 W/m^2 (i.e. more than 60% of the value outside the atmosphere) although it is more normally about 500 W/m^2. When thin cloud is present it usually lies in the range 200–400 W/m^2. In summer in the UK, diffuse sunlight is thought to account for about 50% of the irradiance. However, in winter when there are few clear days, diffuse radiation, mainly from clouds, accounts for a very high proportion of the total irradiation (about 65%). These relative proportions are important, because diffuse radiation cannot be focused. It can be used in most photovoltaic and photochemical systems but cannot be utilised in solar furnaces and other focusing devices. Its angular variation is also fairly isotropic being best collected on a horizontal surface, so collecting surfaces which are tilted to capture more direct radiation fail to collect as much of the diffuse component.

Although not significant in the present context, it should be noted that a substantial amount of radiation is received at the earth's surface re-radiated from water vapour and carbon dioxide. This lies mainly in the range 4×10^3 to 120×10^3 nm and is present during both day and night. Its intensity lies within the range 200–300 W/m^2 and although its heating effect is small (source temperature about 280 K on average), it has the effect of reducing the rate of cooling of the earth at night.

The spectrum of light incident on the earth's outer atmosphere is sometimes referred to as the AM0 (air mass zero) spectrum. Because of the various loss processes referred to above, this is different to the AM1 (air mass one) spectrum which is the theoretical spectrum at ground level on a clear day with the sun at the zenith (1000 W/m^2).

6.3 Radiation data for the UK and other countries

The horizontal surface irradiation for three meteorological stations in the UK is shown schematically in figure 6.2. There is a clear reduction in insolation on moving from south to north and higher levels are measured in the west of the country. This is further illustrated in the maps of average solar radiation per day shown for December and June in figure 6.3. The most significant feature of both figures, however, is the very strong seasonal variation in the available energy. This illustrates the fact that although the UK summer irradiance is not dramatically lower than that of many equatorial areas, the winter level most certainly is. This is illustrated in table 6.1 below adapted from a similar table in reference [4] which shows that the ratio of insolation on a horizontal surface (in MJ/m^2 per day) between midsummer and midwinter in the UK is more than five times that of some other countries.

Table 6.1 Insolation (MJ/m^2/day) in various countries at different times of the year.

	Mean	Summer	Winter	Ratio Summer: Winter
UK	9	18	1.7	11
Central USA	19	26	11	2.4
South France	15	24	5	4.8
Israel	22	31	11	2.9
Australia	20	23	13	1.8
Japan	13	17	7	2.4

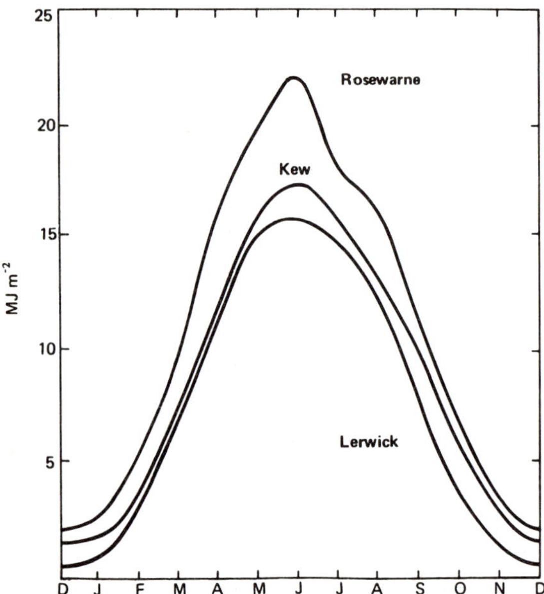

Figure 6.2 Annual variation of mean daily totals of direct and diffuse insolation on a horizontal surface at Rosewarne (Cornwall), Kew (London) and Lerwick (Orkney). (From reference [4] courtesy HMSO.)

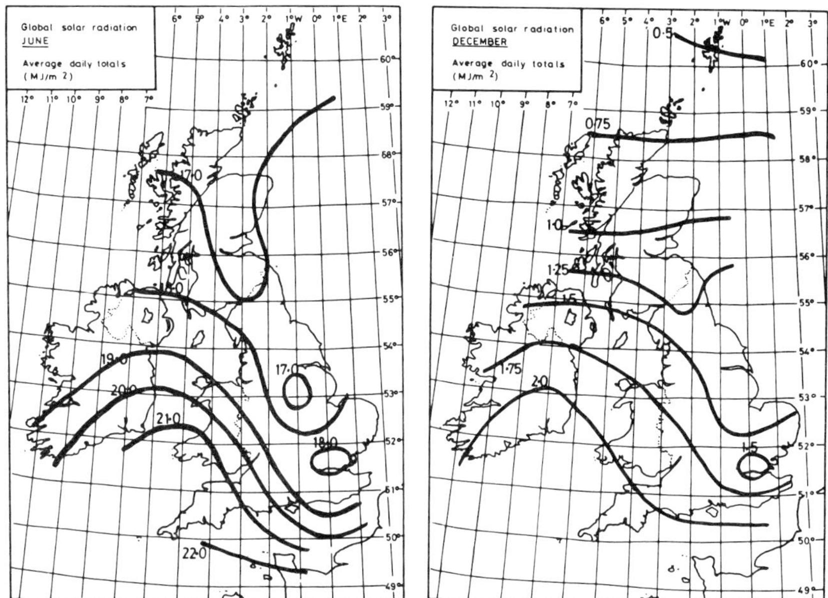

Figure 6.3 Monthly mean global irradiation for June and December (MJ/m^2). (From reference [3] courtesy International Solar Energy Society.)

The highest irradiances on a world basis are found in the desert areas just north and south of the equator. The highest annual mean (horizontal surface) value is about 300 W/m^2 around the Red Sea compared with a value of about 100 W/m^2 in the UK. The world variation is shown in figure 6.4. Regions of lower value near the equator are those where the movement of global air masses leads to relatively cloudy conditions.

The diurnal variation of the irradiation at Kew in June and December for average days is shown in figure 6.5. If the same plots are drawn for cloudless days in June and dull days in December, the variation is nearly two orders of magnitude. There is, of course, very great variability from one day to another and detailed analysis shows that this variation is very much greater in winter. These daily figures average the data over a period of time. There will also be, on days of some cloud, very considerable short term variation.

Most of the data discussed here are for irradiation on a horizontal plane. Inclined planes will receive radiation directly from the sun, diffusely from the sky and as reflected radiation from the ground and surroundings. It is very difficult to optimise the angle for all seasons, weather conditions and locations. Clearly, the maximum energy is captured if the collector follows the sun on clear days and remains horizonal in cloudy conditions, but such tracking systems are too expensive for most applications. The direction in which the surface is pointing and the time of the year are both important. For example in midsummer a horizontal surface collects most energy, followed by east- and west- and southeast- and southwest-facing surfaces; from September to April a south-facing vertical surface collects most radiation. It is likely that the best overall compromise for collecting the maximum energy over the whole year is to hold the collector at an angle to the horizontal of about 45° in a southerly direction in UK conditions. The optimum will clearly vary from one latitude to another and with differing proportions of direct and diffuse radiation. A detailed discussion of these points is given in reference [3].

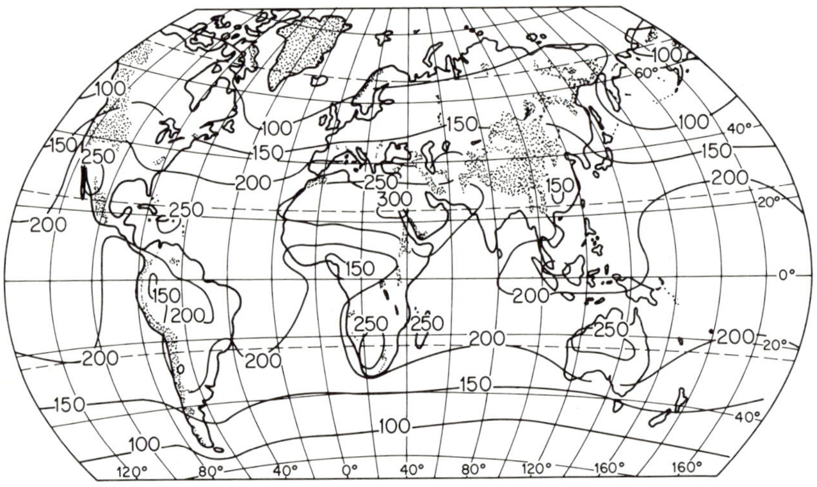

Figure 6.4 Annual mean irradiance on a horizontal plane at the surface of the Earth (W/m² averaged over 24 hours). (Redrawn from reference [3] courtesy International Solar Energy Society.)

6.4 Problems in utilising terrestrial solar energy for centralised electricity generation in the UK

Apart from the generally poor climatic conditions in the UK and the consequent need for even greater reductions in cost than in sunny parts of the world before solar electricity could be competitive with other forms of generation, there are several factors specific to the UK which need to be recognised. The dramatic difference between the irradiation received in summer and winter has already been discussed; this is particularly serious as far as the utilisation of solar energy in the UK is concerned because peak electricity demand in winter is almost twice as high as in summer. It is sometimes suggested that interseasonal storage might be employed to overcome this difficulty, but using current methods this would be prohibitively expensive. In some parts of the world, particularly the USA, air conditioning loads lead to a situation in which demand is greatest at the sunniest time of the year and thus the difficulties inherent in this mismatch between demand and supply do not arise.

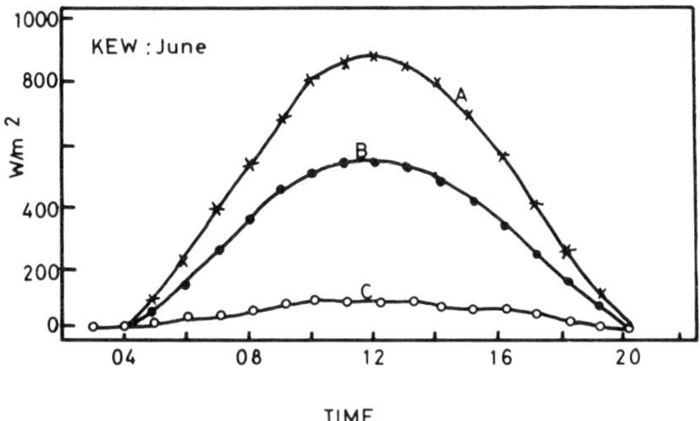

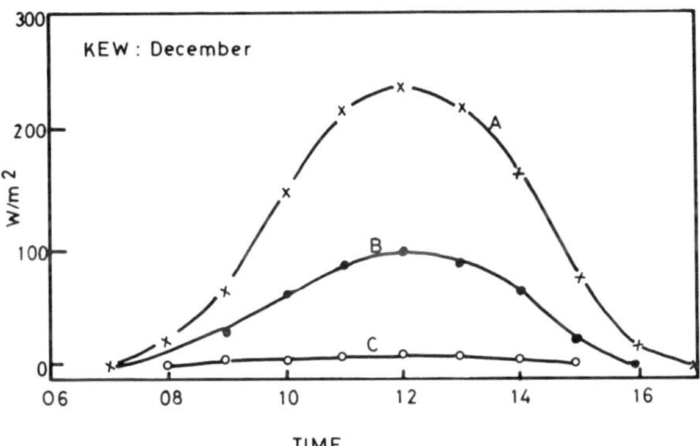

Figure 6.5 Diurnal variation of the mean irradiance on a horizontal surface at Kew for three classes of day for June and December. (a) Average of 2% of days with highest daily totals (b) average of all days and (c) average of 2% of days with lowest daily totals. (From reference [3] courtesy International Solar Energy Society.)

The relatively high proportion of cloudy days giving rise to a high ratio of diffuse to direct radiation has two consequences: firstly, devices which concentrate the solar energy by focusing are unlikely to be attractive in the UK, and secondly the passage of clouds across the sun means that for arrays of collectors in one location, quite sudden variations in electrical output may occur. For very large arrays, or more particularly for collectors dispersed in a number of different geographical locations, the effect will be less serious but it is clear that such variations can occur on a shorter time scale than for wind or wave energy and greater problems of integration into the grid could be presented as a result of this (see §2.12).

Although direct solar conversion is probably one of the most environmentally benign ways of producing electricity, competition for land use in a highly populated country such as the UK would almost certainly present difficulties for centralised, large-scale generation. For a photovoltaic system of 10% efficiency overall, for example, an average power production of 1 GW would require a surface area of 100 km^2 to be entirely covered with collectors and 200–300 km^2 would probably be needed for the total site area. Although this is small compared with the total grassland and cropped area of England, Wales and Scotland (about 225 000 km^2) and arable area (about 65 000 km^2) [4], much of this land in southern England is agriculturally very productive and there may be difficulty in justifying its release for solar electricity production on a large scale. To do so would certainly be costly. Unlike wind energy, use for solar collection and conversion would almost certainly preclude the use of the land for other purposes. Many advocates of solar energy would argue that large-scale generation should not be the primary aim of this technology but that it offers an opportunity to generate at a community level. In this case, land constraints may be less onerous.

Despite these somewhat pessimistic comments about the UK, the use of solar energy for generating electricity elsewhere in the world shows considerable promise. In the rest of this chapter, the available techniques will be discussed and progress reviewed.

6.5 Photovoltaic conversion

6.5.1 Principle of operation of photovoltaic devices

In order to understand the way that a solar cell works, it is necessary to review the physics of semiconductors. In the simplest picture of an atom, the positively charged nucleus has a number of electrons 'orbiting' about it. Quantum theory predicts that electrons are only able to take up certain well-defined orbits corresponding to well-defined energy levels. If energy is absorbed by the atoms—by absorbing a photon, for example—electrons may, in principle, be promoted from one energy level to another. When atoms are joined together to form a solid, the discrete energy levels referred to above are smeared out, and a number of bands of allowable energies are created. The differences between metals, semiconductors and insulators are explained by the size of the gaps between these bands and the number of electrons in them (figure 6.6).

If attention is focused on the two most energetic of these electron bands—the so-called valence and conduction bands—many of the properties of semiconductors can be readily understood. A photon has an energy which is inversely proportional to its wavelength (blue light is more energetic than red light). If a crystalline material has an energy gap, E_g, between its valence band and conduction band (see figure 6.6) which is much greater than the energy of an incident photon, no energy is transferred and the material may be optically transparent. If the energy of the incoming photon is greater than the band gap, some of the energy is absorbed in promoting an electron from the valence to the unoccupied conduction band. In an analogous way, thermal energy is also able to excite electrons from one band to another. In both cases an empty state or 'hole' is left in the valence band. Electrons in the conduction band and holes in the valence band are sufficiently energetic that they will flow as an electrical current when an external voltage is applied to the semiconductor and thus the absorption of light (or the addition of heat) will lead to rapid changes in the electrical conductivity of such materials. In the case of light absorption, this phenomenon is known as photoconductivity.

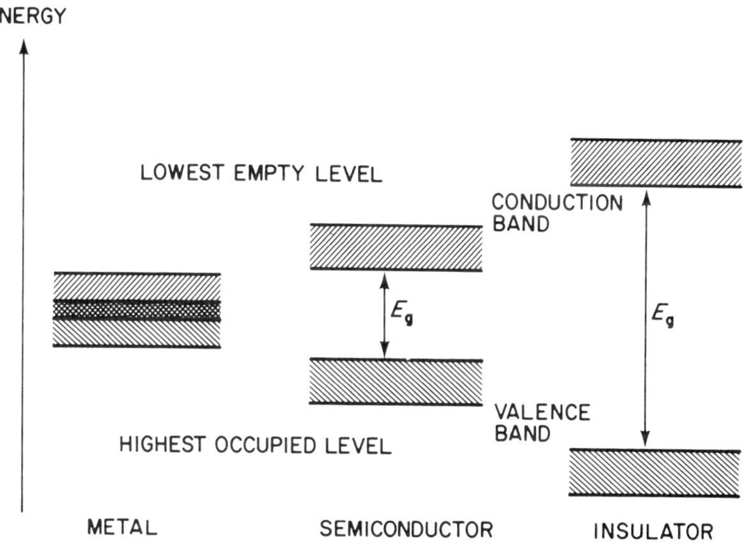

Figure 6.6 Energy bands in a solid at 0 K.

Such semiconductors are referred to as intrinsic semiconductors. The process can be assisted by the addition of suitable impurity atoms and these so-called extrinsic semiconductors are used in solar cells. Silicon, for example, is an intrinsic semiconductor, each atom in its crystal lattice having four valence band electrons and each atoms being bonded to each of its nearest neighbours by 'sharing' these electrons between the atoms. This is illustrated in two dimensions in figure 6.7. If a small number of the silicon atoms are replaced by phosphorus which has five valence band electrons, the electron which is not needed for bonding will be taken into the conduction band and the material will conduct electricity. Such a semiconductor with an excess of negatively charged electrons, is called an n-type material. Conversely, a silicon crystal doped with boron which can only supply three bonding electrons will leave a vacant state or a hole in the valence band. Because it contains an excess of positively charged holes, such an extrinsic semiconductor is termed p-type. In any semiconductor, there is a tendency for holes and electrons to encounter each other and the electron recombines with the hole and is again bound into the structure of the crystal.

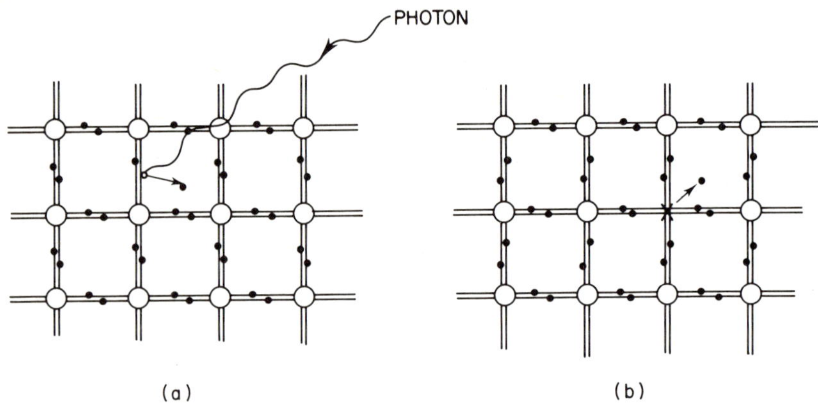

Figure 6.7 Doping in crystalline silicon. (a) In two dimensions silicon atoms (O) form a square array. Each atom shares 2 electrons (●) with each neighbour. When a photon of sufficient energy is available, an electron will be promoted to a higher energy state (conduction band) leaving behind a vacant state or hole (o). (b) n-type silicon is formed by substituting a phosphorous atom for a silicon atom. The P atom (X) has an extra bonding electron which is surplus to requirements. This increases the electrical conductivity of the silicon. When boron is added to silicon the reverse is true; there are then insufficient electrons to satisfy all the bonds and a hole is produced, again increasing the conductivity. In this case the material is p-type.

In an isolated p–n junction, excess electrons in the n-region will tend to diffuse across the junction. Eventually this diffusion will be opposed by ionised donors and acceptors and no further net diffusion will occur. The fixed ionic charges set up an electrostatic field in the so-called depletion region. This is illustrated in figure 6.8(a) and 6.8(b). The physics can be seen most clearly in the energy band representation (figure 6.8(c)). The effect of the depletion region is to align the energy bands in the way shown. Then any free electrons and holes produced by the absorption of photons will be separated by the internal field—electrons moving 'downhill' to the n-side and holes moving 'uphill' to the p-side. Figure 6.9 shows the whole solar cell schematically.

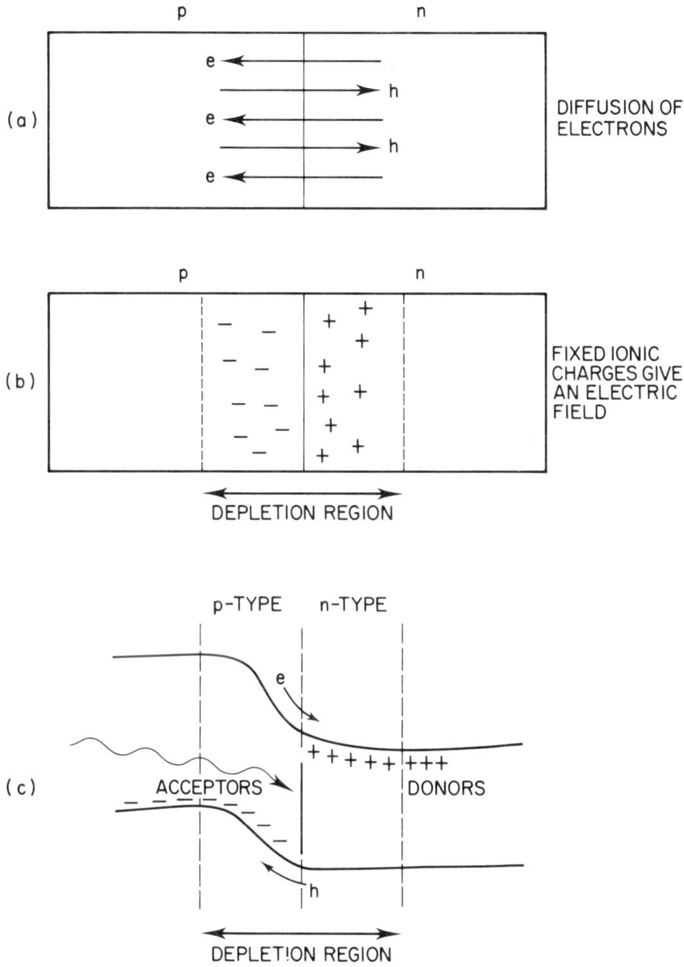

Figure 6.8 An isolated p–n junction. (a) and (b) show an electrostatic field being set up in the so-called depletion zone, (c) shows an energy band representation of the process.

If such cells are illuminated by monochromatic radiation of the correct wavelength, the efficiency for converting the light energy into electricity can be very high. It will be restricted only by such factors as reflection of light from the top surface, masking of the surface by the electrical contact (thus reducing the incident light on the cell), absorption of light in the layer of semiconductor through which it passes to reach the junction and the recombination of holes and electrons before they have been separated by the junction potential [5], [6]. Much research and development has gone into minimising each of these losses. Anti-reflection coatings are applied to the top surface of the cell (although these do not respond with equal efficiency for all wavelengths). Surface contacts in the form of a fine grid of deposited conductor may cover 10% of the surface and are designed to strike a balance between such losses and high resistance for the device.

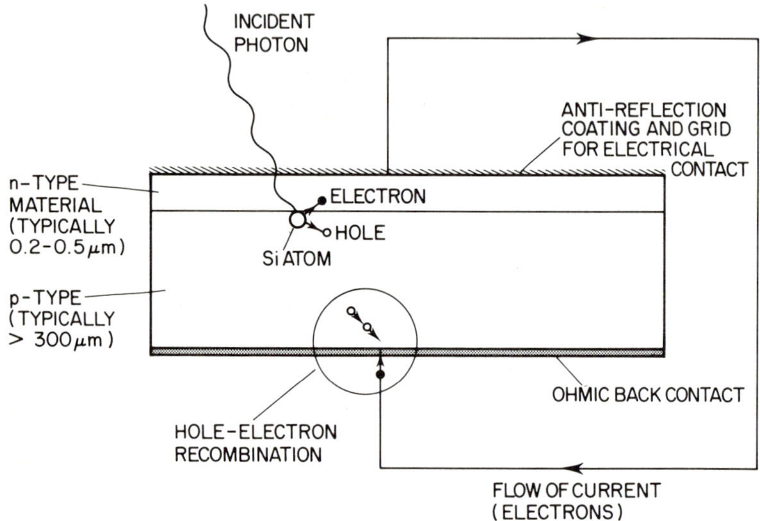

Figure 6.9 Schematic diagram of a silicon solar cell. Photons of the right energy (wavelength) penetrate to the junction where they encounter a silicon atom. An electron is liberated which drifts into the n-type region. The hole formed drifts through the p-type material until recombination takes place with electrons flowing round the external circuit.

The depth of the junction and the level of doping strongly affect the diffusion length of the holes and electrons. This is a measure of the distance that they can expect to travel before recombining. To some degree it depends upon the purity of the material because vacancies, impurity atoms and surface conditions can influence the rate of recombination. However, the doping level cannot be set too low or the open circuit voltage will drop and the series resistance of the device will become too high. A compromise has to be achieved. A similar trade-off exists for optimising the thickness of n-type material to ensure that for the wavelengths of interest, absorption occurs close to the boundary. A layer of semiconductor which is too thick will be wasteful in material and again increase the resistance [5].

The performance of a solar cell, even when illuminated by monochromatic light, is reduced by three further factors. Firstly, the open circuit voltage (typically about 0.6 V for a silicon cell) is always less than the energy gap. Secondly, for a typical current–voltage characteristic (as illustrated in figure 6.10) the maximum power is always less than the product of the short circuit current, I_s, and the open circuit voltage, V_{oc}. Thirdly, power losses occur because of the series resistance in the cell. A measure of these last two factors is offered by the so-called fill-factor which is defined as the ratio of the maximum power to the product of I_s and V_{oc}.

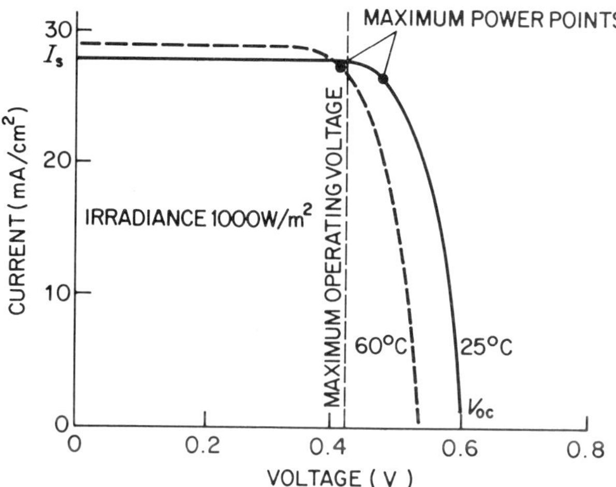

Figure 6.10 Effect of temperature change on performance of a solar cell. (Courtesy Mr F Treble and reproduced from reference [8] with permission of the IEE.)

In practice, of course, solar cells are rarely used to convert monochromatic radiation into electricity. The major loss of efficiency in real conditions arises from the fact that photons with energy less than the band gap are unable to create an electron–hole pair, whilst those with much greater energy tend to waste as heat the excess energy beyond that required to promote the electron across the band gap. For AM0 or AM1 radiation, for example, there is an optimum size of band gap which will maximise the efficiency of the cell. If a material is chosen with a band gap which is too great, a high open circuit voltage can be established at the terminals but relatively few electron–hole pairs will be created. Conversely, if the band gap is too small, high currents but low voltages will result. To achieve maximum power, an optimum band gap for AM1 radiation is believed to be near 1.5 eV. This corresponds to a theoretical efficiency (neglecting the loss processes described above) of about 25%. The usable range of band gaps is in the range 1–2 eV. The position of several of the commonly used solar cell materials on the maximum efficiency curve is shown in figure 6.11. The various loss mechanisms discussed briefly here, have been reviewed in detail by Wolf [7].

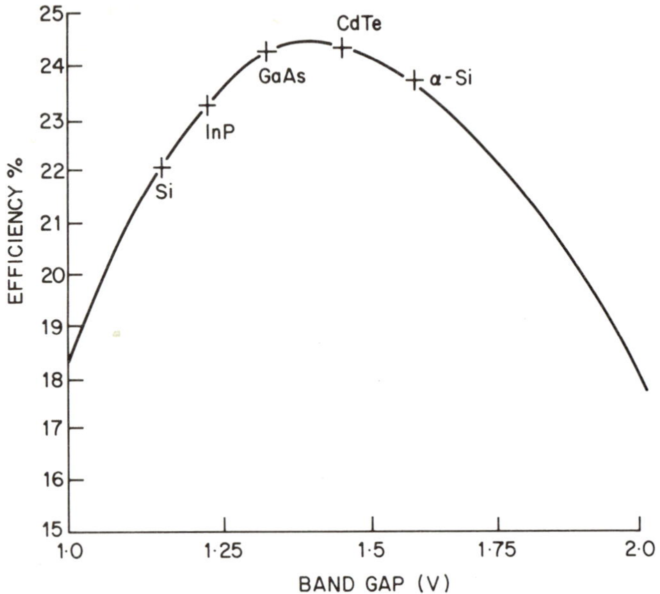

Figure 6.11 Theoretical maximum efficiency of a solar cell as a function of band gap.

The discussion above has assumed that the device is operating at (or near) room temperature. Some semiconductors such as crystalline silicon have conductivities that are fairly strongly temperature dependent and this is reflected in the behaviour of cells fabricated from them. The broken line in figure 6.10 shows the effect of increasing temperature from 25 to 60 °C. There is a small rise in the short circuit current but a distinct drop in the open circuit voltage. Thus the maximum power obtained is reduced. For a typical silicon solar cell, the temperature coefficient for maximum power is about 0.5 percent per degree C. Temperature dependence is very important in designing concentrator cells in which radiation is focused by a trough, dish or Fresnel lens onto a small area of cell. The reason why GaAs is successfully used in concentrator devices is because its temperature coefficient is much lower than for many other semiconductors.

Finally, it should be pointed out that solar cells based on p–n junctions are not the only systems by which a photovoltage can be produced. In the Schottky barrier [5], [8], transfer of electrons takes place from an n-type semiconductor (or holes from a p-type) to a metal layer deposited on it. This is illustrated in figure 6.12(c). On illumination, electrons flow into the semiconductor and holes into the metal thus generating a voltage in an analogous way to that shown in figure 6.8. At present, these devices do not appear to offer as much potential as the conventional p–n junction. Semiconductor–liquid systems will be discussed in the section on photogalvanic devices later in this chapter. Other devices such as p–i–n and MIS-junctions are shown schematically for one semiconducting material (α–Si) in figures 6.12(b) and (d) taken from reference [9]. The advantages of the MIS cell, which it is thought may eventually be competitive with p–n junction technology, have recently been reviewed by Cheek and Mertens [10].

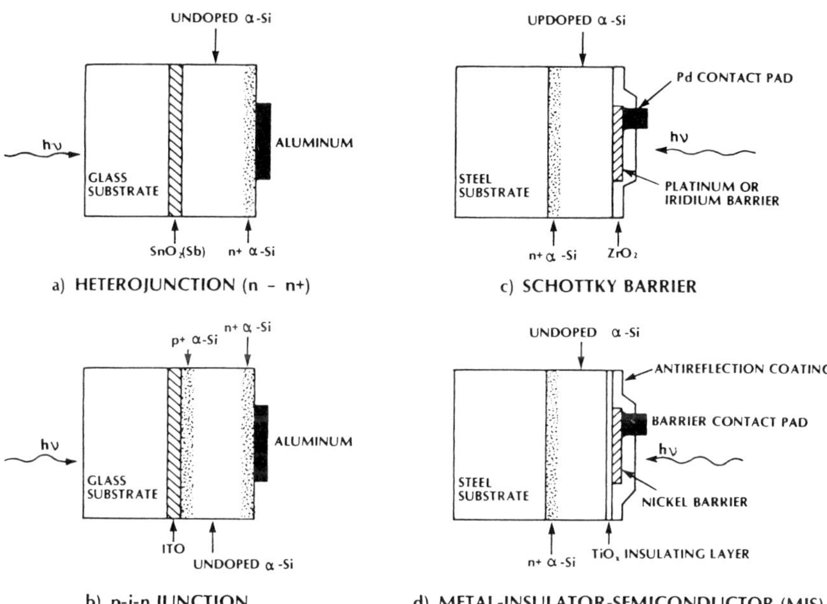

Figure 6.12 Schematic diagram of various amorphous silicon (α-Si) solar cell structures (diagrams not to scale). (Reproduced from reference [9] courtesy Dr H K Charles and *Solar Energy*.)

6.5.2 Types of solar cell

In this section, the technology of various types of cell using crystalline silicon, amorphous silicon, GaAs, copper sulphide and several more exotic alternatives will be discussed. Some of the main parameters and advantages and disadvantages of the cell types are summarised in table 6.2 adapted from a similar table given in reference [11].

Table 6.2 Photovoltaic cell materials options (adapted from reference [11] with permission of the author and the IEE).

Type of solar cell	Silicon			Gallium arsenide Ga Al.As	Cadmium sulphide Cds–Cu$_2$S
	Monocrystal	Polycrystal	Amorphous		
Maximum theoretical efficiency (% AMl)	22%	c. 20%	?15%	24%	12–16%
Typical commercial efficiency (% AMl)	12–14%	6.5%	—	12–19%	8.5%
Achieved laboratory efficiency (% AMl)	18%	10%	5%	23%	8–12%
Band gap (eV)	1.1	0.7–1.0	0.6–0.9	1.4–2.2	1.2–2.4
Advantages	Excellent silicon material availability Established technology		Cost potential	Temperature/ concentration properties Efficiency	Cost potential
Disadvant-ages	Cost	Brittleness restricts area	Efficiency	Material availability Cost	Stability Efficiency Cadmium availability

Crystalline silicon. Solar cells made from this material have been widely used in space applications where reliability and efficiency are of prime importance. Crystalline silicon has a band gap of 1.12 eV and thus has a theoretical limit to efficiency in AM1 illumination of about 22%. In practice about 18% is the upper limit for 'perfect' crystals. The major problems are to find ways of reducing costs so that silicon cells can become competitive for large-scale terrestrial applications.

Crystalline silicon solar cells are in two forms. They are either nearly perfect single crystals or a polysilicon aggregate of much smaller crystallites. Both varieties will now be considered.

The starting point for manufacture is semiconductor grade silicon which is purer and more expensive than might be justified for solar cell technology. This is used, however, because of its wide availability, owing to its use in the electronics industry. It is prepared by the reaction of silica and carbon to give metallurgical grade silicon, formation of SiHCl$_3$ by reaction with HCl followed by fractional distillation and reduction. This is known as the Siemens process. In order to obtain cheaper Si, which would nonetheless be of acceptable quality for solar cells, several other processes are being explored; these include increasing the purity of metallurgical grade silicon by zone refining, the use of gas blowing methods or sodium reduction of fluorosilicate. At the moment, the most favoured methods seem to be the production of silane from metallurgical grade silicon followed by pyrolysis, which is being developed by Union Carbide [12] and the reduction of silicon tetrachloride by zinc vapour (Battelle Columbus Laboratories) [13]. There are, in fact, several processes which could achieve the important goal of reducing the cost of the starting material.

Until recently, commercial collectors have been produced from the Czochralski (Cz) process in which a seed crystal is rotated whilst being removed from a crucible of molten silicon doped with boron. Perfect single p-type crystals up to 150 mm in diameter can be produced at a rate of up to 10 cm/h by this automated (but nonetheless expensive) process. The crystal then needs to be cut into thin wafers (250 microns thick) several hundred at a time. One face of each wafer is made n-type by conventional gaseous diffusion of phosphorous atoms, solid–solid diffusion, glow discharge deposition or ion implantation. Contacts are then applied by evaporation, electrolysis, photolithography or silk-screen processes. Advanced wafering techniques have assumed particular importance since much material can be lost during this part of the process. A good description of how the process may be automated and the potential for cost reductions has been given by Whale [14].

Several other processes which avoid ingot growth and wafering are being developed. These include edge-defined film-fed growth (EFG) pioneered by the Tyco Corporation in the USA, illustrated in figure 6.13 (a) taken from reference [8]. In this process a graphite die rests in a pool of molten silicon which rises through a narrow slit in the die and forms a layer of liquid silicon. A seed crystal is dipped into this liquid and drawn upwards so that the interface between the crystal and the liquid remains just above the die. In this way a ribbon can be formed which can be chopped without significant loss of silicon. Variations on this process in which the ribbon is pulled horizontally or downwards are being studied in Japan and France. Other techniques [15], [16] are being pioneered in the USA by Westinghouse (the dendritic growth process shown in figure 6.13(c)) and in France (the LEP process) [17]. This latter process (figure 6.13(e)) involves pulling a graphite paper (as substrate) down through a bath of molten silicon. Other possible methods are also illustrated in figure 6.13. Another polycrystalline film method has been developed by AEG. Which of the many processes, if any, will result in cost effective production of silicon cells is not yet clear. Progress in the various steps in moving from silicon to the complete cell have been recently reviewed by Charles and Ariotedjo [10], Smith [15] and Van Overstraeten [18].

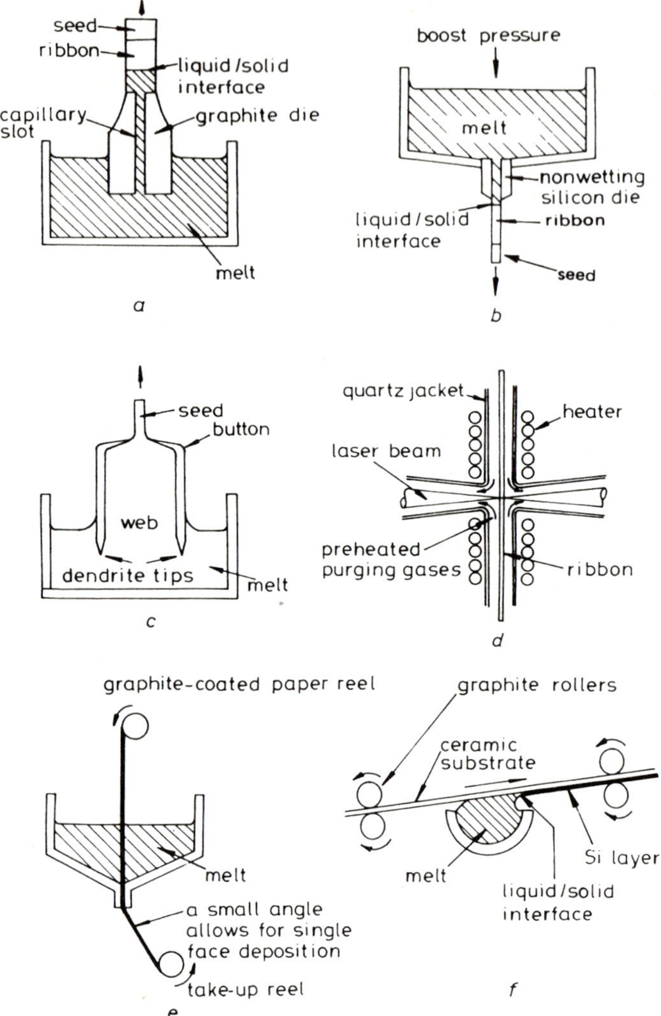

Figure 6.13 Silicon ribbon techniques. (a) EFG and CAST; (b) inverted Stepanov and pendant drop growth; (c) dendritic web; (d) ribbon-to-ribbon; (e) silicon on graphite-coated paper; (f) silicon on ceramic. (Courtesy Mr F C Treble and reproduced from reference [8] with the permission of the IEE.)

These processes frequently produce much poorer crystals than the Cz method. For many years it was believed that the grain boundaries in polycrystalline silicon would radically reduce efficiency. This has not proved to be the case, however. Practical experience seems to suggest that whatever the technique used for production—ribbon processes, fusion, dipping, coating etc—polycrystalline silicon devices are formed with acceptable efficiencies of around 10% as long as grain boundary walls are aligned perpendicular to the layer. In fact, it has been argued that far from degrading the cell by introducing traps and short circuits, grain boundaries segregate impurities from the silicon. It may be that there is a critical surface to area ratio for the crystallites and this is being explored.

Amorphous silicon. Amorphous silicon ($\alpha - Si$) does not have its atoms arranged on a regular crystal lattice but instead they are, like a glass, positioned at random in the solid, like a 'frozen' liquid. It is only in the last four or five years that such materials have been taken seriously as potential rivals to the more conventional systems.

For a long time, it did not seem likely that controlled doping could be achieved in these amorphous semiconductors and it was presumed that addition of a dopant such as phosphorus just provided extra electrons which satisfied the dangling bonds in the material and did not contribute towards the conductivity. However, Spear and his co-workers at the University of Dundee were able to obtain controllable doping of silicon formed by a glow discharge from silane (SiH_4), by incorporating phosphine or diborane in the gas and formed the first p–n junction from amorphous silicon [19]. Solar cells with routine efficiencies of up to 6% have now been produced by RCA Energy Conversion Devices Inc.and Xerox in the USA and by Japanese workers (e.g. [20]). It may be that the material formed is really an alloy of silicon and hydrogen and the hydrogen donates electrons to 'tie up' the dangling bonds in the solid.

The advantage of using amorphous silicon layers to form solar cells is that processes such as sputtering and glow discharge can be automated and used to cover large areas of substrate thinly, thus economising on material. Furthermore the effective band gap for amorphous silicon (the band edges are not sharp for an amorphous material but smeared out) is about $1.6\,eV$. This is therefore close to the optimum shown in figure 6.11. On the other hand the irregular nature of the solid means that resistances are much higher and there are probably many more sites at which recombination of holes and electrons can occur, thus potentially reducing the highest efficiency that can be reached.

Although α-Si solar cells are regarded as being very promising and are indeed being manufactured on production lines for use in watches and calculators, their use as power producing cells for utilities depends in the long term on whether this reduction in the charge-carrier lifetime can be overcome. In Japan, amorphous silicon is now regarded as sufficiently promising for the photovoltaics element of the Japanese 'Sunshine Programme' to be oriented towards this form of cell. In the USA, Energy Conversion Devices Inc. are planning to market rooftop systems based on amorphous silicon formed from silane and silicon fluoride.

A good discussion of both the scientific and technical progress in developing high efficiency solar cells from amorphous silicon has been given by Carlson [21]. Charles and Ariotedjo [9] give a full discussion of devices. Table 6.3 adapted from the latter reference shows the technical parameters for a number of α-Si cells of the types illustrated in figure 6.12.

Table 6.3 Present status of amorphous silicon (α-Si) solar cells (from [9]).

Heterojunction cell	
Cell structure	ITO/α-Si†
Short circuit current density (mA/cm^2)	10.0
Open circuit voltage (V)	0.43
Efficiency (%)	<1.5
Antireflection coating	none
Illumination intensity (mW/cm^2) [(airmass number)]	(AM1)
p-i-n junction cell (figure 6.12(b))	
Cell structure	Pt–SiO$_2$/p-i-n α-Si
Short circuit current density (mA/cm^2)	7.00
Open circuit voltage (V)	0.77
Efficiency (%)	3.30
Antireflection coating	none
Illumination intensity (mW/cm^2) [(airmass number)]	(AM1)
Schottky barrier cell (figure 6.12(c))	
Cell structure	Ir/α-Si
Short circuit current density (mA/cm^2)	14.50
Open circuit voltage (V)	0.75
Efficiency (%)	6.00
Antireflection coating	ZrO$_2$
Illumination intensity (mW/cm^2) [(airmass number)]	(AM1)
Metal–insulator–semiconductor cell (figure 6.12(d))	
Cell structure	Ni/TiO$_x$/α-si
Short circuit current density (mA/cm^2)	7.90
Open circuit voltage (V)	0.68
Efficiency (%)	4.80
Antireflection coating	ZrO$_2$
Illumination intensity (mW/cm^2) [(airmass number)]	(AM1)

† not specified

Gallium arsenide. The advantages of GaAs as a material for forming solar cells are that it has a band gap (1.43 eV) close to optimum, absorbs sunlight in a very thin layer (2–5 μm) and has an ability to perform satisfactorily at elevated temperatures. The disadvantages are that polycrystalline material does not seem to be successful and single-crystal materials are very expensive.

Cells incorporate a p-type layer of an alloy of GaAs with Al on an n-type GaAs substrate [5]. The process currently used to produce nearly all GaAs cells is to form a transparent wide band gap 'window' of (GaAl)As containing some zinc on the substrate, and zinc diffusion forms an intermediate layer of p-type GaAs about a micron thick. The layer of the aluminium alloy is grown by a process known as liquid phase epitaxy on the GaAs single crystal. This heterostructure improves the efficiency of charge collection and reduces the internal resistance of the cell. Cells with efficiencies in AMl light as high as 22% have been achieved for non-production cells. The structure of such a cell is shown schematically in figure 6.14.

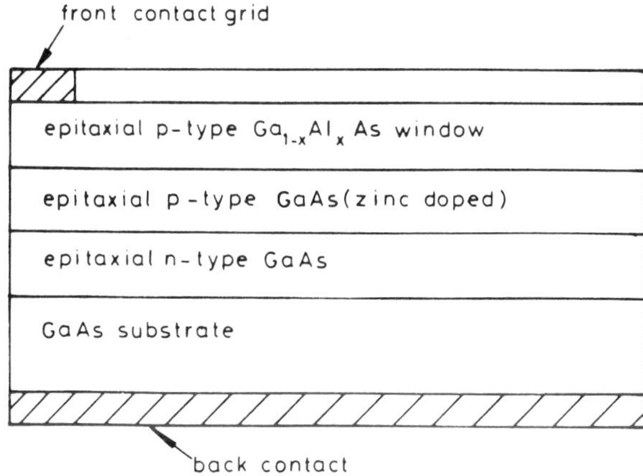

Figure 6.14 GaAs solar cell. (Courtesy Mr F C Treble and reproduced from reference [8] with the permission of the IEE.)

The wider band gap means that the temperature of GaAs can be raised to quite high values without markedly affecting the efficiency (few electrons are thermally excited across the band gap to produce a leakage current). This has given rise to the idea of using a lens to concentrate sunlight on the cell. Such concentrators must, of course, contain a suitable heat sink. Incident radiation intensities of up to 2000 suns have been tried and efficiencies have remained close to 20%. Normally, however, concentrations range from a few hundred down to 50. Concentration allows the size of cell to be reduced but is an extra cost element. The size of the concentrator must be the same as a normal solar array in order to collect the same energy and, because focusing is only possible for direct sunlight, steering towards the sun is also required. A review of concentrator technology has recently been given by Boes and Shafer [22].

Despite the increasing use of concentrator, multi-junctions or tandem cells (in which light is split by a dichroic mirror or prism such that different wavelengths fall on different cells of different band gap) and the high efficiencies (approaching 30%) achieved by such complex systems, the use of GaAs seems likely to be confined to specialist applications unless major innovations can automate production and reduce costs of materials and processing. GaAs cells have been used in hybrid thermal-photovoltaic systems [23].

An excellent review of some of the exotic systems which have been developed using GaAs technology is given by Wilson [24].

Cuprous sulphide–cadmium sulphide. This lightweight solar cell is based on a heterojunction between n-type CdS and p-type Cu_2S giving an energy gap of 1.2 eV. Fabrication relies on thin film techniques and single crystals are not used. The cells are among the least understood scientifically and much progress has been empirical.

The methods most usually employed for forming these cells are evaporation, sputtering and spraying. CdS can be evaporated onto a hot Zn-coated copper or molybdenum substrate from a quartz or carbon crucible at 1000 °C, it can be sputtered from a CdS target to form thinner and more uniform films or sprayed onto a substrate at about 350 °C [25]. In the spray method aluminium chloride is usually added to the spray solution in order to achieve doping and evaporation is usually performed in the presence of some cadmium halide. The Cu_2S layer is added either by a reaction of a copper salt with the surface of the CdS involving dipping in cuprous chloride or by direct deposition by evaporation, sputtering or spraying.

The maximum efficiency for these cells, achieved over many years of development, is about 10%. Production cells have been formed with efficiencies of 6–7%. Unfortunately, these cells suffer significant degradation and encapsulation in non-oxidising conditions is essential. The main forms of deterioration seem to be oxidation of the cuprous sulphide and a photosensitive electronic degradation (reversible in the dark) reducing voltage and electrochemical degradation. This is probably due to decomposition of the cuprous sulphide and subsequent diffusion of copper ions into the cells.

Recently there has been interest in using a Zn–CdS base layer and forming the cells using solid state reactions to give more uniform Cu_2S layers, albeit more slowly. The relatively low efficiency and degradation of the cells are likely to continue to be major reasons why these solar cells find difficulty in competing commercially with silicon. The technology and performance of these cells has been reviewed by Hill and Williams [26].

Other types of cell. Indium phosphide is similar in many ways to GaAs but some prospects exist for the use of thin-film technologies to reduce cost. Progress does not seem, however, to have been very rapid in substantiating these hopes. Ternary systems such as $CuInSe_2$–CdS have also been proposed and these seem to show considerable promise, with efficiencies of nearly 10% [27].

Another approach being tried is to combine a narrow band gap p-type material (e.g. Si, GaAs, InP or CdTe) as an absorber layer, with a wide band gap window layer formed from n-type materials such as CdS, ZnCdS, ZnSe, ZnO or indium tin oxide (ITO) alloys. This latter material combined with CdTe looks the most promising, having achieved efficiencies approaching 10%.

Other amorphous materials have also been the subject of claims as yet unsubstantiated. It has been suggested that amorphous semiconductors may be formed by 'modifying' glasses such as SiO_2, Si_3N_4, As_2Se_3 and the so-called STAG glasses (glasses made from selenium, tellurium, arsenic and germanium) with up to 20% of transition metals such as tungsten and nickel [28]. Little scientific evidence is available about the band structure of such materials or about the lifetimes and mechanisms for recombination of electrons and holes. It is not clear how seriously these ideas are now being pursued.

One proposed method for concentrating solar radiation into photovoltaic cells is by the use of fluorescent collectors [29], [30]. The idea is to absorb the incident radiation in a thin slab of plastic containing a fluorescent dye. The dye is chosen to emit radiation matching the band gap of the photovoltaic cell being used and the emitted radiation is channelled by total internal reflections within the plastic slab to a band of cell material along the edge of the slab. The major problems involved in this method are:

(1) The difficulty of synthesising dyes with all the required properties in the right wavelength region, including fluorescent emission matched to the solar cell's band gap.

(2) The identification of inexpensive plastic materials with high transparency, particularly in the near infra-red regions, as carriers of the dyes.

High efficiencies can in theory be obtained by forming multi-junctions of cells with different band gaps which utilise different parts of the spectrum with high efficiency [5]. These are technically quite difficult to produce and prototype devices have only recently been built. Another method in which sunlight is concentrated to heat a radiator and the infra-red radiation which is emitted is absorbed by a silicon cell has been claimed to achieve efficiencies of nearly 30% [31]. This is illustrated in figure 6.15. It is not yet clear whether multi-junctions or thermo-photovoltaic devices will be economically attractive, except for specialist applications where efficiency is of paramount importance.

Finally, a long-standing dream has been to develop an organic cell of higher efficiency than nature's own (the chloroplast). Various dyes and complex organic molecules have been tried and some have achieved efficiencies of 0.1% or more [32]. However, it is evident that there is a very long way to go before a successful cell of this type emerges.

A number of advanced solar cell concepts have recently been reviewed by Feucht [30].

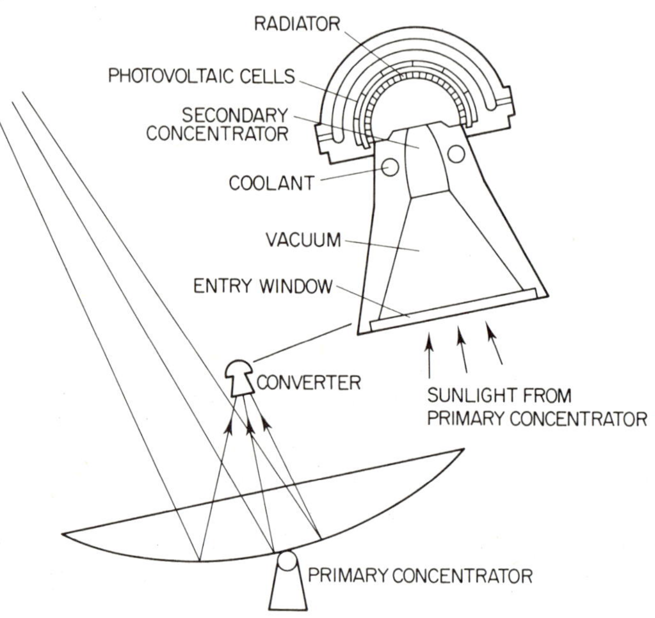

Figure 6.15 Possible configuration of a complete thermo-photovoltaic (TPV) system, in which two stages of sunlight concentration would be employed and (shown enlarged) a converter subsystem, in which the concentrated sunlight would heat a radiator. Visible light would thus be changed to infra-red energy, absorbed by the single-crystal silicon cell and converted to electricity. Because the cell is more responsive to infra-red energy than it is to visible light, conversion efficiencies for TPV cells could be higher than those for conventional photovoltaic arrays. (Courtesy Electric Power Research Institute.)

6.5.3 *Photovoltaic systems*

It is sometimes forgotten that producing the photovoltaic cell is only part of the story in developing a complete system or array. The cells have to be protected against environmental damage. In particular moisture, corrosion, abrasion, ice, temperature changes, deposition of dust and other debris may lower efficiency or cause failure. Some of these potential problems are overcome by encapsulating the cell in plastic or glass and this produces a considerable extra element in the cost of photovoltaic energy.

Individual cells must be linked together in the form of a module in order to give a worthwhile output for power applications. A single silicon cell in bright sunlight (1000 W/m^2) gives an open circuit voltage of about 0.6 V and a short circuit current of up to 350 A/m^2. Individual cells must thus be connected in series and parallel in order to match the source with the load and provide the desired current and voltage relationship. A typical array of solar cells is shown in figure 6.16.

Figure 6.16 60 kW photovoltaic array at Mount Laguna, USA. (Courtesy Mr R D W Scott, BP Solar Systems.)

Table 6.4 Approximate percentage breakdown of costs for a photovoltaic installation (1979).

Panel costs	Encapsulation	30%
	Silicon	20%
	Manufacture	16%
	Sundries, profits etc	34%
Total system	Panel	50%
	Batteries (24h)	3%
	Civil engineering	10%
	Converter	7%
	Installation	5%
	Transport	7%
	Sundries, profits etc	11%

Forming an array of such modules involves further costs and 50% or more of the system costs for small systems is accounted for by the array support, wiring and power conditioning, civil engineering, transport, land costs and other sundries. In particular, the output from photovoltaic cells is DC whilst most utilities generate AC. Efforts are now being made to reduce the costs of the array subsystem [33]. Table 6.4 shows a cost breakdown of a typical silicon cell array. Treble [8] has given an excellent account of the various designs and other factors which must be considered in forming modules and arrays.

6.5.4 *Future prospects and economics*

It has been estimated that in 1979, world spending on photovoltaics amounted to $200M of which about $120M came from the US DoE. This is likely to be cut-back substantially by the Reagan administration in the US in the early 1980s but it is believed that the growing participation of industry will offset some of the effects of this cut in funding.

The long term goal of the US government has been to ensure that photovoltaic systems supply a 'significant' amount of electrical energy to the nation by 2000 and President Carter set a target of a 20% share. The US DoE is attempting to reduce costs for the solar cell module from the 1980 figure of about $6–10 per peak watt to $2.80 in 1982, $0.70 in 1986 and $0.15–0.5 in 1990. This is illustrated in figure 6.17. These module costs imply system costs of about $8 per peak watt in 1982, $1.6–2.2 by 1986 and about $1.2 in 1990. At these levels it is believed that the market for solar cells will be sustained, moving progressively from the higher cost, low volume, remote applications currently being considered, through intermediate size and cost applications to the high volume utility market in the 1990s. In the late 1970s and early 1980s a $98M federal photovoltaics utilisation programme (FPUP) was set up to provide 'lift-off' for the US market and $1.5B was authorised for commercialisation. At the time of writing (1982) it seems likely that these programmes will not be pursued by the Reagan administration. Detailed targets have been set in the USA for overcoming the technical problems associated with the leading cell types. These include reducing the cost of production of polysilicon material and of the formation of large crystalline sheets as well as seeking cheaper methods of cell encapsulation and module fabrication. In 1981, projects were underway to construct a 100 kW installation in Utah, 240 kW and 200 kW schemes in Mississippi, a 600 kW system in Washington D.C. and a 350 kW installation in Saudi Arabia. A full description of the US federal programme has been given by Forney [34] and in a US DOE publication [35].

Government funded programmes in Europe are substantial [17], [36] but are dwarfed by the US effort. Germany has been spending about $12M p.a., France about $8M in a rapidly expanding programme and the EEC has been spending $5M p.a. on R and D. In Japan, a major programme, mainly on amorphous silicon technology, is being pursued. About £1.5M of government money was spent in 1980 with considerable industrial funding. Programmes elsewhere, including the UK, are very modest.

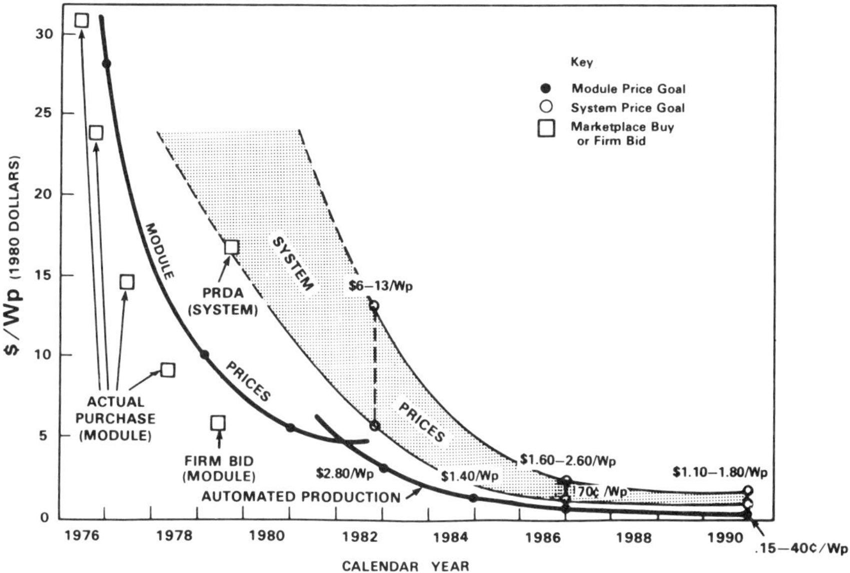

Figure 6.17 Recent and projected solar photovoltaic equipment prices. (Courtesy US Department of Energy.)

Industrial interest and involvement has been considerable [17], Solarex (minority interest held by AMOCO), Solar Power Inc. (involved with Exxon), Sensor Technology, Atlantic Richfield and Motorola are the major companies. Smaller companies such as Tyco Corporation who developed the edge defined growth method, Solar Energy Systems and Photon Power are linked with other major oil companies (Mobil, Shell and Total respectively). Most of these companies are concentrating their efforts on silicon in its various forms but Photon Power and Solar Energy Systems are developing CdS cells. In Europe, Radiotechnique in France, Lucas BP Solar Systems Ltd and Solapak in the UK and AEG Telefunken in Germany are among the leaders. Several American companies including Solarex, Solar Power and Sensor Technology have been developing commercial links in Europe, particularly in France. In Japan, Sanyo, Mitsubishi and Fuji are now routinely producing amorphous silicon solar cells for calculators.

The world development of photovoltaics in 1980 was hindered by a shortage of photovoltaic modules and of silicon wafers in particular. For some years world-wide production doubled, leading to an estimated output sufficient to supply about 2 MW in 1979. The result of this was that the encouraging reduction in prices did not continue to fall markedly at the end of the 1970s. Some sources, however, believe that the US targets are still being met. It would appear that finding applications for small stations (up to 50–100 W) will not prove difficult. The transition to medium (up to several kilowatts) and large stations may be much more difficult.

Success will depend primarily on whether the US cost targets can be met. In an independent study by the American Physical Society, who looked at the ultimate cost breakdown of a large automated production line for various silicon technologies on optimistic assumptions, it was concluded that costs as low as $0.50 per peak watt are doubtful. Durand [17] believes, as do other experts, that $1–2 per peak watt may be more realistic as an attainable goal by 1990.

6.6 Photogalvanic systems

These systems (sometimes referred to as photoelectrochemical systems) are similar in many respects to photovoltaics but instead of using a junction between two semiconductors, electricity (or chemicals) is produced by the irradiation of an electrode/electrolyte system. These cells are still at the research stage and so far only low efficiencies have been achieved. Their particular attraction is their ability in some cases to store energy as a chemical product which regenerates the reactants on subsequent conversion to electricity.

6.6.1 Electrolyte cell

In this type of cell a thin layer of electrolyte is held between a transparent electrode (e.g. doped tin oxide) and a metallic counter electrode (figure 6.18). The electrolyte contains two redox couples A, B and Y, Z where A and Y are electron donating (reducing) molecules and B and Z are electron accepting (oxidising) molecules.

If the free energy is negative, the redox reaction (1) proceeds spontaneously [37]

$$Y + B \rightarrow Z + A. \tag{1}$$

The reaction occurs at separate electrodes as illustrated in figure 6.18(b) and the cell discharges until reaction (1) has reached equilibrium. This is normally well to the right-hand side of the equation.

A is chosen to be a dye with an excited state A^* and on irradiation the photochemical reactions

$$A \xrightarrow{h\nu} A^*$$
$$A^* + Z \rightarrow B + Y \tag{2}$$

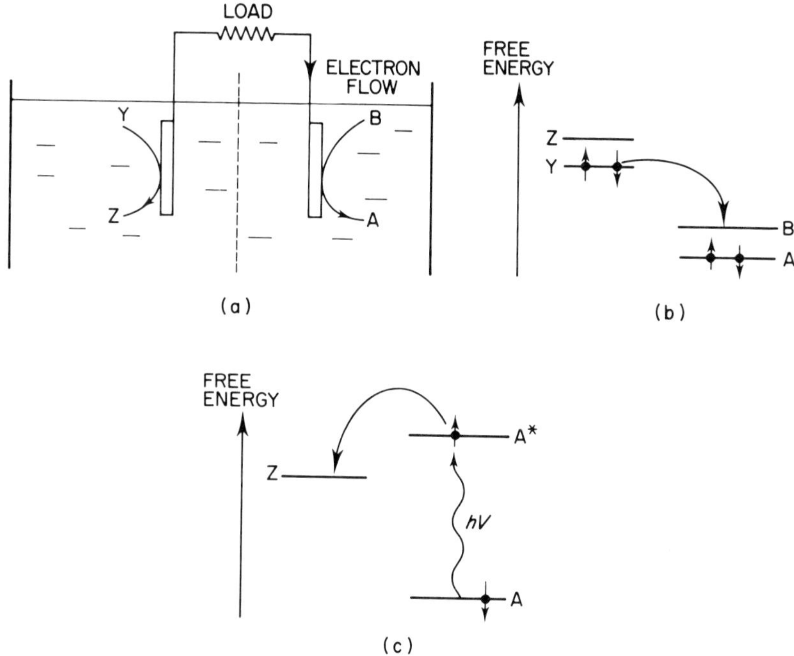

Figure 6.18 Schematic diagram of a photochemically rechargeable storage battery. (a) Electrolyte cell; (b) cell discharge—thermodynamically spontaneous and producing electric power; (c) cell recharge.

take place. The net reactions at the electrodes can thus be summarised as

$$\text{transparent electrode} \qquad B \pm e \rightarrow A$$
$$\text{counter electrode} \qquad Y \rightarrow Z + e.$$

Provided the products Y and B of equation (2) can be separated before they back-react and can be replaced in the left- and right-hand compartments of the cell, photochemical recharging of the battery may be accomplished as shown in figure 6.18(c).

Theoretical calculations [37] suggest that maximum efficiencies of about 18% might be achievable with power outputs of 100 W/m^2. In order to achieve these efficiencies the dye (A) (normally Ru(bpy)$_3^{2+}$, where (bpy) is 2,2′ bipyridene) needs to be rather soluble, the transparent electrode must select the A, B redox couple rather than the Y, Z couple and the rate of the reaction should be slower than the time required for the various species to diffuse to the electrodes. The separation of the excited electron from its associated hole must also be controlled through reaction rates. No chemical system has yet been found which satisfies all these requirements.

6.6.2 *Semiconductor cell*

In this type of cell, the dye A is removed so that light incident on the transparent electrode passes through the electrolyte to be absorbed at a semiconducting counter electrode. This is normally an n-type semiconductor. It should be chemically stable in the electrolyte, have a band gap which is as close as possible to the peak in the solar spectrum and a small electron affinity. As in a conventional Schottky barrier the interface is characterised by a depletion layer at the surface of the semiconductor. Photons of energy greater than the band gap are absorbed in the depletion layer and the electrons and holes created are separated by the internal electric field. Holes drift to the anode surface and recombine at surface states with electrons from the reduced member of a redox couple in the electrolyte. This process can alternatively be viewed as charge injection at the junction followed by oxidation of the lower valence state ion in the redox couple. At the transparent electrode, the reverse reaction takes place with reduction of the higher valence state ion in the electrolyte. The circuit is completed by diffusion of the reduced and oxidised species along the concentration gradients produced by the electrode reactions. To improve performance, a small external voltage is sometimes applied to the cell. The mechanism in these devices is thus similar to that in conventional photovoltaic cells but the photoelectrochemical device has the inherent advantage that the semiconducting material may be polycrystalline or amorphous without significantly impairing the efficiency.

In most cases studied to date, the photocurrent is accompanied by chemical reactions at the electrode. In the case of n-type semiconductors, the reaction is the dissolution of metal ions into solution. For p-type materials, oxidation occurs leading to passivation of the electrode. Various methods have been tried to overcome this problem. One method is to coat the surface with a protective film which does not inhibit the photochemical process, another is to use a process known as spectral sensitisation. This technique involves the addition of dye molecules to the anode surface. It is claimed that by absorbing the incident radiation in the dye, whose energy levels are fine-tuned to the solar spectrum, electron transfer can take place from the dye to the semiconductor. A wide band gap, chemically stable, semiconducting electrode such as TiO_2 or $SrTiO_3$ can then be used without incurring a loss in efficiency due to the mismatch between the semiconductor band gap and the peak intensity in the solar spectrum (TiO_2 is sensitive only to photons of wavelength less than 400 nm, where the solar energy content is less than 4%). Ellis *et al* [38], have suggested using an electrolyte of mixed NaOH and Na_2S in conjunction with a CdS electrode. In this cell, the electrolyte itself is reduced at the cathode and oxidised at the anode resulting in no net chemical change in the system. Efficiencies of up to 9% have been claimed for this cell using etched photoelectrodes.

Although not strictly related to the direct production of electricity, it may be of interest to note that in certain versions of the semiconductor cell, photoelectrolysis can be used to incorporate energy storage by the production of hydrogen. Fujishima and Honda [39] used a single-crystal anode of n-type TiO_2 with a Pt counter-electrode to electrolyse water. In order to split the water molecule into its constituents, a voltage of 1.23 eV is required and practical cells have to employ a semiconductor with a band gap in excess of about 2.2 eV in order to avoid irreversible decomposition of the electrode. The process is thus, again, inherently inefficient because of the mismatch between this value and the energy distribution in the solar spectrum and maximum theroretical efficiencies of about 20% have been calculated. Efficiencies an order of magnitude lower than this have been achieved to date. Much interest has also recently focused on the photo-chemical cleavage of water by photocatalysis. Particularly promising is the work of Gratzel and his co-workers at the Federal Polytechnic in Lausanne [40] who have found that by adding ruthenium to titanium dioxide particles (about 20 nm in diameter) and adding a suitable sensitiser, visible light can be used to split water into hydrogen and oxygen without the application of an external voltage. Much research still needs to be carried out to perfect such systems but they show great potential.

6.7 Satellite solar power stations

The idea of collecting solar energy in space and returning it to earth using a microwave beam was first suggested in 1968 by Glaser [41]. Although regarded by some as closer to science fiction than engineering reality, such an idea has its attractions. Firstly, the full solar irradiation of 1350 W/m² would be available at the collector at all times (except when the sun is eclipsed by the earth) and thus about five times the energy could be collected compared with the best terrestrial sites. Secondly, power could, in principle, be directed to any point on the earth's surface. Thirdly, the zero gravity and high vacuum conditions in space would allow much lighter, lower maintenance structures and collectors. Considerable interest has been shown (mainly in the USA) in the idea and it has now been analysed in some depth. A team from Arthur D Little, Raytheon, Grumman Aerospace and Textron Inc. produced one fairly detailed design and costing in 1976 and several different schemes have also been assessed. An excellent review has been given by Collins [42].

The basic concept typically involves the construction in geosynchronous orbit (35 680 km above the earth's surface) of two 6 km by 5 km aluminium girder structures which would support a complex of silicon solar cells. The DC output would be converted to microwaves and transmitted back to earth at 2.45 GHz from a central antenna about 1 km in diameter. The beam would be reconverted to electricity at a 100 km² receiving station of rectennas and the output of a central facility of this size could be up to 5 GW. The overall efficiency has been estimated to be 5–10%. Other schemes have suggested larger collector areas leading to a peak output of up to 10 GW. Designs have also been suggested which employ thermal electric cycles (Brayton cycle) [43].

In technical terms, there is no one part of the satellite solar power station (SPS) concept which could be singled out as being unrealistic. At the moment the microwave transmission system using klystrons or amplitrons is probably the most technologically demanding component. To ensure high conversion efficiency (about 60%) the beam must be coherent over the area of the transmitter and it has been suggested that this might be achieved by actively aligning and keeping in phase by electronic phase shifters—slotted waveguide arrays up to 20 m square. Control would be achieved by reference beams from the rectenna which would maintain the pin-point accuracy required for hitting the terrestrial target. Vast power transfers would need to be achieved from the collectors through a rotary joint to the antenna. For such massive injections of power only very large and strong utility grids could be expected to absorb the rapid changes in output (5 GW in seconds) which will occur at eclipses or as a result of system failures.

Constructing and maintaining an SPS in orbit presents daunting problems. The transportation of the constructional materials using future generations of space shuttle not only requires a continuing space programme to develop these materials, but the number of launches required could lead to significant pollution of the upper atmosphere from exhaust propellants. The major environmental constraint is likely to be the level of microwave radiation at the receiver and in fringe areas. It is assumed that the beam intensity would be acceptable at 25 mW/cm². The current US exposure standard is 10 mW/cm² and this is 1000 times higher than Soviet standards. Radio interference might be considerable and interference between the beam and the ionosphere could occur. It has been claimed that the SPS would prevent useful radio astronomy measurements being carried out on earth. Finding large areas of land for receiver stations would undoubtedly present difficulties in many parts of the world. This has recently been discussed by Collins and Tomkins [44] who have suggested building artificial islands at sea.

Costings have assumed that space transportation and solar cell cost targets for 1990 are met and range from 3 c/kWh to 10 c/kWh at a load factor of 90%. In 1978, the US DoE procured a 'reference' design from NASA [45] which utilised a 5 km by 10 km by 0.4 km collector area and produced 5 GW(e) on earth. It was estimated that each SPS would cost $15 000M at 1980 prices, which at a load factor of 80% gives a cost of electricity of about 6 c/kWh for a thirty year life assuming US criteria for evaluating the economics.

Funding of SPS studies has emanated almost entirely from the US (although British Aerospace for the European Space Agency and several European universities and research associations have carried out preliminary studies) and has now exceeded $20M. It now seems likely that plans to develop a prototype SPS will, however, be shelved. Some further recent evaluations of the costs of SPS by the US National Research Council have concluded that earlier projected costs were very optimistic. It is clear that only nations such as the USA could contemplate developing the SPS on their own. Most nations including the UK could only obtain power from such as source through massive international collaboration.

6.8 Solar thermal generation

The direct use of the sun's heat has a wide range of possible applications—heating buildings, producing industrial process heat and generating electricity. Its flexibility in areas of high direct solar irradiation is further demonstrated by the scale of electrical applications for which it may be used. These range, in principle, from providing power for remote small-scale use up to plant of several hundred megawatts. The concept was the subject of a British patent as long ago as 1893 but it is only in the last decade that it has been studied seriously for electricity generation.

A solar thermal power generation system consists of up to five elements: a solar concentrator to collect the sunlight, an absorption system to convert the concentrated radiation into a heat fluid, a transfer system for the fluid and a power generation system. There is also an opportunity to incorporate heat storage in the system before the generator. Three basic plant types can be envisaged.

(1) A distributed system in which a large number of receivers each convert the sunlight to thermal energy and transfer the heat to a central generating facility. Sunlight is usually concentrated by a factor of up to 100.

(2) A central system in which sun tracking flat-plate reflectors (heliostats) transfer energy on to a central collector system on a tower (power tower). This has the advantage over (1) of disposing of the need to conduct a working fluid through a pipe system but requires accurate tracking and reflection of the sunlight on to the power plant. High concentration ratios (200–1000) and thus very high temperatures can be achieved.

(3) A hybrid approach combining the features of (1) and (2), using flat-plate reflectors to deflect sunlight on to a parabolic dish which then focuses on to an absorber tube. In this arrangement, tracking requirements

are less onerous than for (1), heat transmission is minimised but the price paid is that the receiver is more costly and reflection losses much higher. Concentrations of about 200 may be achieved.

These types of system are illustrated in figure 6.19. For large-scale electricity generation (more than 1 MW(e)), the power tower central receiver system now seems to be the most favoured arrangement.

One other concept should also be mentioned. It has been suggested that an existing fossil-fuel power plant could be repowered using solar thermal energy. Alternatively a specially designed power station could be operated jointly using coal, oil or gas and solar energy. The fossil fuel allows continuity of output to be achieved. This concept is being studied in the USA [46].

6.8.1 *Distributed collectors*

Because distributed systems only appear to offer advantages over central receiver systems at small sizes they will only briefly be considered here. For a full account of the wide ranging US programme aimed at the development of these devices, the reader is referred to reference [46].

Flat-plate collectors can generally only achieve temperatures of up to about 150 °C because surface thermal losses then become appreciable. There are two basic types of concentrator which can either stand alone or be used as part of a distributed system; these are the linear focusing and point focusing devices. The former are normally equipped with tracking facilities along a single axis whilst the point devices can move along two axes. Linear concentrators can theoretically achieve concentrations of about 100 whilst point concentrations of about 1000 can be obtained.

Many types of linear system have been designed including parabolic troughs, fixed-mirror mobile-receiver collectors, fixed-absorber mobile-mirror collectors and Fresnel lenses. The parabolic trough is the most widely used because it is simpler and cheaper than the others. Its disadvantage is that a large mass must be steered into the sun. The main point focusing device is the parabolic dish in which the sunlight is brought to focus near a central absorber (it is usually left slightly out of focus in order to reduce local hot spots in the absorber).

Some of the discussion below on research, advanced systems and components for centralised use also applies to the tracking collectors.

6.8.2 *The power tower*

The principle of the central receiver system or power tower is illustrated in figure 6.19(a). For a typical power station with a peak output of 10 MW(e), about 2000 heliostats presenting a reflecting area of about $10^5\,\text{m}^2$ are required. Good general reviews of such systems have been given by McNelis [47], [48].

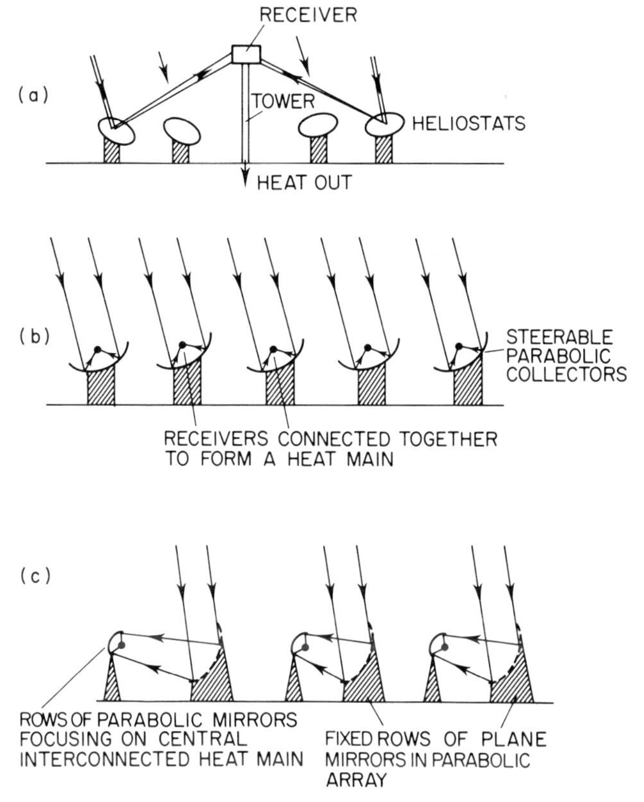

Figure 6.19 Three basic plant types: (a) central receiver or power tower system; (b) distributed system; (c) plane–parabola system.

The heliostats comprise up to 50% of the system cost. Several designs have been suggested and four of these are illustrated in figure 6.20. The reflector consists typically of a base and mounting, structural substrate, a reflective surface and a cover material which gives environmental protection. The assembly is mounted on a pedestal and in the McDonnell Douglas design several types of mounting are possible including equatorial, universal, receiver oriented altazimuth and simple altazimuth. The latter is generally favoured because of its low cost and simplicity, although incorporation of an active optical sensor using phototransistors to maintain orientation is more difficult than for the universal mount. The problems of providing orientation for the McDonnell Douglas design have been discussed in some detail in reference [49]. The reflector material itself must have high specular reflectivity, the ability to be mass produced at low cost and to give mechanical rigidity. Second-surface silvered float glass, bonded or clamped to a steel honeycomb framework has been advocated for this purpose. A major design problem is overcoming the occasionally very high wind loadings which the heliostats encounter.

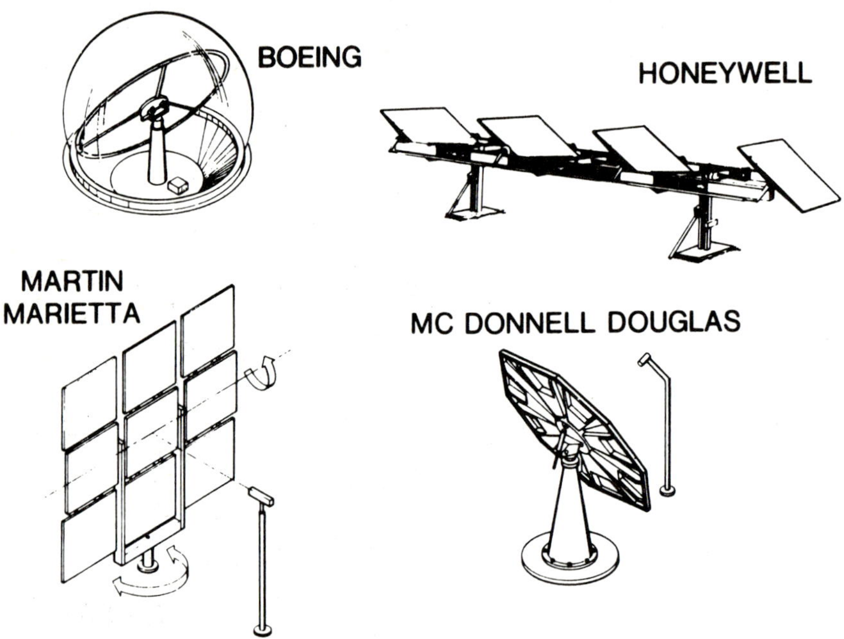

Figure 6.20 Pilot plant heliostat concepts. (Drawing courtesy Sandia Laboratories.)

The receiver consists in most designs of a guyed or lattice steel tower, which for a 10 MW(e) plant needs to be over 200 m high in order to avoid shadowing of one collector by another. The detailed design of the receiver itself depends upon the working fluid which is chosen, but two basic types—externally heated and cavity heated—are used. Certain principles are common in all designs. The collector tube is normally roughened and covered with a suitable temperature resistant matt black surface to enhance radiative capture although in high concentration systems some losses are accepted because the surface can become unacceptably hot. In lower concentration systems (particularly the hybrid type) the receiver may be enclosed in a Pyrex glass evacuated envelope, the internal surface of which has been coated by an evaporated layer of indium oxide to reflect back into the receiver any re-radiated energy. Plasma spraying of chromium oxide on the receiver tube is being tried in an effort to increase absorption at the receiver and careful attention is being paid to minimising heat loss in the absorber support and at joints with the vacuum envelope. Details of receiver design have been discussed by Francia [50].

The configuration of the collector field and receiver is determined partly by the latitude. At the equator the tower is placed at the centre of the field, whilst in the northern hemisphere it is located south of the centre. When the receiver is surrounded by collectors (a 360° field) lower tower height and higher efficiency can be achieved and the design of the receiver is simplified.

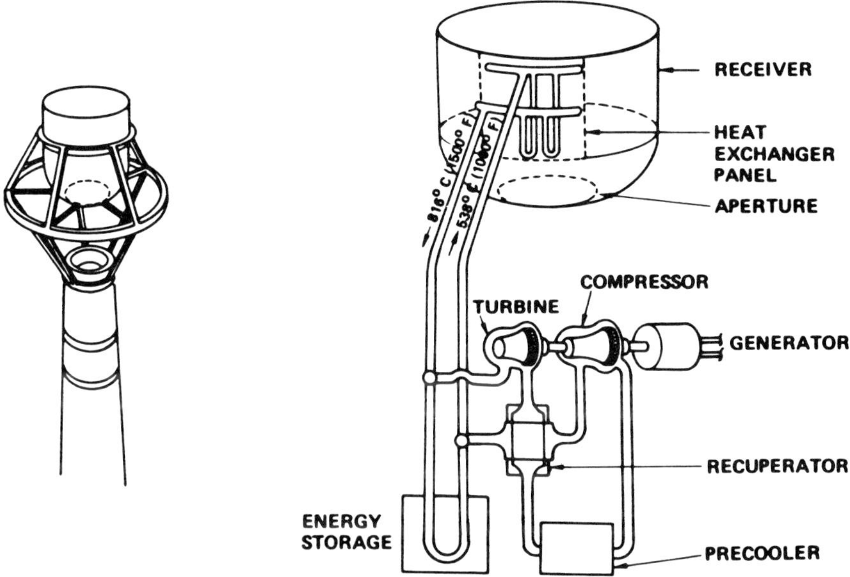

Figure 6.21 Schematic diagram of Boeing receiver and system design.

The most usual working fluid is water/steam operating in a conventional Rankine steam cycle with an upper temperature of about 500°C. An overall plant efficiency of nearly 20% is then expected. However, several advanced systems have been evaluated including the use of sodium or molten salts as the heat transfer fluid. Boeing [51] have analysed the prospect of using a closed cycle helium system operating at over 800°C to generate 100 MW(e) from one site. This is illustrated in figure 6.21. The receiver comprises a lower hemispherical section and an upper cylindrical section which, it is claimed, optimises heat containment and distribution. The input is collected through an aperture in the base of the receiver. Heat transfer panels within the cylindrical section heat the circulating gas, which is under a pressure of about 3.5 MN/m^2 (500 psi). Another study [52] carried out by the consulting engineers Black and Veatch investigated the possibility of using an open Brayton cycle which operates with very high temperature air (about 1000 °C). It has been claimed that these systems have advantages because they can achieve 40–45% efficiency, do not require a cooling tower or large amounts of water (this is particularly important in arid areas) and that gas turbines can be mounted near the top of the tower and can be run by auxiliary fuel when the sun is not shining. Other designs utilise liquid sodium [53] and molten salts [54]. There is no doubt that a number of these cycles offer good prospects, but it is likely that in the early phases of the large-scale solar thermal programme, the well understood lower temperature Rankine cycle will be utilised.

If solar thermal devices are not used in conjunction with gas turbines or to repower conventional power stations, it is important to incorporate some form of storage in the system. This is because clouds passing the sun can, in a matter of seconds, reduce the direct component of sunlight dramatically. The fact that a thermodynamic cycle is used in these systems means however, that to provide the necessary storage should be relatively simple and inexpensive and methods of obtaining heat storage of up to six hours production are being investigated. These include molten salt re-generators in which the latent heat of fusion accompanying a phase change in a mixture of salts (e.g. KCl, LiCl or KNO$_3$, NaNO$_3$) is stored or heat storage in high-temperature oils and rock beds filled with fluid.

6.8.3 *Progress with solar thermal technology*

The major R and D effort in this area has once again been carried out in the USA ([46], [55]). In 1979, over \$100M was committed to solar thermal energy, of which \$24.6M was spent on large power (more than 10 MW(e)) R and D, \$28M on a 10 MW(e) prototype plant at Barstow in California and \$13.5M on 'advanced technology'. The remaining funding was directed mainly towards small systems.

The Barstow project (Solar One) which first produced electricity in April 1982 is a collaborative demonstration project being carried out jointly between the US DoE, the Los Angeles Department of Water and Power, the California Energy Commission and Southern California Edison. A 90 m high power tower is heated by 1800 heliostats, each 40 m^2 with two-axis control to follow the sun. The output from the steam cycle is 10 MW(e) peak with an oil–rock thermal storage facility. The aims of the project are to establish the technical feasibility of solar thermal, electricity production, obtain development, production, operation and maintenance cost data and to confirm cost predictions for the technology. It is also intended 'to determine environmental impact, collect operational data and to further involvement, acceptance and interest by utilities, commercial

groups and the public'. McDonnell Douglas have designed the system, an
aerial view of which is shown in figure 6.22.

Figure 6.22 The Barstow Solar One installation. (Courtesy US Department of
Energy.)

The aim of the US programme is to achieve the design of a 100 MW(e)
central power station within a capital cost target of $1300 per kilowatt (at
1978 prices) for production of a large number of such stations. A design
study for such a commercial scale facility is currently being carried out by
McDonnell Douglas in collaboration with Southern California Edison and
Bechtel. In 1985, it is hoped to be able to generate electricity at 5–6 c/KWh
(again at 1978 prices). To put this target into perspective, the cost of the
prototype power station at Barstow has been estimated to be about $13 000
per kilowatt. In order to achieve significant cost reductions it is clear that
advanced technology will need to be employed and to this end two test bed
systems have been built. The 5 MW(th) solar thermal test facility (STTF) at
Sandia Laboratories in New Mexico, comprises 250 heliostats focusing on a
60 m tower and the 400 kW(th) advanced component test facility (ACTF) at

the Georgia Institute of Technology in Atlanta, which is capable of testing receivers at temperatures in excess of 1200 °C, comprises 550 heliostats.

Research on improving components is placing particular emphasis on high-temperature receiver designs and advanced cycles, of which several have been mentioned. High-temperature receivers are particularly important for uses which involve process heat such as high purity metallurgical processing. Ceramic, heat pipe, falling molten salt and chemical decomposition (of SO_3) systems are being studied. Improving the absorbtivity has led to studies of retro-reflectance, polarisation and characterisation of surfaces. In particular, co-evaporated Pt/Al_2O_3 on Pt deposited on a quartz substrate, thin-film molybdenum and stabilised organo-metallic absorber systems have shown promise as advanced selective absorber surfaces for intermediate (about 600 °C) temperatures. At higher temperatures, special paints of (Fe–Mn–Cu)O pigments in silicate binders have performed well. Collector system studies have included work on composites of silver on glass, silver on plastic films, anodised aluminium and studies of silver corrosion mechanisms. Many of the novel systems have been tried on the test bed facilities.

Work is continuing in the USA on distributed sytems, particularly on high performance, lightweight, low cost parabolic dishes, point focusing Fresnel lenses and liquid-metal heat pipe receivers and heat transport systems.

Many of the research areas mentioned have wide applicability, not just for large-scale electricity generation but also for the other potential uses including total energy systems, small community systems and remote (irrigation) systems. Major demonstration projects are proceeding in each of these areas [44].

France and Italy have an impressive record of scientific studies and demonstrations of solar thermal systems. Francia designed a small central receiver station in 1965 for the development of high-temperture receivers on which he is a leading expert [56]. His current installation at St Ilaria near Genoa (prototype for the ACTF in the USA) is rated at 100 kW(th). In 1952, Trombe constructed a solar furnace at Mont Louis in the Pyrenees which consisted of 540 back-silvered mirrors reflecting light on to a 10 m diameter parabolic concentrator containing 3500 plane mirrors. The furnace was used for metallurgical studies. A further furnace designed by Trombe and his co-workers in the late 1960s was built at Odeillo in the Pyrenees, again for high-temperature technology. Of similar design to the Mont Louis system, it comprised 63 heliostats each composed of 180 mirrors reflecting onto a paraboloid concentrator consisting of 9500 plane mirrors. The output is estimated to be 1 MW(th). In 1975, a high-temperature electricity producing project (Themis) was initiated. Originally the output was planned to be 10 MW(e) but has subsequently been de-rated to 2.5 MW(e). The Themis plant has 200 heliostats focusing on to a cavity mounted on a 101 m high central tower. The heat is transferred by a molten salt mixture (sodium nitrite, sodium nitrate and potassium nitrate) at 450 °C, which is used to provide storage, and thence to the steam generator. Because of a shortage of water for cooling, dry cooling towers are used. Electricité de France, who will operate Themis, hoped to couple the station to its grid in early 1982. Recently delays have been incurred because of wind damage to some of the mirrors.

The EEC have funded a 1 MW(e) system (Eurelios), sited in Sicily which has been supplying electricity to ENEL, the Italian state utility, since May 1981. The heliostat field comprises 182 heliostats covering 7800 m², half constructed by Cethel, a French company and half by Messerschmitt-Bölkow-Blohm (MBB) in Germany. The receiver is a 4.5 m aperture cavity-type system on a 55 m high tower. A steam cycle is used to generate electricity; a top temperature of over 50 °C being achieved. A small amount of heat storage (about 30 minutes at reduced output) has been designed into the system. The system is fully discussed in reference [57].

An IEA sponsored project supplies electricity to a Spanish utility from a site in Almeria [53]. Two 500 kW(e) systems are now operating—one a distributed collector design and the other a central receiver system. The aims are to demonstrate the feasibility of both types of plant using existing technology and assess technical performance and economics. The distributed plant uses two types of parabolic trough design (Acurex and MAN). The receiver for each module type has a selective surface (chromium black) and a glass envelope. A high-temperature oil is used for heat transfer with an upper working temperature of 295 °C and this oil also provides some storage. Electricity is generated using an organic Rankine cycle with toluene as the working fluid. The central system which first provided power to the grid in late 1981, uses Martin Marietta heliostats with an area of 6150 m² and the collector field has been designed to cover an area of 300 m by 300 m. The receiver tower is 45 m high and a temperature of 530 °C is achieved in the 6 m wide cavity-type receiver. A steam cycle with an upper temperature of 500 °C is used to generate electricity. Storage is provided by a two-tank sodium system.

Finally, major projects are proceeding in Japan [58]. Two 1 MW(e) systems were completed in 1981. These are shown in figure 6.23. The first of these is a hybrid plane-parabolic system in which a bank of flat-plate collectors track the sun and a small number of parabolic trough reflectors collect this solar radiation and further concentrate it. Nearly 2500 flat-plate reflectors and 124 parabolic reflectors are employed. The absorber is a horizontal tube with a vacuum envelope and power is generated from a steam cycle with a superheated temperature of nearly 400 °C. Storage is provided by a combination of accumulator and molten salt regenerator. The accumulator is used to store the heat transfer medium and the LiCl, KCl salt mixture uses the latent heat of fusion to store heat at temperatures close to the superheated steam. The other system is a more conventional power tower plant and uses 807 heliostats, each 4 m × 4 m, reflecting solar energy to a tower-suspended cavity-type receiver. Electricity is again generated by a steam cycle but using saturated steam at 250 °C. Because superheated steam is used in this plant, only a pressurised water heated storage tank is installed and there is no molten salt regenerator system.

Figure 6.23 An aerial view of the Japanese solar thermal power plant prototypes. On the right-hand side of the photograph is the power tower system and on the left-hand side a plane-parabola system.

Estimates of costs are still not well established but for a number of the projects described in this section estimated costs have been given by McNelis [48]. As prototypes, there are, of course, considerable variations in cost but they appear to be mainly in the range £4000–6000 per kW(e) (at 1978 prices). Assuming a load factor of 25%, a 25 year life, a discount rate of 5% p.a. and 2% p.a. operation and maintenance costs, this is equivalent to a cost of electricity of 20 p/kWh. It should be noted, however, that the figure is sensitive to load factor which may be much improved if storage of several hours duration is designed into the system. Nonetheless it seems that even for favourable locations, costs would need to be reduced in production by a considerable factor to be competitive with conventional power stations. As mentioned above, this indeed is the aim of the programme in the USA. A summary of the main power tower projects is given in table 6.5.

Table 6.5 A summary of the major (more than 500 kW(e)) solar thermal stations in operation or construction.

Location/ project name	Sponsor/ contractor	Power output (MW(e))	Date of completion	Cost estimate (1978£M)
Sicily (Adrano) Eurelios	EEC Ansaldo/Cethel/ ENEL/MBB	1.0	1981	4–5
France (Targasonne) Themis	CNRS/EdF Cethel	2.5	1982	8–9
Spain (Almeria)	IEA/ International Consortium	0.5	1981	~7
Japan (Nio)	(i) Sunshine Project/ EPDC/Mitsubishi (central receiver)	1.0	1981	3.6 [49]
	(ii) Sunshine Project/ EPDC/Hitachi (plane-parabolic)	1.0	1981	not known
USA (Barstow, California)	DoE/S Cal. Edison/ LA Water and Power/ California Energy Commission/McDonnell Douglas	10.0	1982	65–70
USA (Albuquerque)	DoE/Sandia Labs/Martin Marietta	5 (thermal)	1977	11.5

6.9 The solar chimney

In 1978, it was suggested by Professor Schlaich of Stuttgart University that power could be produced by a hybrid solar/wind system. His concept is shown in figure 6.24. The solar chimney system consists of a central tower or chimney surrounded by a solar collector area in the form of a

transparent plastic 'greenhouse'. As air in the 'greenhouse' is heated it
rises up the chimney and turns an air turbine. Air is sucked in from outside
the collection area, to replenish the supply of air to be heated.

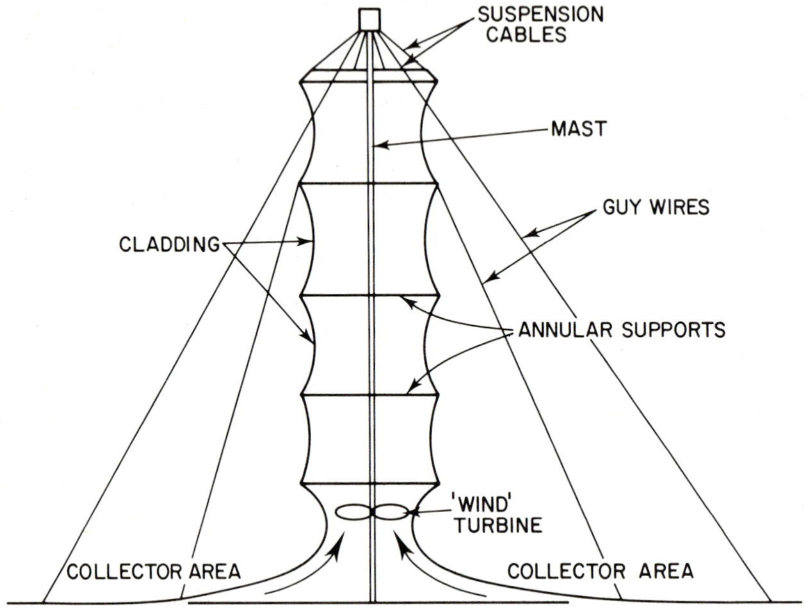

Figure 6.24 The solar chimney concept.

In 1982, a 100 kW prototype of this concept was completed in Manzanares in Spain and is now being tested. It has a 200 m high, 10 m
wide central tower with a plastic collecting area extending to a radius of
125 m. The chimney is constructed from corrugated steel, attached to
annular rings and is supported by guy wires. The 'wind' turbine is
housed in the base of the chimney and consists of four 5 m, variable pitch
blades.

The plant at Manzanares is being used to study plastic materials for the
collector area (including methods of disposing of rain water) and the
turbine system. Professor Schlaich claims that the prototype is performing
well and is interested in the possibility of building a full-scale prototype
followed by commercial designs generating 100 MW or more. According to
Schlaich, a 10 MW plant would require a collector area of $2.5 \times 10^5 \, m^2$ and
a turbine diameter of 40 m. The chimney would need to be 400 m high and
50 m wide. He believes that 10 MW designs could be constructed for less
than £20 M.

6.10 Solar ponds

A solar pond is a novel device for collecting and storing solar energy thermally in a pond of water typically two or more metres deep. It is most frequently used as a source of low-grade heat for domestic heating but can be used for central power generation. The idea of the solar pond originated from the observation that some lakes with natural salt gradients developed very high temperatures well below the surface. The subject has recently been reviewed by Tabor [59].

The principle is simple. Incident radiation passes through the transparent upper layers and is absorbed in the lower layers. Conduction through surrounding soil is normally low and conduction of heat into the upper layers is likewise usually low for a liquid. Thus total heat transfer can be minimised if convection can be prevented. A method which is used to achieve this, is to dissolve common salt (or $MgCl_2$, which is more soluble) in the lower layers of the liquid thus setting up a vertical concentration gradient. The resulting density gradient can be such that it counter balances any negative density gradient leading to convection. If the solution density is ϱ, the thermal conductivity β, temperature T and depth, z, the above requirement for stability against steady state convection becomes

$$d\varrho/dz > 0 \qquad \text{or} \qquad (\partial\varrho/\partial z)_T > \varrho\beta \, (dT/dz).$$

A similar equation exists for the suppression of oscillations in the concentration gradient but in this case the right-hand side of the above equation is modified by a term involving viscosity, thermal diffusivity and the salt diffusivity. The theoretical temperature and density gradients for a solar pond in equilibrium (a) with salt, (b) without salt and in an unstable configuration (c) are shown in figure 6.25. The unstable case arises from the setting up of a mixing layer at depth due to overheating, or at the surface due to evaporation.

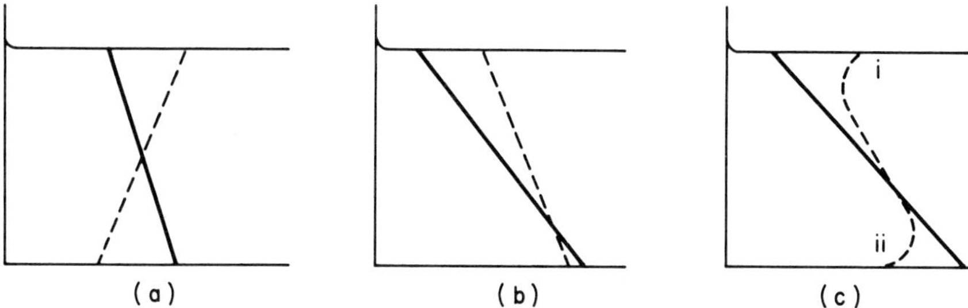

Figure 6.25 Obtaining equilibrium in a solar pond: (a) no salt; (b) stable; (c) unstable. i, is the mixing layer due to evaporation and ii, the mixing layer due to overheating. The dotted line is the temperature gradient and the broken line the density gradient.

Eventually, if left to itself, salt will diffuse upwards and the concentration gradient would even out leading to a homogenous salt solution. This can be prevented by the so-called 'falling pond' method in which solution is continually drawn off at the bottom in order to impose a vertical velocity on the whole pond which is greater than the salt diffusion velocity. The solution withdrawn is flash distilled, the pure salt replaced at the bottom and the water at the top. In this way the salt concentration gradient remains constant in time and only the water moves. Prevention of evaporation from the surface can also be achieved as part of the same process.

The maximum temperature achieved depends on optimising the depth for a given pond area so that the greater losses due to decreased transmittance of the incident radiation balance the reduced heat losses by conduction. Boiling has been achieved in some ponds.

Solar ponds have been studied in detail particularly in Israel, Australia and the USA. In prototype ponds developed in Israel, temperatures of about 90 °C have been achieved in the lower layers and with the insolation levels prevailing (an energy equivalent of about 2000 GWh/year) an energy density of 400 GWh/km^2 can be built up. With a flow rate of 0.1 m/s from a 12 cm wide extraction layer, a low-temperature organic turbine could generate about 5 MW per square km of pond at a 75% load factor. Dirt and wind cooling impair performance and are the main reason why solar collection efficiencies of only about 20% can be achieved for large areas even where edge effects are small.

The costs appear to be highly competitive with diesel oil generation in the most favourable locations and the provision of long term storage is an important advantage. Tabor [60] has estimated that the cost of electricity generated by solar ponds in such conditions is about 13 c/KWh, with some prospect of achieving 5 c/KWh. To be efficient and to generate power for utility use, pond areas clearly need to be very large and this is a drawback inherent in the concept. Furthermore for latitudes of greater than about 40°, the more oblique solar incidence leads to increased reflection losses. It has been suggested that solar ponds could be used for home heating in the UK but electricity generation would be very expensive.

The ponds developed in Israel are sizeable and there are plans to build large-scale plant which, it is believed, will generate electricity competitively. In 1979 a 150 kW power station was opened on the shores of the Dead Sea. A schematic diagram of this plant is shown in figure 6.26. This is a pilot plant for a 5 MW installation due to be completed in 1982 which, it is hoped, will generate electricity at 10–13 c/KWh. Larger schemes of up to 40 MW are planned. Tabor [60] has estimated that the world potential could be in the range 6–32 GW. This could be greater if ideas originally proposed by Assaf [61] for very large schemes in the Dead Sea prove practicable. It was suggested that a solar pond could be 'floated' on a large saline lake by enclosing it in a thermally insulated wall and forming an inverse gradient below the bottom of the normal gradient in the top few metres. The mass of water below the boundary would then act as a huge heat store. Such a scheme could, in principle, supply all the electricity needs of Israel.

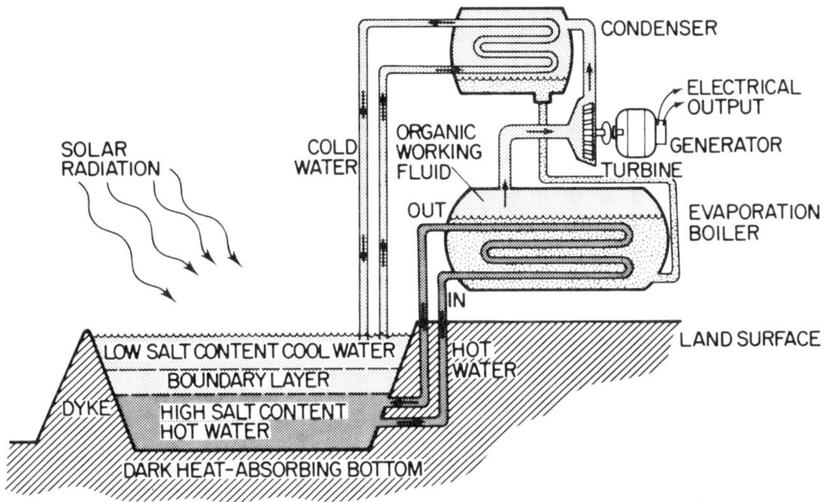

Figure 6.26 The operation of the Dead Sea solar power station. The diagram is not drawn to scale. In practice the cool water is taken from the top 10–15 cm of the pond, which is separated by a 1 m boundary layer from the 2 m deep hot water sink.

References

There is an extensive literature on solar technology and the reader should find little difficulty in obtaining good books and review articles on the subject for deeper study. Solar heating and cooling of buildings is not included in this volume; the interested reader is referred to the recent volume published by the UK International Solar Energy Society (*Solar Energy Today* ed L Jesch) which has good introductory chapters on this subject. Biomass energy is well reviewed in the first three references of this chapter and there are many specialist texts and reviews. A good basic introduction to the physics of solar energy is given in Wilson's book [6] and several of the more recent review articles given in the references on photovoltaic energy ([9], [15], [17] and particularly [8]) provide excellent specialist surveys. For good review articles on solar power satellites and solar ponds the interested reader is referred to [42] and [59] respectively.

1 Watt Committee on Energy 1978 *Teach in on Energy from the Biomass* (London: Watt Committee on Energy)

2 North W J 1980 *Solar Energy* **25** 387–97

3 *Solar Energy, A UK Assessment* 1976 (London: International Solar Energy Society)

4 *Energy Paper* 16 1975 (London: HMSO)

5 Debney B T and Knight J R 1978 *Contemp. Phys.* **19** 25–45

6 Wilson J I B 1979 *Solar Energy* (London: Wykeham Publications)

7 Wolf M 1970 *Proc. 8th IEE Photovoltaics Specialists Conf.* (London: IEE) p360

8 Treble F C 1980 *IEE Proc.* **127** 505–27

9 Charles H K Jr and Ariotedjo A P 1980 *Solar Energy* **24** 329–39

10 Cheek G and Mertens R 1981 *Proc. 15th IEEE Photovoltaics Specialists Conf, Florida* (London: IEEE)

11 Wolfe P R 1979 The developing role of photovoltaic generation *IEE Future Energy Concepts Conf.* (London: IEE) pp26–9

12 Yaws C L *et al* 1979 *Solar Energy* **22** 547

13 Yaws C L *et al* 1980 *Solar Energy* **24** 359

14 Whale A V 1979 Silicon solar array technology and the prospects for cost reduction *Conf. on Photovoltaic Solar Energy Conversion* (UK International Solar Energy Society)

15 Smith J L 1981 *Science* **212** 1472–8

16 Maycock P D 1980 *Proc. 14th IEEE Photovoltaic Specialists Conf., San Diego* (London: IEEE) pp6–12

17 Durand H L 1979 Present status and prospects of photovoltaic solar energy conversion *Conf. on Photovoltaic Solar Energy Conversion* (UK International Solar Energy Society) pp93–8

18 Van Overstraeten R J 1981 *Proc. 15th IEEE Photovoltaic Specialists Conf., Florida* (London: IEEE) pp372–6

19 Spear W E and Le Comber P G 1976 *Appl. Phys. Lett.* **28** 105–7

20 RCA Laboratories 1977 *Report* ERDA Contract EY-76-C-03-1268

21 Carlson D E 1980 *Proc. 14th IEEE Photovoltaic Specialists Conf., San Diego* (London: IEEE) pp291–7

22 Boes E C and Shafer E D 1981 *Proc. 15th IEEE Photovoltaic Specialists Conf., Florida* (London: IEEE) pp 305–22

23 Mertens R 1979 Hybrid thermal-voltaic systems *Conf. on Photovoltaic Solar Energy Conversion* (UK International Solar Energy Society) pp65–72

24 Wilson B L H 1979 Gallium arsenide solar cells for use in concentrated sunlight *Conf. on Photovoltaic Solar Energy Conversion* (UK International Solar Energy Society) pp45–54

25 Hill R 1979 Thin film cuprous sulphide–cadmium sulphide solar cells *Conf. on Photovoltaic Solar Energy Conversion* (UK International Solar Energy Society) pp25–36

26 Hill R and Williams E W 1982 *Proc. Conf. 'Photovoltaics: A New Industry for Britain?'* (UK International Solar Energy Society, Conference Proceeding C30)

27 Leang J and Deb S 1981 *Proc. 15th IEEE Photovoltaic Specialists Conf., Florida* (London: IEEE) pp1016–20

28 Ovshinsky S R 1977 *7th Conf. On Amorphous Semiconductors, Edinburgh* ed W E Spear (Edinburgh: University of Edinburgh Centre for Industrial Consultancy and Liaison) pp519–23

29 Weber W H and Lambe J 1976 *Appl. Opt.* **15** 2299

30 Feucht D L 1981 *Proc. 15th IEEE Photovoltaic Specialists Conf., Florida* (London: IEEE) pp648–53

31 EPRI 1979 *EPRI Journal* April issue 19–22

32 Merrit V Y and Hovel H J 1976 *Appl. Phys. Lett.* **29** 414–5

33 Post H N and Schueler D G 1981 *Proc. 15th IEEE Photovoltaics Specialists Conf., Florida* (London: IEEE) pp718–24

34 Forney R G 1979 *Conf. on Photovoltaic Solar Energy Conversion* (UK International Solar Energy Society) pp81–92

35 *Photovoltaic Energy Systems* 1980 US DoE/CS-0146

36 Van Overstraeten R J 1979 The EEC photovoltaic solar energy R and D programme *Conf. on Photovoltaic Solar Energy Conversion* (UK International Solar Energy Society) pp73–80

37 Albery J W and Archer M D 1977 *Nature* **270** 399–402

38 Ellis B E, Kaiser S W and Wrighton M S 1976 *J. Am. Chem. Soc.* **98** 1635–6

39 Fujishima A and Honda K 1972 *Nature* **238** 37–8

40 Sprintschnik G, Sprintschnik H W, Kirsch P P and Whitten D J 1976 *J. Am. Chem. Soc.* **98** 2337

41 Glaser P E 1968 *Proc. 3rd IECEC* (New York: American Institute of Chemists) pp98–104

42 Collins P Q 1979 Satellite solar power stations—current status and prospects *Proc. IEE Future Energy Concepts Conf.* (London: IEE) pp21–5

43 Gregory D L 1977 *Proc. 12th Intersoc. Energy Conv. Eng. Conf.* (American Nuclear Society) pp1386–90

44 Collins P Q and Tomkins R 1981 Economics of satellite solar power system operation *Proc. IEE Future Energy Concepts Conf.* (London: IEE) pp169–72

45 Woodcock G R 1978 SPS cost considerations *Report* (Boeing Aerospace Company)

46 US DoE 1979 *Solar Thermal Power Systems Annual Technical Progress Report* FY 1978 DoE/ET-0078-T

47 McNelis B 1979 Large scale solar thermal power generation *Proc. IEE Future Energy Concepts Conf.* (London: IEE) pp264–7

48 McNelis B 1978 Solar thermal power generation: a general introduc-

tion *Conf. on Solar Thermal Power Generation* (UK International Solar Energy Society) pp1–9

49 Easton C R, Hallet R W, Gronich S and Gervais R L 1974 Evaluation of central solar tower power plant *Proc. 9th Intersoc. Energy Conv. Eng. Conf., San Francisco* (New York: American Society of Mechanical Engineers) pp271–7

50 Francia G 1968 *Solar Energy* **12** 51–64

51 *EPRI Report* No. ER-629 1978

52 *EPRI Report* No. ER-652 1978

53 Kalt A C 1978 The IEA solar power stations project *Conf. on Solar Thermal Power Generation* (UK International Solar Energy Society) pp57–65

54 Etievant C *et al* 1978 *Proc. DFVLR Symp., Germany* paper 18

55 Selvage C S 1978 The US DoE solar thermal power generation program *Conf. on Solar Thermal Power Generation* (UK International Solar Energy Society) pp66–74

56 Floris R and Francia G 1978 High concentration solar receiver *Conf. on Solar Thermal Power Generation* (UK International Solar Energy Society) pp101–9

57 Anon 1981 *Mod. Power Syst.* **1** 26–30

58 *Solar Thermal Power Generation* (Tokyo: Sunshine Project Operation Bureau, Electric Power Development Company Limited) 1980

59 Tabor H 1981 *Solar Energy* **27** 181–94

60 Tabor H 1980 *Phil. Trans. R. Soc.* A **295** 423–33

61 Assaf G 1976 *Solar Energy* **18** 294–9

Concluding remarks

There is a clearly defined potential for alternative energy sources in both the developed and developing world as reliance on conventional generation brings with it increasing uncertainty, and development in much of the world creates a rapidly growing thirst for energy. In the next few years, several of the alternative sources of electricity discussed in the book will be capable of being evaluated fully—technically, economically and in terms of their environmental impact. If these hurdles can be overcome they will have the ability to make a valuable and growing contribution in reducing world reliance on fossil fuels.

It would be naive to suggest that technical and economic factors alone will determine the degree to which alternative energy sources will be exploited either in the developed world or in less developed countries. It is likely that political and social developments will be at least as important. As relative newcomers to the scene, in the sense that electricity utilities have only recently begun to take the alternative sources seriously, alternative sources have much ground to make up. To some degree, the speed with which they find a place in the plans of governments and utilities, will also depend upon the degree to which difficulties with other technologies are encountered.

This book has not set out, however, with the primary aim of looking at the broad context in which alternative sources might be developed. Instead it has tried to provide a clear guide to their technical and commercial development. It is always dangerous to attempt to encapsulate the arguments of a whole book in a short summary. However, with the clearly stated qualification that this should be read in conjunction with the text, an attempt has been made to summarise the state of development of the various technologies in table 7.1. It is clear that there are no outright winners. Several of them, however, are now reaching a stage where technical maturity allied to reducing costs means that, where resources are available and where competing energy costs are high, they are likely to make a significant contribution to electricity production.

Table 7.1 A summary of the present status of the alternative energy sources discussed in this book. The more * in each entry the more promising the technology.

Source	Technical maturity	Requirement for capital per module	Requirement for mature industrial base to use the technology	Present cost of 'good' sites cf. oil	Environmental acceptability	Ability to provide 'firm' power	World resources	UK resources	Promise in UK
Wind	***	***	**	***	**	*	**	***	***
Wave	*	**	*	**	***	*	*	**	**
OTEC	**	**	*	**	****	****	*		
Geothermal steam or hot water	****	***	**	****	**	****	**	*	*
Geothermal hot rock	*	**	**	***	***	****	**?	**?	***
Tidal	****	*	**	***	***	**	*	**	***
Photovoltaics	***	***	***	**	****	*	***	*	*
Solar thermal	***	**	**	**	***	**	***		
Solar ponds	***	***	**	**	****	***	*		

Index